"By 2050, a quarter of the world's population will live in Africa. This important book is a must-read for everyone committed to addressing the tremendous challenge of feeding the rapidly growing African population under climate change".

Jonathan Verschuuren, *Professor of International and European Environmental Law, Tilburg Sustainability Centre and Tilburg Law School; Extraordinary Professor, North West University, Potchefstroom, South Africa; Marie Sklodowska-Curie Alumnus, Australian Centre for Climate and Environmental Law, University of Sydney; Extraordinary Professor, Global Environmental Law Centre, University of the Western Cape, South Africa*

Agricultural and Climate Change Adaptation Law in Africa

This book presents an analysis of climate change and agricultural laws in Kenya, Nigeria, and South Africa in order to determine whether they adequately addressed the concept of agricultural adaptation.

Agriculture is one of the sectors of the economy that is contributing to climate change, and at the same time the sector is heavily impacted by climate change. Therefore agricultural adaptation is required. Focusing on three countries, this book provides a novel, comparative examination of how and to what extent the law promotes agriculture-focused adaptation in these countries. The role of the law in addressing issues such as water management strategies, soil conservation methods, and crop production methods is discussed. This book identifies gaps in the regulatory frameworks for agricultural adaptation and highlights the lack of adaptive capacity of African agriculture due to weak or non-existing legal frameworks. It discusses ways to remedy these gaps through specific on-farm adaptation strategies, legislative amendments to consolidate all relevant national climate change-related policies and laws with agricultural policies and laws that have relevant provisions on adaptation as medium-term solutions, and the development of a specific framework law for agriculture-focused adaptation, incorporating essential agricultural adaptation strategies, could perhaps be enacted as long-term solutions to the regulatory gaps.

This book will be of great interest to students and scholars of law, climate change, food and agriculture, sustainable development, and African studies.

Habib Sani Usman is an Assistant Professor at the School of Law, American University of Nigeria, and a Visiting Researcher at the Centre of African Legal Studies, Bayreuth University, Germany. He has 20 years' experience as a Solicitor and Advocate of the Supreme Court of Nigeria. He holds a doctorate degree (LLD) in International Aspects of Law, with specialisation in agricultural and climate change adaptation laws, from the North West University, Potchefstroom, South Africa.

Earthscan Food and Agriculture

Globalisation and Livelihood Transformations in the Indonesian Seaweed Industry
Edited by Zannie Langford

Principles of Sustainable Aquaculture
Promoting Social, Economic and Environmental Resilience, 2nd edition
Stuart W. Bunting

The Spatial Organization of Urban Agriculture in the Global South
Food Security and Sustainable Cities
Ada Górna

Digital Communication for Agricultural and Rural Development
Participatory Practices in a Post-COVID Age
Edited by Ataharul Chowdhury and Gordon Gow

Regenerative Farming and Sustainable Diets
Human, Animal and Planetary Health
Edited by Joyce D'Silva and Carol McKenna

Sustainable Cacao Cultivation in Latin America
Edited by Luz Cecilia García, Naga Raju Maddela, Freddy Zambrano Gavilanes and Carolina Aguilar Duarte

Food Policy in the United States
An Introduction, 3rd edition
Parke Wilde

Agricultural and Climate Change Adaptation Law in Africa
Reflections from Kenya, Nigeria, and South Africa
Habib Sani Usman

For more information about this series, please visit: www.routledge.com/books/series/ECEFA/

Agricultural and Climate Change Adaptation Law in Africa

Reflections from Kenya, Nigeria, and South Africa

Habib Sani Usman

Routledge
Taylor & Francis Group
LONDON AND NEW YORK

earthscan
from Routledge

First published 2025
by Routledge
4 Park Square, Milton Park, Abingdon, Oxon OX14 4RN

and by Routledge
605 Third Avenue, New York, NY 10158

Routledge is an imprint of the Taylor & Francis Group, an informa business

British Library Cataloguing-in-Publication Data
A catalogue record for this book is available from the British Library

Library of Congress Cataloging-in-Publication Data
Names: Usman, Habib Sani, author.
Title: Agricultural and climate change adaptation : reflections from Kenya, Nigeria, and South Africa / Habib Sani Usman.
Other titles: Earthscan food and agriculture.
Description: Abingdon, Oxon [UK] ; New York, NY : Routledge, 2025. | Series: Earthscan food and agriculture | Includes index. |
Identifiers: LCCN 2024046227 (print) | LCCN 2024046228 (ebook) | ISBN 9781032881577 (hardback) | ISBN 9781032881591 (paperback) | ISBN 9781003536437 (ebook)
Subjects: LCSH: Agricultural laws and legislation—Environmental aspects—Kenya. | Agricultural laws and legislation—Environmental aspects—Nigeria. | Agricultural laws and legislation—Environmental aspects—South Africa. | Climatic changes—Law and legislation—Kenya. | Climatic changes—Law and legislation—Nigeria. | Climatic changes—Law and legislation—South Africa. | Sustainable agriculture—Environmental aspects—Kenya. | Sustainable agriculture—Environmental aspects—Nigeria. | Sustainable agriculture—Environmental aspects—South Africa.
Classification: LCC KQC768 .U86 2025 (print) | LCC KQC768 (ebook) | DDC 343.6707/6—dc23/eng/20241028
LC record available at https://lccn.loc.gov/2024046227
LC ebook record available at https://lccn.loc.gov/2024046228

ISBN: 978-1-032-88157-7 (hbk)
ISBN: 978-1-032-88159-1 (pbk)
ISBN: 978-1-003-53643-7 (ebk)

DOI: 10.4324/9781003536437

Typeset in Optima
by codeMantra

For my late mother Aminatu Isma'il, and late Hon. Kadi Habib A. Tijjani. You are always remembered for your wise counselling and prayers.

Contents

Foreword xv
Acknowledgements xvii
List of abbreviations xix
List of statutes xxiii

1 Introduction 1
1.1 Background 1
1.2 Agriculture and climate change 2
1.3 A brief mapping of agriculture and climate change at country levels 3
1.3.1 Kenya 3
1.3.2 Nigeria 4
1.3.3 South Africa 5
1.4 Adaptation in the context of climate change 6
1.4.1 Forms of adaptations 7
1.4.2 Need for climate change adaptation 8
1.4.3 Stages of adaptation in the context of climate change 10
1.4.4 Mainstreaming in the context of climate change adaptation 11
1.4.5 Challenges for adaptation in the context of climate change 12
1.5 Adaptation in the context of agriculture 13
1.5.1 Forms of adaptation in the context of African agriculture 13
1.5.1.1 On-farm adaptation strategies 14
1.5.1.2 Technological solutions 14
1.5.1.3 Insurance and financial assistance 15
1.5.1.4 Climate-smart agriculture 16
1.5.2 Levels and plans for adaptation in the context of agriculture 17

1.5.3 Challenges for adaptation in the context of agriculture 20
1.6 Conclusion 20
References 43

2 Climate Change Adaptation, Agriculture, and the Law 55
2.1 Introduction 55
2.2 The Law in the Context of Climate Change Adaptation 56
2.3 Principles of Climate Change Adaptation Laws 58
2.3.1 The United Nations Framework Convention on Climate Change 59
2.3.2 The Kyoto Protocol 61
2.3.3 The Paris Agreement 61
2.4 Agricultural Adaptation Laws 62
2.4.1 Laws for controlling agricultural production 64
2.4.1.1 Fertiliser application and chemical-use Laws 65
2.4.1.2 Livestock production Laws 66
2.4.2 Laws for promoting agricultural production 67
2.4.2.1 Water-management Laws 67
2.4.2.2 Land-use and soil-conservation Laws 69
2.4.2.3 Agricultural insurance Law 71
2.5 Conclusion 72
References 84

3 Kenyan policy and legal framework for adaptation and agriculture 90
3.1 Introduction 90
3.2 Regulatory framework for climate change adaptation 90
3.2.1 The Constitution of Kenya 91
3.2.2 Policies for adaptation in Kenya 93
3.2.2.1 National Environment Action Plan 93
3.2.2.2 National Climate Change Response Strategy 94
3.2.2.3 National Climate Change Action Plan 95
3.2.2.4 National Adaptation Plan 96
3.2.3 Adaptation-related Laws 97
3.2.3.1 The Environment Management and Coordination Act of 1999 97
3.2.3.2 Climate Change Act of 2016 99
3.3 Regulatory framework for agriculture 101
3.3.1 Kenyan Climate-Smart Agriculture Strategy of 2017 102

3.3.2 Agricultural policies 103
3.3.2.1 National Land Policy 103
3.3.2.2 Draft National Livestock Policy of Kenya 103
3.3.2.3 Draft National Agricultural Insurance Policy 104
3.3.3 Agriculture-related Laws 105
3.3.3.1 Water Act 106
3.3.3.2 National Drought Management Authority Act 107
3.3.3.3 Kenyan Agriculture and Livestock Research Act 108
3.3.3.4 Crops Act of 2013 108
3.3.3.5 Land Act of 2012 109
3.3.3.6 Agriculture and Food Authority Act 110
3.4 Gaps in the Kenyan policy and legal frameworks for climate change adaptation and agriculture 111
3.5 Conclusion 113
References 126

4 Nigeria: The policy and legal framework for adaptation and agriculture 130
4.1 Introduction 130
4.2 Regulatory framework for climate change adaptation in Nigeria 131
4.2.1 The constitution 131
4.2.2 Adaptation Policies in Nigeria 132
4.2.2.1 National Adaptation Strategy and Plan of Action for Climate Change in Nigeria 132
4.2.2.2 Nigeria Climate Change Policy Response and Strategy 133
4.2.2.3 National Policy on Environmental Conservation 135
4.2.2.4 National Adaptation Plan 136
4.2.2.5 Nigeria's Nationally Determined Contributions 137
4.2.3 Adaptation-related Laws 139
4.2.3.1 National Environmental Standards and Regulations Enforcement Agency Act 139
4.2.3.2 Climate Change Act 2021 140
4.2.4 CSA Strategy 142
4.3 Regulatory framework for agriculture 143
4.3.1 Agricultural-based Policies 143

4.3.1.1 Nigerian Agricultural Policy of 2001 144
4.3.1.2 National Agricultural Resilience Framework of 2014 146
4.3.1.3 National Water Policy 147
4.3.1.4 National Drought Plan 2018 149
4.3.1.5 National Policy on Erosion and Flood Control of 2006 151
4.3.2 Agriculture-based Laws 151
4.3.2.1 Water Resources Act 152
4.3.2.2 River Basins Development Authority Act 153
4.3.2.3 Land Use Act of 1978 154
4.3.2.4 NESEFC Regulations of 2011 155
4.3.2.5 Nigerian Agricultural Insurance Corporation Act 156
4.4 Deficiencies in the Policy and legal frameworks for adaptation and agriculture in Nigeria 157
4.5 Conclusion 158
References 174

5 South Africa: The policy and legal framework for adaptation and agriculture 178
5.1 Introduction 178
5.2 Regulatory framework for adaptation in South Africa 179
5.2.1 The Constitution 181
5.2.2 Adaptation-related Policies 183
5.2.2.1 National Climate Change Response Strategy 183
5.2.2.2 National Climate Change Response Policy White Paper 184
5.2.2.3 National Climate Change Adaptation Strategy 2019 186
5.2.2.4 National Development Plan 2030 187
5.2.2.5 National Determined Contribution 187
5.2.3 Adaptation-related Laws 188
5.2.3.1 National Environmental Management Act 188
5.2.3.2 Disaster Management Act 190
5.2.3.3 Carbon Tax Act 191
5.2.3.4 Climate Change Act 192
5.3 Regulatory framework for agriculture 194
5.3.1 Agricultural Policies 195

5.3.1.1 Draft Climate-Smart Agriculture Strategic Framework 2018 195
5.3.1.2 Climate Change Sector Plan for Agriculture, Forestry, and Fisheries 196
5.3.1.3 Second National Action Programme for South Africa to Combat Desertification, Land Degradation, and the Effects of Droughts 197
5.3.2 Agriculture-related Laws 198
5.3.2.1 Water Services Act 198
5.3.2.2 National Water Act 199
5.3.2.3 Conservation of Agricultural Resources Act 202
5.3.2.4 Spatial Planning and Land Use Management Act 204
5.4 Weaknesses of the South African regulatory frameworks for climate change adaptation and agriculture 205
5.5 Conclusion 207
References 225

6 Conclusion 229
6.1 General background 229
6.2 Summary of the analysis and findings 230
6.2.1 Climate change, agriculture, and adaptation 230
6.2.2 Climate change adaptation, agriculture, and the Law 232
6.2.3 Assessment of the Kenyan, Nigerian, and South African regulatory frameworks for climate change adaptation and agriculture 233
6.2.3.1 Regulatory frameworks for adaptation: Kenya, Nigeria, and South Africa 233
6.2.3.2 Kenyan, Nigerian, and South African regulatory frameworks for agriculture 236
6.3 Regulatory gaps 240
6.4 Recommendations 244
6.4.1 Short-term solutions 244
6.4.2 Medium-term solutions 246
6.4.3 Long-term solutions 247
6.5 Future research 247

Index 253

Foreword

This book makes a crucial finding: although agriculture has been the backbone of the African continent and is poised to remain one of the most vital economic activities for Africa's future, most countries in Africa lack laws that address both short-term and long-term adaptability to climatic changes, despite agriculture's clear vulnerability to these changes. This raises an important question: what can be done? Habib examines the gaps in agricultural adaptability within the legal frameworks of Kenya, Nigeria, and South Africa, using these countries as examples to illustrate broader issues across the continent. This book is essential reading for policymakers seeking to develop agriculture-focused adaptation laws and policies, as well as for those interested in the complex dynamics of climate governance in Africa.

Brewster Caiphas Soyapi
Professor of Environmental Law, Sustainability Governance, and African Judicial Environmentalism – Faculty of Law, North West University, Potchefstroom, South Africa; Deputy Director of Africa, Global Network for Human Rights and the Environment

Acknowledgements

My unreserved gratitude goes to the Almighty, creator of the heavens and earth, who gave me the ability, courage, and wisdom to undertake this research.

I am heavily indebted to Professor Louis J. Kotzé of the Faculty of Law, North West University, Potchefstroom, South Africa who served as my mentor and promoter for this research. Your comments on each chapter did not only give me insight, but also provoked me to think critically and to view issues from an entirely different perspective.

I am grateful for the insightful and critical feedback received from the anonymous reviewers assigned by the Routledge editorial team. Special thanks also go to numerous experts and colleagues, who I will not name here, who made this study possible.

I must also acknowledge the useful support rendered by the American University of Nigeria. In particular, I appreciate the advice offered by Professors Lawal Ahmadu and Yusuf M. Yusuf and the flexibility to travel allowed by the Interim President Professor Attahir B. Yusuf.

To my father Muhammad Sani Usman, my wife Khadija Uthman, and our daughters Amina and Halima: Thank you all for your prayers, patience, and encouragement for the entire duration of this research.

I also acknowledge the support rendered by other extended family members, relatives, friends, and well-wishers whose names could not be listed due to space and time constraints.

Abbreviations

°C	Degree Celsius
AFA	Agriculture and Food Authority Act
AFDIA	Agricultural and Fishery Disaster Insurance Act
AMCEN	African Ministerial Conference on the Environment
AU	African Union
BRS	Basel, Rotterdam, and Stockholm
C40	Cities Climate Leadership Group
CAF	Cancun Adaptation Framework
CARA	Conservation of Agricultural Resources Act
CBA	Community-based adaptation
CBD	Convention on Biological Diversity
CBDR&RC	Common But Differentiated Responsibility and Respective Capabilities
CCAQM	Climate Change and Air Quality Management
CCP	Cities for Climate Protection
CCSPAFF	Climate Change Sector Plan for Agriculture, Forestry and Fisheries
CEEPA	Centre for Environmental Economics and Policy in Africa
CEN-CENELEC	European Committee for Standardization
CFRN	Constitution of the Federal Republic of Nigeria
CO_2	Carbon dioxide
CO_2-eq	Carbon dioxide equivalents
COP	Conferences of the Parties
CRSA	Constitution of the Republic of South Africa
CSA	Climate-Smart Agriculture
CTA	Carbon Tax Act
DALRRD	Department of Agriculture, Land Reform and Rural Development
DEFF	Department of Environment, Forestry and Fisheries
DFFE	Department of Forestry, Fisheries and the Environment
DFID	Department for International Development
DMA	Disaster Management Act

DWS	Department of Water and Sanitation
EAC	East African Commission
ECOWAS	Economic Community for West African States
EDE	Ending Drought Emergency strategies
EIA	Environmental Impact Assessment
EMCA	Environment Management and Coordination Act
EU	European Union
FAO	Food and Agricultural Organization
GCAA	Global Climate Action Agenda
GCF	Green Climate Fund
GDP	Gross Domestic Product
GEF	Global Environmental Facility
GHG	Greenhouse Gases
Gt	Giga Tonnes
HPG	Humanitarian Policy Group
ICCPM	International Code of Conduct on Pesticide Management
ICJ	International Court of Justice
ICLEI	International Council for Local Environmental Initiatives
IDPs	Integrated Development Plans
IMCCC	Inter-Ministerial Committee on Climate Change
INECE	International Network for Environmental Compliance and Enforcement
IPCC	Intergovernmental Panel on Climate Change
IUCN	International Union for Conservation of Nature
KALRA	Kenyan Agriculture and Livestock Research Act
KJWA	Koronivia Joint Work on Agriculture
LDCs	Least Developed Countries
LDCF	Least Developed Countries Fund
LULUCF	Land Use, Land Use Change, and Forestry
MDGs	Millennium Development Goals
MEA	Multilateral Environmental Agreement
MtCO2e	Million Tons of Carbon Dioxide Equivalent
NAIC	Nigerian Agricultural Insurance Corporation
NAPs	National Adaptation Plans
NAP-CDLDD	National Action Programme for South Africa to Combat Desertification, Land Degradation and the effects of Drought
NARF	National Agricultural Resilience Framework
NAS	National Adaptation Strategy
NASPA-CCN	National Adaptation Strategy and Plan of Action for Climate Change in Nigeria
NCC	National Council on Climate Change
NCCAP	National Climate Change Action Plan
NCCAS	National Climate Change Adaptation Strategy
NCCC	National Climate Change Council

NCCPRS	Nigeria Climate Change Policy Response and Strategy
NCCRP	National Climate Change Response White Paper
NCCRS	National Climate Change Response Strategy
NCSASF	National Climate-Smart Agriculture Strategic Framework
NDCs	Nationally Determined Contributions
NDMA	National Drought Management Authority Act
NDP	National Development Plan
NEAP	National Environment Action Plan
NEEDS	National Environmental, Economic and Development Study
NEMA	National Environment Management Authority
NER	National Environmental Regulations
NESEFC	National Environmental Soil Erosion and Flood Control
NESREA	National Environmental Standards and Regulations Enforcement Agency
NESREA Act	National Environmental Standards and Regulations Enforcement Agency Act
NGOs	Non-Governmental Organisations
NPEC	National Policy on Environmental Conservation
NPEFC	National Policy on Erosion and Flood Control
NWA	National Water Act
NWRP	National Water Resources Policy
NWRS	National Water Resources Strategy
OECD	Organization for Economic Co-operation and Development
OPEC	Organization of Petroleum Exporting Countries
PCBs	Polychlorinated Biphenyls
PCC	Presidential Climate Commission
PROACT	Pro-Resilience Action
PTD	Public Trust Doctrine
RACNNR	Revised African Convention on Conservation of Nature and Natural Resources
RBDA	River Basins Development Authority Act
RECs	Regional Economic Communities
REN 21	Renewable Energy Policy Network for the 21st Century
SADC	South African Development Commission
SAIIA	South African Institute of International Affairs
SAN	Sustainable Agriculture Network
SCCF	Special Climate Change Fund
SDGs	Sustainable Development Goals
SEA	Strategic Environment Assessment
SIDs	Small Islands Developing States
SPULMA	Spatial Planning and Land Use Management
Stell LR	Stellenbosch Law Review
TMNs	Transnational Municipal Networks
TNCs	Transnational Corporations

UNCCD	United Nations Convention to Combat Desertification
UNCED	United Nations Conference on Environment and Development
UNCTAD	United Nations Conference on Trade and Development
UNDP	United Nations Development Program
UNECE	United Nations Economic Commission for Europe
UNEP	United Nations Environment Program
UNFCCC	United Nations Framework Convention on Climate Change
UNO	United Nations Organization
USD	United States dollars
VGGT	Voluntary Guidelines for Responsible Governance of Tenure of Land, Fisheries and Forestry in the Context of National Food Security
VGSSM	Voluntary Guidelines for Sustainable Soil Management
WIM	Warsaw International Mechanism for Loss and Damage
WRMRP	Water Resources Management Reform Program
WSA	Water Service Act
WSSD	World Summit on Sustainable Development
WTO	World Trade Organization
WWF	World Wide Fund for Nature

Statutes

Azerbaijan

Agrarian Insurance Law 1617-VQ of the Republic of Azerbaijan of 2019

Germany

Berlin Rules on Water Resources 71 of 2004

Kyrgyz Republic

Kyrgyz Law on Soil Fertility and Protection of Agricultural Lands 165 of 2012

Kenya

Agriculture and Food Authority Act 13 of 2013
Climate Change Act 11 of 2016
Constitution of Kenya 2010
Crops Act 16 of 2013
Environment Management and Coordination Act 8 of 1999
Environmental (Impact Assessment and Audit) Regulations, Legal Notice 101 of 2003
Government Lands Act 280 of 2010
Insurance Act 487 of 1984
Kenya Agriculture and Livestock Research Act 17 of 2013
Land Act 6 of 2012
Land Consolidation Act 283 of 2012
Land Control Act 302 of 2012
Land Registration Act 3 of 2012
Land Titles Act 282 of 2010
National Drought Management Authority Act 4 of 2016
Registration of Titles Law of 6 of 2010
Riverbanks, Lake Shores and Sea Shore Management Regulations 387 of 2009

Water Act 43 of 2016
Water Institute Act 372A of 2001
Wetlands, Riverbanks, Lake Shores and Sea Shore Management Regulations 19 of 2009

Nigeria

Climate Change Act of 2021
Constitution of the Federal Republic of Nigeria 1999 (as amended)
Land Use Act of 1978
National Environmental (Soil Erosion and Flood Control) Regulations 39 of Lagos 2011
National Environmental (Soil Erosion and Flood Control) Regulations 39 of 2011
National Environmental Standards and Regulations Enforcement Agency (Establishment) Act 25 of 2007
Nigerian Agricultural Insurance Corporation Act N89 of 1993
River Basins Development Authority Act R9 of 2004
Water Resources Act 101 of 1993

Republic of Korea

Agricultural and Fishery Disaster Insurance Act 9477 of 2009

South Africa

Carbon Tax Act 15 of 2019
Climate Change Bill 580 of 2018
Constitution of the Republic of South Africa 1996
Conservation of Agricultural Resources Act 43 of 1983
Draft Conservation of Agricultural Resources Amendment Bill 673 of 2013
Disaster Management Act 57 of 2002
Local Government: Municipal Systems Act 32 of 2000
National Water Act 36 of 1998
National Environmental Management Act 107 of 1998
Spatial Planning and Land Use Management Act 16 of 2013
Water Services Act 108 of 1997

Uganda

Ugandan Cattle Grazing Act Cap 42 of 1945

United Kingdom

Climate Change Act c27 of 2008

International and regional instruments

Directive 2000/60/EC establishing a framework for the community action in the field of water policy
Basel Convention on the Control of Transboundary Movement of Hazardous Wastes and their Disposal of 1989
Convention on Biological Diversity of 1992
International Code of Conduct for the Effective Use and Management of Fertilizers of 2019
International Code of Conduct on Pesticide Management of 2013
Rotterdam Convention on the Prior Informed Consent Procedure for Certain Hazardous Chemicals and Pesticides in International Trade of 1998
Stockholm Convention on Persistent Organic Pollutants of 2001
United Nations Economic Commission for Europe Convention on the Protection and Use of Transboundary Watercourses and International Lakes of 1992
Revised African Convention on Conservation of Nature and Natural Resources of 2003
SADC Protocol on Shared Water System of 1995
Water Framework Directive 2000/60/EC of the European Union
Voluntary Guidelines for Responsible Governance of Tenure of Land, Fisheries and Forestry in the Context of National Food Security of 2012
Voluntary Guidelines for Sustainable Soil Management of 2017
World Soil Charter of 2015
United Nations Framework Convention on Climate Change of 1992
Paris Agreement of 2015

1 Introduction

1.1 Background

Climate change is mostly caused by the indiscriminate emission of greenhouse gases (GHGs) especially carbon dioxide (CO_2), methane (CH_4), and nitrous oxide (N_2O), emanating from certain human activities such as the exploitation of mineral resources and agricultural practices, especially crops and livestock production.[1] In the same regard, the impacts of climate change, such as heat waves, rising sea levels, flooding, droughts, and desertification, are partly responsible for the loss of arable land and the reduction of agricultural yields (crop failure and loss of livestock). Agriculture happens to be the pillar of the economy in most African countries especially Kenya, Nigeria, and South Africa, serving as the source of gross domestic product (GDP) and employment. According to Mendelsohn et al.,[2] agriculture contributed about 21% of the GDP of African countries in 2000, and due to climate change, the agriculture-based GDP of African countries will likely reduce by up to 4% by 2100. Essentially, a loss of, or a reduction in, agricultural yields could exacerbate other existing socio-economic problems and related humanitarian disasters such as famine and displacement of people in African countries.[3]

The impacts of climate change on agriculture are expected to multiply, thus exceeding the natural abilities of human and natural systems to cope.[4] Climate adaptation measures that seek to address the impacts of climate change are necessary in order to safeguard agriculture against the impacts of climate change and to protect the environment from agriculture-induced climate change. Adaptation can involve measures and strategies aimed at building adaptive capacity, increasing resilience, and reducing the vulnerability of human and natural systems to climate change. Adaptation measures and strategies in the context of agriculture in African and developing countries cannot be effective without a well-developed governance infrastructure that also includes effective legal and regulatory standards.

In light of the above considerations, this book seeks to investigate how agriculture-focused adaptation laws in Kenya, Nigeria, and South Africa could be developed. In order to achieve this, the book will:

DOI: 10.4324/9781003536437-1

- Briefly evaluate the theoretical framework of adaptation and agriculture.
- Explore how adaptation in the context of agriculture intersects with the law.
- Investigate, for the purpose of context, the international and African regional legal framework and briefly explore other soft-law instruments related to adaptation and agriculture.
- Examine the adaptation laws and policies of Kenya, Nigeria, and South Africa in order to determine the extent to which they provide for adaptation in the context of agriculture.
- Make recommendations to fill the gaps in these laws, where they exist, so that they could better cater for adaptation in the context of agriculture.

In light of the above considerations, this book seeks to answer the following central research question that will act as a guide to the overall study: to what extent do Kenyan, Nigerian, and South African laws provide for climate change adaptation in the context of the agricultural sector, and where there are gaps, how could these be addressed?[5]

To answer this question, this book adopts a doctrinal legal research that employs a descriptive and analytical study of discourse, where data are gathered through physical and electronic libraries. The research methodology and design adopted for this book are appropriate because literature and academic scholarship on agricultural adaptation laws in Africa (and in specific countries) are lacking. Most of the available literature deals with risks and vulnerabilities associated with climate change adaptation, or policy and legal barriers to climate change adaptation.[6]

Hence, it is relevant to outline how climate change and agriculture affect each other, as the basis for subsequently understanding how the law promotes adaptation in a broad context and in the context of agriculture.

1.2 Agriculture and climate change

Agriculture contributes to climate change, while climate change affects agriculture. For example, some methods associated with the production of crops, especially the application of industrial fertilisers such as ammonia (NH_3) and (NH_4), as well as non-industrial fertilisers (such as manure derived from fermentative digestion by cows, horses, and sheep), to boost soil nutrients qualify as sources of GHG emission, leading to climate change.[7] Likewise, unconventional tillage farming methods[8] and other farming activities such as the clearing of forests to provide farmland[9] are also potential causes of GHG emission and climate change.[10] Other than the clearing of forests and unconventional tillage practices, the production of high nitrogen-based crops, such as maize and sorghum, and the manufacturing of manures, especially from rice swamps,[11] contribute to the emission of nitrous oxide gases (also a form of GHG) and climate change.[12] Similarly, the rearing of livestock contributes to the emission of methane (a form of GHG), which leads to climate change.[13]

Cows, for example, emit methane through a process called enteric fermentation.[14] According to Glatzle,[15] livestock-rearing accounts for 18% of global GHG emissions, and this is more than the GHG emissions from the transportation sector.[16]

Agriculture is vulnerable to climate change because the sector is heavily dependent on biophysical and ecological (climate-sensitive) resources, particularly conducive temperature, water, soil moisture, and fertile land.[17] Changes in climatic conditions, such as higher temperatures, for instance, could affect the developmental stages of crops, as it takes longer for crops to mature and in some cases lead to lower crop yields.[18] Likewise, water in sufficient and correct quantities at the correct time is required to boost crops and livestock productivity.[19] The types of water required by crops and livestock include rainwater,[20] water flowing in streams and rivers (blue water),[21] and water that already exists in the soil (ground water or soil moisture).[22] Limited supply of all forms of water especially in climate hotspot areas of Africa[23] means that only certain types of crops such as millet, sorghum, groundnuts, cotton, and maize can grow successfully.[24] In addition to crop production, livestock production is also vulnerable to the impacts of climate change.[25] For example, the ability of animals to adjust their metabolic rate is affected by temperature increase, and this deficiency is a factor that is responsible for the reduction of productivity in meat, eggs, and milk.[26]

Although climate change is a global problem, its consequences are mostly felt at the local or country level.[27]

1.3 A brief mapping of agriculture and climate change at country levels

1.3.1 Kenya

Agriculture plays a prominent role in the socio-economic development of Kenya. According to reliable statistics, more than 80% of the rural population in Kenya are supported by the agricultural sector, while about 75% of total employment in urban areas is provided by agriculture.[28] More specifically, Vision 2030 of the Second Medium Term Plan (Kenyan blueprint for national development plan) has described agriculture "as central to the economic pillar for guiding Kenya to middle income status".[29] In 2020, for instance, agriculture contributed up to 30% of the GDP of Kenya, which was about 2,478.3 billion Kenyan Shillings (equivalent to 21.2 million USD).[30] In addition, the agricultural sector in Kenya supplies over 65% of raw materials for agro-industries domestically and for export.[31] Agricultural products in Kenya include crops such as coffee, maize, cotton, and cashew nuts in addition to livestock, especially cattle, camel, donkeys, sheep, and goats.[32] These products mostly rely on conducive temperature and rain-fed water for sustenance, which has been confirmed to be highly vulnerable to climate change and variability, which for instance, manifest as an increase or fluctuating rainfall pattern.[33] It is on record that "Kenya has experienced increased temperature of 1.0°C in 2020 and it is expected to further increase to 2.3°C by 2050".[34] Consequently, a decline in crop productivity, especially post-harvest losses, and low yields in Kenyan livestock systems should be anticipated.[35] This could further lead to a "loss of income from the agriculture sector, poverty, food shortage and malnutrition".[36]

Even though Kenya is not among the major contributors of GHG emission in Africa, the GHG emission profile of Kenya is rising at an alarming rate. For example, in 2010, it was 73 million tons of carbon dioxide equivalent ($MtCO_2e$), and this figure is expected to rise to almost 143 $MtCO_2e$ by 2030.[37] About one-third of these emissions are derived from agriculture, particularly through methane emissions from livestock and practices related to land use.[38] Against this background, adaptation measures are required in Kenya in order to address the impacts of climate change.

1.3.2 Nigeria

Nigeria is a developing country endowed with mineral and natural resources,[39] including petroleum and agricultural products, as its main source of economic growth and development.[40] In 2021, for instance, the petroleum sector contributed 70% of budget revenues, and 90% of total export value in Nigeria,[41] while agriculture accounted for about 25% of the country's GDP.[42] In addition to providing food for humans and feeds for animals in the country, the agricultural sector is serving as a means of livelihood for about 70% of Nigerians, especially those living in rural areas.[43]

However, the exploitation of petroleum products and agricultural activities are the underlying reasons for the emission of GHGs in Nigeria.[44] About 60% of Nigeria's GHG emissions in 2020, for instance, accrued from the petroleum (fossil fuel) sector, while the rest of the GHG emissions emanated from the agricultural sector, through inefficient methods such as shifting cultivation[45] and indiscriminate bush burning associated with the farming of crops (mainly maize, rice, and millet) and the overconcentration or overgrazing livestock in a particular location as a result of uneven distribution and inequitable allocation of land.[46] This alarming rate of GHG emissions is partly responsible for temperature increase and a decrease in rainfall in the country, among other problems.

Currently, there is an already unevenly distributed rainfall pattern in Nigeria (where the northern part of the country has 7.5 cm of rainfall annually and the southern part of the country has 15 cm of rainfall annually), and these could be exacerbated by temperature increases.[47] By implication, less rainfall leads to reduced river flow, excessive droughts, and desertification in Northern Nigeria,[48] and more rainfall could mean more flooding, storm surges, and the destruction of farms located in coastal areas of Southern Nigeria.[49]

By extension, these fluctuating trends in temperature and precipitation could have dire consequences on agricultural practices in Nigeria, especially crops and livestock production.[50] It has been predicted that climate change could lead to a decline in agricultural productivity between 10 and 25% in Nigeria by 2080.[51] This loss in agricultural productivity could also lead to a decline in Nigeria's GDP which, in monetary terms, is estimated between N15 trillion (USD 100 billion) and N69 trillion (USD 460 billion).[52] Hence, adaptation is required in order to safeguard Nigeria's natural resources and economic development against the impacts of climate change.

1.3.3 South Africa

South Africa is situated on a latitudinal range of 20–35°S and is composed of a variety of agro-ecological zones consisting of sub-humid and wet areas in the eastern coastal region, as well as desert and semi-desert areas in the north-western region.[53] As a result of its geography and location, South Africa is already highly susceptible to a low annual precipitation rate, temperature rise, and sea level rise.[54] Climate change owing to GHG-intensive activities in the country, especially coal mining (which satisfies the bulk of South Africa's energy needs[55]), and to some extent agricultural practices, further aggravates low annual precipitation rate, temperature rise, and sea-level rise in South Africa.[56]

Climate change in South Africa usually manifests through a reduction in rainfall patterns, higher temperatures, or water scarcity and droughts.[57] As a result of climate change, South Africa's average annual rainfall is just 450 mm compared to the global average annual rainfall rate of 860 mm.[58] Furthermore, only 10% of South Africa's land mass receives 450 mm rainfall, with more than half of that being utilised by the agricultural sector.[59] With less rainfall and less water availability,[60] it means that more irrigation services are required to support crops and livestock production.[61] By implication, irrigation services require more carbon uptake, and this could exacerbate climate change.[62] As already noted above, climate change could only mean more water shortages and droughts – with dire consequences for agriculture, which serves as a major source of employment, food, and raw materials for industries in South Africa.[63] The contribution of agriculture to South Africa's economy could be attributed to the dualistic nature of the agricultural system that exists in the country. A highly developed commercial-based farming method consisting of well-diversified crops and livestock co-exists with a communal base and small-scale subsistence farming (with an average farm size of 0.5–2 ha).[64] Combined, both forms of agricultural systems satisfy South Africa's domestic and export needs.[65] For example, the two forms of agricultural system contributed about 2.53% of South Africa's GDP in 2020.[66] Despite the role of agriculture in South Africa's economic growth and development, there are about 14 million people in the country who are considered to be vulnerable to food insecurity, unemployment, and poverty.[67] These problems are further compounded by the impacts of climate change on small-scale and poor farmers in South Africa, whose adaptive capacity is weak.[68]

Consequently, climate change adaptation is urgent and necessary in order to safeguard ecosystems and to enhance the socio-economic development of poor and vulnerable people in South Africa, just as in Kenya and Nigeria.[69] Hence the question: what is adaptation, and in what ways could it be facilitated in the context of agriculture?

1.4 Adaptation in the context of climate change

According to the Oxford English Dictionary, adaptation includes "action or process of fitting or suiting one thing to another, or the process of changing something to fit into a new situation".[70] In technical terms, adaptation has been construed to mean any form of "adjustment that is made in order to enhance the viability of social and economic activities and to reduce their vulnerability to climate change".[71] This form of adjustment could consist of some multi-sectoral efforts and actions aimed at curbing, coping with, and preparing for the current and anticipated impacts of climate change on human systems[72] and natural systems.[73] These multi-sectoral efforts and actions are illustrated by the construction of ocean walls or coastal defences as part of a reaction to, or in anticipation of manifestations, of climate change impacts such as sea-level rise and flooding.[74] Other examples of multi-sectoral efforts and actions intended to adjust the impact of climate change in the specific, but narrowed, context of agriculture include introducing drought, flood, and saline-tolerant crops; and improving fisheries, livestock breeding, and farming techniques.[75] These specific multi-sectoral measures and actions in support of climate change adaptation are, however, seen through the lens of politics. For example, the planning and implementation of flood control mechanisms such as hydro sacks in a community affected by climate change-induced flooding is shaped by the dynamics of politics, just as the distribution of relief materials and resources to the victims of other epidemics and disasters. Likewise, the governance of adaptation measures and actions especially the negotiation and signing of Multilateral Environmental Agreements (MEAs) such as the Paris Agreement, the enactment of national climate laws (for instance, *the Nigerian Climate Change Act* and *the South African Climate Change Act*, as subsequently discussed in Chapters 4 and 5 of this book), and the formation of national policies on adaptation are motivated by political considerations.[76] This opinion is shared by Eriksen and others who consider adaptation as "a socio-political process that mediates how individuals and communities deal with multiple and concurrent environmental and social changes". [77] They contended that:

> what counts as 'adaptive' is always political and contested. What is seen as positive adaptation to one group of people may be seen as mal-adaptation to another, and political processes determine which view is considered more important at different scales and to different constituencies.[78]

Pelling[79] also observed that climate change adaptation actions, just like other kinds of measures which societies partake in responding to epidemics such as earthquakes and famine, should be considered as a social process that is often shaped by multiple factors occurring simultaneously.[80] For the purpose of illustration, climate change impact reports such as those prepared by the Intergovernmental Panel on Climate Change (IPCC), for instance, are considered to be heavily politically motivated, especially when those reports are expected to be relied upon as the basis for justifying certain types of adaptation measures in areas affected by the impacts of climate change.

According to the IPCC, adaptation is, or should at least be, a continuous and well-coordinated process, instead of a one-time action, given the peculiarity of climate change as a continuously and varying phenomenon capable of increasing in terms of frequency and intensity.[81] In the same manner, the United Nations Convention Framework on Climate Change (UNFCCC) regards climate change adaptation as "actions taken to help communities and ecosystems to cope with changing climate condition".[82] In this regard, it is important to note that adaptation practices vary according to location, context, and timing.[83]

1.4.1 Forms of adaptations

Depending on its manifestations, timings, goal, and motive of implementation, adaptation could either be anticipatory/proactive, or reactive, or autonomous and planned.[84]

Anticipatory adaptation takes place before the actual impacts of climate change occur, for instance, building levees or storm water drainage structures to prevent flooding.[85] Reactive adaptation is the type of adaptation that takes place after the initial impacts of climate change have occurred, for example, when levees, sea walls, or storm water drainage structures are constructed to prevent further flooding.[86] As noted by Parry,[87] autonomous adaptation is a type of adaptation that:

> [d]oes not constitute a conscious response to climatic stimuli, but is triggered by ecological changes in natural systems and by market or welfare changes in human systems.

This is, for instance, illustrated by improved forecasting for farming, extreme events, and disaster management; improved water management for drinking and agriculture; and improved food security through crop diversification.[88] Therefore, autonomous adaptation could be well suited for the agricultural sector as it allows farmers to "adjust their behaviour in accordance with changing climate conditions".[89] Planned adaptation according to the IPCC is:

> [a] type of adaptation that occurs as the result of a deliberate policy decision, based on awareness that conditions have changed or are about to change and that action is required to return to, maintain, or achieve a desired state.[90]

Planned adaptation is applied as a strategy to cope with anticipated impact of climate change in the future.[91] In the context of agriculture, planned adaptation is often used in building adaptive capacity against climate change-related risks through, for instance, the development and implementation of planned initiatives such as the use of new drought-resistant crop varieties, the use of technologies to boost the efficiency of irrigation systems, or adopting agroforestry approaches.[92]

In this book and with reference to the agricultural sector, except where specifically stated otherwise, climate change adaptation refers to anticipatory and planned adaptation.[93] Anticipatory and planned adaptation is chosen as the focus of this book because adapting to the impending impacts of climate change well ahead of time could be more cost-effective with significant advantages for national development.[94] For example, resorting to anticipatory and planned adaptation could enable governments and organisations to make efficient use of scarce resources by focusing on other sectors, rather than spending on building separate institutions and processes to support reactive adaptation.[95] Also, unlike other types of adaptation, anticipatory and planned adaptation are already accommodated by the international legal order and therefore require further modifications only to suit the specific needs of regions or individual countries that are adversely affected by climate change.[96]

1.4.2 Need for climate change adaptation

Adaptation is required in the short and longer term to cope with the impacts of climate change[97] or to adjust to the adverse impact of climate change on human and natural systems.[98] In order to cope with, or adjust to the impacts of climate change, adaptation strategies must be tailored towards reducing the vulnerability of systems to the impacts of climate change, boosting the resilience of systems against the impacts of climate change, and enhancing the ability of systems to adapt against impending and reoccurring impacts of climate change.[99]

Reducing vulnerability means initiating a process that can reduce the impending risk of climate change impacts on human and natural systems,[100] or measures that can protect ecosystems against the anticipated impact of climate change.[101] For example, coastal wetlands designed as protection against rising sea levels could serve in reducing the vulnerability of coastal communities to the negative impacts of climate change.[102] In discussing adaptation, the term vulnerability takes into consideration the specific nature and circumstances of a system (human or natural) such as geographic location, economic advancement, and institutional and regulatory capacity that makes it prone to the impacts of climate change.[103] In order to reduce the vulnerability of systems to the impacts of climate change, adaptation decisions must focus on measures that are aimed at protecting the environment and improving socio-economic well-being.[104] For example, the management and conservation of forests,[105] erosion control, conservation of water-source areas, and building insulation against heat are adaptation measures that can reduce the vulnerability of the environment against the increased pressure of climate change while improving the socio-economic well-being of the people concerned.[106]

Boosting resilience entails taking measures that are aimed at strengthening "the ability of human and natural systems to absorb shocks or ride out changes".[107] This could also mean "maintaining the essence and integrity of a system or process".[108] Enhancing the adaptive capacity of systems to adapt to the impacts of climate change could mean taking steps to increase the robustness of adaptation infrastructures such as improving societal awareness on the potential risk created by climate change impacts[109] and/or promoting actions that will make human systems learn how to adapt.[110] Resilience explains the ability of vulnerable systems to deal with adverse or beneficial changes arising from the impacts of climate change.[111] For this reason, Klein et al.[112] are of the opinion that "a system is vulnerable because it is not resilient; and it [a system] is not resilient because it is vulnerable".[113] Hence, adaptation measures and action are embarked upon in order to facilitate the development of techniques that can help increase the resilience of human and natural systems against the impacts of climate change. The resilience of human and natural systems can be increased, for instance, "through resource conservation measures or through actions that enable specific populations to recover from loss".[114] These measures would not be possible if affected communities do not have the adaptive capacity to implement them.

Adaptive capacity relates to the interplay of social, economic, and political actions with livelihood systems that influence the ability to cope with climate change and variability.[115] Low or weak adaptive capacity in African and developing countries is attributed to an array of factors, which include deficiency in institutional, technological, and equity factors. In addition, these factors interact with each other to "influence vulnerability and shape outcomes from climate change".[116] Jobbins et al.[117] further observe that the ability of African countries to adapt to climate change is inhibited by weak and underdeveloped legal frameworks and institutions.

Hence, the objectives of an effective adaptation strategy are to reduce the risk of exposure to a stress event (vulnerability), increase the ability of systems to cope with a stress event (resilience),[118] and enhance the ability of systems to adapt to the impacts of climate change (adaptive capacity).[119]

It must, however, be emphasised that adaptation approaches aimed at increasing resilience could be the same globally, but adaptation approaches tailored towards reducing vulnerability and enhancing adaptive capacity vary according to geographic location. Vulnerability and adaptive capacity are characteristics of human and natural systems as such they are dynamic and multi-dimensional, and are influenced by complex interactions among social, economic, and environmental factors. Further, decisions on reducing vulnerability and increasing adaptive capacity are shaped by the operating context within which decisions are made (for example, existing natural resource quality and non-climate stressors, government policy, and programmes), access to effective adaptation options, and the individual capability to take adaptive action.[120] On this basis, the stages of adaptation planning and implementation in the context of climate change are discussed.

1.4.3 Stages of adaptation in the context of climate change

Adaptation measures can be implemented at the individual level, at the community level, as well as at the institutional or governmental level.[121] Examples of an adaptation measure implemented at the individual level include diversification of livelihood means and the storage of essential needs such as food and water by inhabitants of communities affected by droughts or floodings. In particular, the storage of essential items with low perishability can serve as an effective measure against climate change-induced power outage.[122]

Adaptation at the community level is based on communal knowledge-sharing and practices, joint actions by community members who "pool risk and resources" for coping with, and integrating the impacts of climate change on livelihood means.[123] Communal knowledge-sharing and practices are demonstrated by the installation of hazard alerts or early warning systems (observation and monitoring technologies), which can forecast the occurrence of extreme weather events such as storms or sea-level rise due to climate change with a view of proactively "altering the exposure of a system"[124] or adjusting against those events.[125] The construction of a flood defence system such as the placing of hydro sacks and sandbags, building of dikes, and the use of water-extraction machines to reduce the vulnerability of systems against the impacts of climate change could also be added to the list of examples of adaptation strategies that could be implemented at the community level.[126] Likewise, migration could be an example of an adaptation strategy that could be implemented by communities that are affected by climate change-induced drought.[127] In other words, migration is embarked upon voluntarily as a matter of choice and as a way of diversifying income, or out of necessity as a coping strategy and as a means of boosting resilience against natural disasters and risks such as climate change.[128] For example, the heavy flooding recorded along the Zambezi River in central Mozambique in 2001, 2007, and 2008 caused migration.[129] Again, following the droughts in Burkina Faso and Mali in the 1970s, about one million people were reported to have migrated to urban centres and across national borders in search of better livelihoods.[130] In this regard, migration could serve as a complementary strategy to adaptation in reducing vulnerability against the impacts of climate change in Africa.

Adaptation strategies at the institutional or governmental level are response measures that have been mainstreamed into national development policies and are supported by legal, technical, financial, and managerial actions that are implemented as short-term, medium-term, or long-term plans.[131]

Short-term adaptation plans are implemented as immediate responses to prevent existing, immediate, or impending impacts of climate change such as flooding and heat waves, from reoccurring.[132] In the words of Persson and Klein,[133] some adaptation strategies such as flood control and cooler homes qualify as short-term plans because (as urgent measures) their implementation is expected to endure one to ten years.[134]

Medium-term adaptation plans are implemented "after assessing the risks of the impacts of climate change that may occur in the medium term".[135] However, unlike short-term adaptation plans, medium-term adaptation plans are intended to last for a period of 10 to 30 years.[136] Medium-term adaptation plans are proactive in nature, requiring gradual development of other modalities such as institutional arrangements and the management of basic information on the nature of (or anticipated) severity of climate change impact and vulnerability on a specific community or sector of the economy.[137]

Where an adaptation strategy is implemented and is envisioned to last for more than 30 years but does not exceed 100 years, it is described as a long-term adaptation plan.[138] Many factors such as societal changes and policy considerations determine the need for a long-term adaptation plan.[139] Long-term adaptation plans are expected to result in longer-lasting behavioural changes.[140] Just like the medium-term adaptation plan, a long-term plan is implemented after assessing the risks of the impacts of climate change on human and natural systems.[141]

It is noteworthy that the success of either short-term, medium-term, or long-term adaptation plans, especially when initiated at the institutional or governmental level, is dependent on how they are integrated with other developmental plans.[142] In the context of adaptation governance, the process of integration is technically described as mainstreaming.[143]

1.4.4 Mainstreaming in the context of climate change adaptation

Although mainstreaming is used interchangeably with terms such as climate policy integration, embedding, aligning, etc., the differences in name do not altogether alter the core objectives of incorporating climate change dimensions into development planning across all sectors of the economy.[144] According to the IPCC, climate change considerations (especially adaptation actions) should be mainstreamed into:

> [e]xisting regulatory programs and structures for water resources, agricultural productivity, forest productivity, and coastal management, at all levels of government, from the international level to local level.[145]

Three dimensions to mainstreaming climate change concerns with development have been identified, and these include horizontal, vertical, and international dimensions.[146] Mainstreaming takes a horizontal dimension when cross-sectoral climate change measures are done by different policy sectors of the economy or different departments of the government.[147] Where mainstreaming is applied by different levels of government, it is often described as being vertical.[148] The integration of climate change adaptation policies, for instance, by the Ministry for Agriculture into the Ministry for Water Resources at the national level to local government level, is an illustration of vertical mainstreaming.[149] Mainstreaming of climate change concerns with development goals takes an international dimension when done between transnational corporations or multi-lateral bodies.[150] These dimensions to mainstreaming make it less problematic to decide where and when to situate mainstreaming activities.

There are several benefits associated with mainstreaming. Through mainstreaming, adaptation measures are properly aligned with other development programmes, and this by implication will minimise conflicting development goals.[151] Also, mainstreaming could lead to the proper consolidation of financial resources and the minimisation of loss, as adaptation and development actions are jointly implemented.[152] Likewise, mainstreaming as a process will foster co-operation among government ministries, departments, and agencies, and it could also facilitate the integration of long-term adaptation plans with development plans.[153] In spite of the importance and effectiveness of mainstreaming climate change adaptation, there are certain challenges that inhibit the success of adaptation measures in an African and developing-country context.

1.4.5 Challenges for adaptation in the context of climate change

In the words of Hettiarachchi and Ardakanian,[154] there could be times when it may not be possible to physically adapt to the impacts of climate change, and even if it is physically possible to adapt, there may be certain challenges. These challenges to adaptation strategies can emanate from economic, social, and ecological constraints, which can vary among economic sectors, communities, regions, and countries.[155] Economic constraints, especially poverty, may hinder the implementation of simple adaptation strategies such as the building of cooler homes or diversification to alternative livelihood means (migration) in communities affected by drought.[156] Social constraints to adaptation are those barriers that are "often regarded as behavioural, cultural, or policy specific obstacles (such as governance challenges) which may not be coped with".[157] For example, relocating people from areas affected by climate-induced flood may be inhibited by cultural obstacles because the inhabitants of the area may have sentiments.[158] Social constraints to adaptation are usually caused by low sensitisation of climate change risk and its associated impacts.[159]

According to Mandryk et al.,[160] economic constraints and social constraints to adaptation, particularly governance challenges (institutional impediments), are the major causes of unsuccessful adaptation measures in most African and developing countries. Baker et al.[161] believe that institutional impediments are a result of fragmentation, which could lead to weak coordination of adaptation measures. For example, the exercise of power and authority may be affected by inter-agency rivalry, where institutions tasked with disaster risk reduction co-exist with other institutions established for reducing climate risk and promoting adaptation.[162] In this instance, there may be an array of actors grappling to perform an almost similar role.[163]

Ecological constraints are physical limitations to adaptation, such as those "largely associated with the natural environment, ranging from ecosystem thresholds to geographical and geological limitations"[164] that might not be solved with ease. Ecological constraints could, for instance, manifest as "the inability to prevent intolerable risks to objectives and/or needs of water and land managements".[165] These challenges affecting adaptation in a broad context also affect adaptation in the context of agriculture as explained below.

1.5 Adaptation in the context of agriculture

The IPCC observes that without adaptation, major crops in tropical and temperate regions will be affected by the prevailing local temperature increases of 2°C or more resulting from climate change.[166] With adaptation, farmers can achieve food security, stable income, and guaranteed means of livelihood in the wake of fluctuating socio-economic and climatic conditions, such as droughts and floods.[167]

Adaptation in the context of agriculture is achieved through synergies of actions and measures.[168] These synergies of actions could include, for instance, the adoption of hybrid and diversified crop varieties and livestock breeds, cover cropping, conservation tillage practices, and water-retention techniques in areas prone to water scarcity.[169] According to Blesh and Wolf[170] and Li et al.[171] adaptation in the context of agriculture is to a considerable extent influenced by the farmers' perception of climate change risk and vulnerability, and the farmers' decisions and adaptive capacity.[172] In other words, adaptation in the context of agriculture is motivated by the level of planning and implementation, as well as the variety of actors concerned.

1.5.1 Forms of adaptation in the context of African agriculture

Adaptation in the context of agriculture is implemented based on the vulnerability of the products involved (for instance, crops and livestock) and the resources that support them (such as water and soil) to climate change impacts. In addition, adaptation in agriculture is implemented based on the location involved, while also considering associated economic, political, and institutional circumstances.[173] Also, the type of adaptation that is implemented in the context of agriculture depends on the stakeholders involved.[174] These forms of agricultural adaptation include on-farm adaptation strategy, technological solutions,[175] insurance, and financial management techniques.[176]

1.5.1.1 On-farm adaptation strategies

On-farm adaptation strategy is any farm management practice that has been adopted in order to boost agricultural yield against the impacts of climate change.[177] On-farm adaptation strategies consist of altering crop planting dates to avoid extreme climatic events, the diversification of crop types that have an early maturity period, or crop varieties that are tolerant to droughts, floodings, and pests.[178] In the same way, adopting improved livestock varieties to accommodate climate variability and associated economic risk is an example of on-farm adaptation strategies.[179] Equally, changing location for the production of crops and livestock, changing the topography of land,[180] and using other practices such as alternative fallow[181] in order to improve soil moisture content that has been affected by climate change impact are also forms of on-farm adaptation strategy. According to Klassen and Gilpen,[182] the practice of irrigation in a bid to improve soil moisture content that has depreciated as a result of drought is also an on-farm adaptation strategy.[183]

An on-farm adaptation strategy also includes making changes in the timings of farm operations to manage the duration of growing seasons for crops in the event of droughts, decreased soil moisture and extreme temperature.[184] Changes in the timing of farming operations can be done by scheduling farming activities such as planting, spraying,[185] and harvesting of crops.[186] It has been observed that changing the timings of farming operations as an on-farm adaptation strategy is associated with numerous benefits including boosting crop yields, averting heat stresses and avoiding soil moisture deficiencies.[187]

Other types of on-farm adaptation strategy are conservation agricultural practices, and these include the use of cover crops and conservation tillage.[188] Cover crops are planted either before or after harvest (or residues from harvests) in a bid to protect the soil from exposure.[189] Examples of cover crops include alley crops, while that of residues from harvests are corn stalks or wheat stubble.[190] Cover crops have been accepted by farmers for their soil moisture-retention capabilities and the reduction of wind, water, and soil erosion.[191] Conservation tillage involves placing seeds in narrow holes instead of placing seeds on seedbeds.[192] Conant[193] argues that farm system resilience and the adaptive capacity of farmers against droughts are enhanced through conservation tillage.[194] In addition to their soil moisture-retention capability, cover crops and conservation tillage have been found to be useful in controlling soil disturbances particularly erosion (generated by water run-off) and water imbalance, as deficiencies emanating from the impacts of climate change.[195] In particular, conservation tillage has been noted for reducing carbon losses that normally happen because of ploughing.[196]

1.5.1.2 Technological solutions

Technological solutions are physical measures that are necessary for facilitating adaptation in the agricultural sector. For example, the installation of environmental control equipment and facilities (as physical measures) is required to avoid damage caused by heat on livestock barns.[197] Technological solutions could also involve other physical measures such as the desalinisation of seawater, delivery of freshwater, erection of blue-green algae control fences, and the installation or augmentation of power generation equipment for water-purification plants.[198] These solutions are essential in adapting to the impacts of climate change on agriculture.

Other technological solutions that can be deployed to address the impacts of climate change on agriculture include the construction of integrated drainages, reservoirs, and recharge areas in a bid to make up for the shortfall in blue water.[199] Cloning and the use of genetic engineering technology to develop new hybrid crop varieties and animal breeds that can withstand the impacts of climate change – especially unsuitable temperature, droughts or lack of soil moisture – are also technological solutions that could facilitate adaptation in the context of agriculture.[200]

Similarly, technology can be used to develop information systems that can forecast weather events and predict the implications of changing climatic conditions on planting and harvesting of crops, and this is also an adaptation strategy in the context of agriculture.[201] By using information services such as early warning systems, farmers can make proper long-term decisions on the timing of their farm operations (such as type and choice of crop varieties to plant, when to use irrigation, and when to seek for financial assistance or insurance), taking into consideration future climatic changes and variability.[202] For example, through information systems, the potential risks associated with El Niño-Southern Oscillation on agricultural products were predicted and accordingly assessed within a reasonable time.[203] The use of information systems as a form of agricultural adaptation also has the potential of addressing anticipated flooding, soil-moisture deficiencies, and droughts resulting from low precipitation.[204]

1.5.1.3 Insurance and financial assistance

Insurance is described as "a contractual transaction that guarantees financial protection against potentially large loss in return for a premium".[205] Insurance and financial management techniques are also forms of agricultural adaptation strategies that can help farmers address problems of low crop yield or losses due to the impact of climate change.[206] In other words, insurance can be used as a form of adaptation strategy to enhance the adaptive capacity of agricultural practices, boost resilience, or overcome the vulnerability of the impacts of climate change on agricultural products.[207]

Depending on the magnitude of risk posed to agricultural practices by the impacts of climate change, farmers opt for either the indemnity-based insurance or the index-based insurance as a form of adaptation to crop and livestock production.[208] For crop production, an indemnity-based insurance or an index-based insurance provides cover against physical loss or damage resulting from disasters such as climate change-induced pests, diseases, fire, windstorm, excessive rainfall, and droughts.[209] In terms of livestock production, an indemnity-based insurance or an index-based insurance serves as relief against the vagaries of climate change impacts that manifest as droughts, pests, diseases, and especially invasive alien species.[210]

In addition to insurance for crops and livestock, other financial support incentives such as subsidies, grants, research and development aid, and infrastructure exist to support farmers in adapting to the impacts of climate change at the individual or household level, community level or national level.[211] For example, the Presidential Employment Stimulus Initiative (PESI) for subsistence producers operated by the Department of Agriculture, Land Reform and Rural Development exists in South Africa.[212] In Nigeria, the Central Bank offers grants and subsidies to farmers through an intervention scheme known as the Anchor Borrowers' Programme.[213] There are financial support incentives also for farmers in Kenya that are similar to those of South Africa and Nigeria. For example, the National Agricultural Value Chain Development Project, which is a World Bank-assisted programme, supports small-scale Kenyan farmers involved in dairy, poultry, fruits (banana, mango, and avocado), vegetables (tomato and potato), coffee, cotton, and cashew nut farming, by providing them with:

> [n]ew opportunities for maximizing finance and private sector investments in the nine value chains through a range of value chain investments and other enabling initiatives such as improved subsidy targeting, through e-vouchers and operationalizing warehouse receipt financing.[214]

All the aforementioned forms of adaptation in the context of agriculture are collectively described as climate-smart agriculture (CSA). In other words, any strategy aimed at boosting agricultural yield, food security, and the sequestration of GHG emissions is attained through CSA.

1.5.1.4 Climate-smart agriculture

The concept of CSA emerged following the outcome of the First Global Conference on Agriculture, Food Security and Climate Change held at The Hague, in 2010.[215] The Food and Agricultural Organisation (FAO) summarised CSA as a means of:

> [m]aximizing global agricultural productivity, increasing the resilience of agricultural systems in the face of climate change, and concerns to minimize greenhouse gas emissions from agriculture.[216]

An agricultural system is considered to be climate-smart when it adopts efficient on-farm adaptation methods (such as conservation agriculture and seed varieties that are drought tolerant and early maturing) in combination with the most suited technologies for irrigation and tillage that are not carbon intensive.[217] Additionally, CSA provides for innovative means of consolidating agriculture, development, and climate change under a common sustainable development agenda.[218] The objectives of CSA are to:

> [i]ncrease agricultural productivity and incomes; adapt and build resilience to climate change; and minimize greenhouse gas emissions where possible.[219]

Stakeholders at the local, national, and international levels rely on CSA as a suitable strategy for aligning agricultural practices with climate change and development.[220] Currently, CSA is being resorted to by African and developing countries that rely on agriculture as a source of income but are facing the challenges of climate change. For example, Kenya, Nigeria, and South Africa adopt a CSA strategy in addressing the impacts of climate change on agriculture and in curtailing agriculture-induced GHG emissions.[221] In strict terms, agricultural adaptation measures including CSA strategies are guided by a well-coordinated planning process that is implemented at different levels and stages and by different actors.

1.5.2 Levels and plans for adaptation in the context of agriculture

The planning and implementation of adaptation in the context of agriculture is done either at the individual level, the communal level, or at the national level. At the individual (or household) level, adaptation in the context of agriculture is implemented through independent agricultural activities developed by small-scale farmers based on "local climatic, economic and social conditions".[222] Some examples of individual-level agricultural adaptation are the use of simple irrigation methods such as sprinkler method,[223] use of alternate crop varieties, adequate storage for crops, shelter for animals, and diversification of livestock and breeding methods in adapting to climatic change and climate variability.[224]

Migration, which has been previously explained as a form of adaptation,[225] could also be implemented at the individual level in the context of agriculture,[226] meaning that the impacts of climate change on wild and remote areas where livestock-rearing communities exist could be addressed through migration.[227] For pastoralists, migration is not only a means to manage their livestock effectively by accessing scarce natural resources (grazing land and water sources) in response to the impacts of climate change and variability[228] but also a means of achieving positive livelihood outcomes.[229] As a matter of necessity, as well as an important resilience strategy, pastoralists can migrate to other areas that are less vulnerable to the adverse impact of climate change in order to safeguard their lifestyles, means of livelihoods, and assets.[230] In Nigeria, for example, the impacts of climate change force pastoralists to migrate from the northern part of the country to the southern part of the country in search of pasture and water.[231] Likewise, in some parts of East Africa, such as Kenya, extreme weather events, flooding or decrease in water availability, lack of fodder for livestock, and diminished soil quality[232] as manifestations of climate change are responsible for migration.[233] Precisely, about 15 million people whose livelihoods depend on livestock had to migrate to other areas not affected by following the occurrence of climate change-induced droughts in East Africa in 2021, which led to the loss of livestock.[234]

In a nutshell, individual agricultural adaptation strategies especially migration could not only help farmers cope with vulnerability on agricultural products but could also lead to optimal production of livestock and even crops in confirmed climate change hotspot regions and regions susceptible to climate variability.[235]

At the community level, the planning and implementation of adaptation in the context of agriculture, which is often described as a "community-based adaptation (CBA)"[236] is carried out by stakeholders including farmers' unions and affiliated groups (such as animal health and seed-distribution companies).[237] The concept of CBA is based on the premise that local communities have the skills, experience, local knowledge, and networks to undertake locally appropriate activities that increase resilience and reduce vulnerability to a range of factors including climate change.[238] In other words, CBA in the context of agriculture is seen as "a decentralized approach that promotes collaboration and coordination between different actors" who are more familiar with the underlying challenges of climate change on agriculture in the community than non-community members. Advocates of CBA claim that it is the most effective mechanism for identifying, assisting, and implementing community-based development activities, research, and policy in regions where adaptive capacity is as dependent on current livelihood opportunities as climatic changes.[239]

Adaptation in the context of agriculture at the national level is planned and implemented by government departments and agencies for agriculture.[240] Agro-forestry techniques including the planting of trees and bushes to reduce soil, water erosion, and increase carbon sequestration in the soil serve as examples of agricultural adaptation usually implemented at the national level.[241] As the focal point for facilitating adaptation at the national level, government departments and agencies establish research institutions and design and implement policies, programmes, and operational decisions and other incentives such as funding to respond to climate change.[242] At the national level of adaptation planning, governments are also responsible for removing barriers against adaptation (generally) and motivating individuals and organisations to act proactively.[243] To actualise this, national governments in African and developing countries have launched National Adaptation Plans (NAPs) in line with global requirements for climate change governance.[244] Specifically, NAPs (explained further in Chapters 3, 4, and 5) represent the central mechanism by which nations can integrate their adaptation planning and action in different sectors (including agriculture) in a coherent way.[245]

It is also the responsibility of national governments, while planning and implementing agricultural adaptation at the national level, to provide direction to individuals and communities, through, for instance, promoting information sharing, building adaptive capacity, and mainstreaming adaptation efforts.[246] National governments also create adaptation platforms that consolidate initiatives and actions by individuals and communities for addressing the impacts of climate change on agriculture.[247] To achieve optimum adaptation in the agricultural sector, adaptation at all the levels must be implemented through a short-term, medium-term, or long-term plan.[248]

A short-term adaptation plan to agricultural adaptation strategy is important, as it will safeguard farmers against sudden-onset disasters resulting from the impacts of climate change, such as pathogens that contaminate soil water.[249] Short-term adaptation plans are mostly implemented at the individual level sometimes by corporations such as seed-producing companies, rather than by governments. For example, seed-producing companies present drought-resistant and faster-maturing crop varieties or animal feeds that can survive climate change-related conditions as short-term adaptation measures in the context of agriculture.[250]

A medium-term adaptation plan for agriculture is adopted following a risk assessment exercise of the impacts of climate change on agriculture that may occur in the medium term.[251] Agricultural adjustment by means of selling off livestock is an example of a medium-term adaptation plan for agriculture.[252] Other examples of a medium-term adaptation plan for agriculture, according to Arcidiacono,[253] include the use of technical solutions to improve ventilation and cooling systems in animal shelters, selecting crop varieties that can adapt to unexpected lengthy growing seasons, or choosing crop varieties that can withstand water scarcity and varying temperature conditions. Unlike short-term adaptation plans, medium-term adaptation plans could be carried out by both individual farmers and national governments.

Just like in any sector of the economy, a long-term adaptation plan in the context of agriculture is prompted by many factors such as societal changes and policy considerations.[254] Ajani[255] mentions making structural changes in the type of cropping and animal breeding through the development of crop and livestock technology that can adapt to climate change stress, or the expansion of water-storage systems as examples of a long-term adaptation plan in the agricultural sector. Likewise, the improvement of forecasting mechanisms and reinforcement of institutional and decision-making structures are part of long-term adaptation plans in the context of agriculture.[256] Long-term adaptation plans especially in the agricultural sector in an African and developing country context are generally implemented by governments. However, long-term adaptation plans including short-term and medium-term adaptation plans in the context of agriculture are mostly faced with challenges, which could be physical (infrastructural deficits)[257] or non-physical (socio-economic challenges).[258]

1.5.3 Challenges for adaptation in the context of agriculture

The barriers to adaptation in the context of agriculture are mostly noted by how farmers respond to them.[259] In other words, socio-economic factors rather than infrastructural factors are the main challenges to adaptation in the context of agriculture.[260] Socio-economic factors in particular are derived from cultural values and religious ideologies coupled with lack of access to financing and other critical agricultural resources such as land.[261] It follows that the absence of "organisational stability" in some African and developing countries with more vulnerability and weak adaptive capacity is also a socio-economic barrier to successful adaptation practices.[262]

In addition, the IPCC identified obstacles of adaptation, in the context of agriculture in African and developing countries, from a socio-economic point of view to include transaction cost, behavioural obstacles, uncertainty, market failure, distributional issues, and coordination failures.[263] Transaction cost could mean the cost associated with sourcing information on projected impacts of climate change and the cost of adjusting to the impacts of climate change.[264] For example, the cost of implementation of water-conservation techniques, crop-preservation methods, planting drought-resistant crops, and so on, are quite expensive if they are to be implemented at the individual level.[265] Behavioural obstacles are described as "significant matters related to decisions taken without considering prevailing social norms".[266] Uncertainty as a socio-economic impediment to agricultural adaptation is explained as unforeseen issues related to future developments and the types of technologies involved. Externalities such as moral hazards and information asymmetries[267] as well as lack of sufficient economic incentives could constitute market failure for adaptation at the individual level or community level.[268] Distributional issues are seen as socio-economic barriers to adaptation when cost-effective projects are chosen or when decisions are made without taking into account the net benefits and impacts on equity.[269] Finally, coordination failures as socio-economic barriers to adaptation in the context of agriculture means taking decisions using insufficient knowledge or resources by a stakeholder at different levels of adaptation. These barriers against agricultural adaptation remain insurmountable with the absence of well-developed governance infrastructure and effective legal mechanisms.[270] Consequently, an agricultural adaptation strategy is only considered effective and comprehensive when it is supported by laws and policies that strive "to contribute towards the global efforts of combating climate change and facilitate approaches that support low-carbon and climate-resilient development".[271]

1.6 Conclusion

This introductory chapter argued that climate change and agriculture are intricately linked, with each having devastating impacts on the other. Agricultural practices such as the production of crops and livestock are some of the drivers of climate change. As a result of climate change, there will be changes and/or increased variability in rainfall patterns, which will endanger arable lands and result in the reduction of agricultural yields (crop failure and loss of livestock). A loss of, or a reduction in, agricultural yields could exacerbate other existing socio-economic problems and related humanitarian disasters such as famine and displacement of people. Due to the importance of agriculture in Africa and its vulnerability to the impacts of climate change and variability, an adaptation strategy that promotes a low-carbon pathway, and at the same time increases agricultural production, is desired. In essence, adaptation strategies supported by a robust regulatory framework are required to protect agricultural practices, especially in African countries. The next chapter will probe the role of law (hard law and soft law) from an international and regional perspective as standard-setting mechanisms for actors and entities responsible for adaptation, with a view to establishing the context and purpose in discussing the role of national laws of African countries particularly Kenya, Nigeria, and South Africa in facilitating adaptation in a broad context and in the context of agriculture.

Notes

1 Human-induced climate change otherwise known as anthropogenic climate change also emanates from among others fossil fuel burning and land use changes. See UNFCCC 2007 Climate Change Conference. https://unfccc.int/event/bonn-climate-change-conference-may-2007.
2 R Mendelsohn, A Dinar and A Dalfelt 2000 *Climate change impacts on African agriculture*. Preliminary analysis prepared for the World Bank, 2.
3 K Warner, M Hamza, A Oliver-Smith, F Renaud and A Julca 2010 Climate change, environmental degradation and migration. *Natural Hazards,* (55), 689–715.
4 CM Godde, RB Boone, AJ Ash, K Waha, LL Sloat, PK Thornton and M Herrero 2020 Global rangeland production systems and livelihoods at threat under climate change and variability. *Environmental Research Letters*, 15(4), 044021.
5 In investigating the extent to which Kenyan, Nigerian, and South African laws provide for climate change adaptation in the context of the agricultural sector, the scope of this book delimits cases law analysis. In addition, this book excludes the following:

- Loss and damage, resilience, climate change mitigation, and climate change justice will not be discussed in this book. Although these are also strategies for addressing the impacts of climate change, these broad areas require different in-depth analysis that falls outside the scope of this book.
- International climate change law regime could be diverse and intricate. Therefore, a discussion of international climate change law in this book is limited to the climate change-specific provisions of the UNFCCC, notably COPs, the Kyoto Protocol, and Paris Agreement dealing with climate change adaptation. Other environmental conservation treaties are excluded from the focus of this book, except those which support the objectives of adaptation such as the Convention on Biological Diversity (CBD).
- Associated aspects such as international trade law and trade treaties will not be discussed.

- Given that soft law connotes very broad and endless list of examples, any reference to soft laws in this book will be limited to pre-legislation policies. Other examples of soft law such as standards, best practices, industry codes, environmental stewardship, and reporting schemes except where relevant and specifically required will not be discussed.

6 For example, see: WB Foden and BE Young 2016 *IUCN SSC guidelines for assessing species' vulnerability to climate change*. Version 1.0. Occasional Paper of the IUCN Species Survival Commission No. 59. IUCN Species Survival Commission, 114; P Antwi-Agyei, AJ Dougill and LC Stringer 2013 *Barriers to climate change adaptation in sub-Saharan Africa*: Evidence from northeast Ghana and systematic literature review. *Centre for climate change economics and policy working paper*, 154.

7 In 2005, industrial fertilisers were reportedly confirmed to be responsible for about 58% of global anthropogenic emissions. This is much compared to the net flux of carbon dioxide (CO_2) emissions (other than emissions from the use of fossil fuel). See SJ Vermeulen, BM Campbell and JS Ingram 2012 Climate change and food systems. *Annual Review of Environment and Resources*, 37(1), 195–222. According to the IPCC, the global warming potential (GWP) value of fertilisers is 23 times higher than CO_2. Similarly, the application of industrial fertilisers such as N_2O accounts for about 60% of global anthropogenic emissions. See IPCC 2007 https://www.ipcc.ch/site/assets/uploads/2018/03/ar4_wg2_full_report.pdf; PK Thornton and M Herrero 2014 Climate change adaptation in mixed crop–livestock systems in developing countries. *Global Food Security*, 3(2), 99–107; Reid et al. 2007 Vulnerability and adaptation to climate risks in Ontario agriculture. *Mitigation and Adaptation Strategies for Global Change*, 12, 609–637.

8 Some examples of unconventional tillage include the practice of "removing weeds, in order to avoid water and nutrient competition, and to incorporate fertilisers and manure into the soil". See AM Palese, N Vignozzi, G Celano, AE Agnelli, M Pagliai and C Xiloyannis 2014 Influence of soil management on soil physical characteristics and water storage in a mature rainfed olive orchard. *Soil and Tillage Research*, (144), 96–109; LA Ribera, FM Hons and JW Richardson 2004 An economic comparison between conventional and no-tillage farming systems in Burleson County, Texas. *Agronomy Journal*, 96(2), 415–424.

9 It has been estimated that about 25% of CO_2 emissions are as a result of deforestation. See RA Houghton 2005 Tropical deforestation as a source of greenhouse gas emissions. In *Tropical deforestation and climate change*, Jul 1, 13.

10 AB Guerrero and E Muñoz 2018 Life cycle assessment of second-generation ethanol derived from banana agricultural waste: Environmental impacts and energy balance. *Journal of Cleaner Production*, 174, 710–717.

11 Manure produced from rice swamps without oxygen (in an anaerobic state) will decompose and add oxygen, and this as a process is responsible for the emission of methane into the atmosphere. In like manner, manure stored as "liquid or slurry in lagoons, ponds, tanks or pits" is also responsible for the emission of methane into the atmosphere. See Global Methane Initiative 2011 Global Methane Emissions and Mitigation Opportunities. https://www.globalmethane.org/documents/analysis_fs_en.pdf.

12 JM Huho and RC Kosonei 2014 Understanding extreme climatic events for economic development in Kenya. *IOSR Journal of Environmental Science, Toxicology and Food Technology*, 8(2), 14–24.

13 MJ Angelo 2009 Corn, carbon, and conservation, rethinking US agricultural policy in a changing global environment. *George Mason Law Review*, 17, 593; C Ifejika Speranza 2010 Resilient adaptation to climate change in African Agriculture Studies, 54, 3.

14 Enteric fermentation occurs when animals such as cows digest microbes, and in the process, the fermented feed produces methane as a by-product. A Thorpe 2009 Enteric fermentation and ruminant eructation: The role (and control?) of methane in the climate change debate. *Climatic Change*, 93(3), 407–431.
15 A Glatzle 2014 Questioning key conclusions of FAO publications 'Livestock's Long Shadow' (2006) appearing again in 'Tackling Climate Change Through Livestock' (2013). *Pastoralism: Research, Policy and Practice*, 4(1), 1–6.
16 See also Reid et al. 2007 Vulnerability and adaptation to climate risks in Ontario agriculture. *Mitigation and Adaptation Strategies for Global Change*, 12, 609–637.
17 JM McGuire, LW Morton, JG Arbuckle Jr, and AD Cast 2015 Farmer identities and responses to the social–biophysical environment. *Journal of Rural Studies*, 39.145–155.
18 JR Porter and MA Semenov 2005 Crop responses to climatic variation. *Philosophical Transactions of the Royal Society B: Biological Sciences*, 360(1463), 2021–2035; JM Antle, SM Capalbo, ET Elliott and KH Paustian 2004 Adaptation, spatial heterogeneity, and the vulnerability of agricultural systems to climate change and CO2 fertilization: An integrated assessment approach. *Climatic Change*, 64(3), 289–315.
19 Gannon et al. 2018 Policy brief climate risks to hydropower supply in Eastern and Southern Africa. Grantham Research Institute on Climate Change and the Environment (LSE), 1–13.
20 Rainwater is described as "liquid in the form of droplets that have condensed from atmospheric water vapour and then becomes heavy enough to fall under gravity". Wikipedia. https://en.wikipedia.org/wiki/Rain.
21 Blue water has been explained as the amount of rainwater that does not "infiltrate into the soil or does not penetrate beyond the root zone, but runs off from the surface of the soil and forms surface storage such as lakes, reservoirs, rivers, wetlands, etc., or subsurface storages such as aquifers". See M Falkenmark and J Rockström 2006 The new blue and green water paradigm: Breaking new ground for water resources planning and management. *Journal of Water Resources Planning and Management*, 132(3), 129–132.1: Rockström et al. 2009 Future water availability for global food production: The potential of green water for increasing resilience to global change. *Water Resources Research*, 45(7), 45
22 In the words of Ringersma et al. remnants of rainwater that infiltrate the soil and is held in the soil, which is subsequently made available to plants, are described as soil moisture. See J Ringersma, NH Batjes and D Dent 2003 Green water: Definitions and data for assessment (No. 2003/2). *ISRIC-World Soil Information*, 1–50. Soil moisture is further explained by Falkenmark as the "invisible water in the soil, which is often forgotten in water resource management". It should be noted that both rainwater and blue water are "converted as soil moisture before being consumed by crops". See M Falkenmark 1997 Meeting water requirements of an expanding world population. *Philosophical Transactions of the Royal Society of London. Series B: Biological Sciences*, 352(1356), 929–936.
23 A climate change hotspot has been defined as a "location within a region in which land use is determined by the interplay of spatial planning, climatic conditions and other factors". According to the IPCC, Africa is considered as climate hotspot because of its "geographic location and geologic variations". See DA Malak, K McGlade, D Pascual and E Pla 2017 *Adapting to climate change: An assessment of vulnerability and risks to human security in the Western Mediterranean Basin*. Springer, 13–61; RK Pachauri, MR Allen, VR Barros, J Broome, W Cramer, R Christ and JP van Ypserle 2014 Climate change 2014: Synthesis report. *Contribution of Working Groups I, II, and III to the fifth assessment report of the Intergovernmental Panel on Climate Change*, 151.

24 For example, a decrease in crop production in 1999/2000 and 2005/2006 in East Africa is an indication of limited water agriculture in Africa. JM Huho and RC Kosonei 2014 Understanding extreme climatic events for economic development in Kenya. *IOSR Journal of Environmental Science, Toxicology and Food Technology,* 8(2), 14–24.

25 The impacts of climate change on livestock could be direct (when it affects animal productivity and health) and indirect (where animal fodder is affected). For example, about 20 to 60% of losses in animal fodder were recorded in temperate regions of Africa because of reduced precipitation in 2003. See V Gitz, A Meybeck, L Lipper, CD Young, and S Braatz 2016 Climate change and food security: Risks and responses. *Food and Agriculture Organization of the United Nations Report,* 110(2), 3–36.

26 E Nesamvuni et al. 2012 "Effects of climate change on dairy cattle, South Africa." *African Journal of Agricultural Research,* 7(26), 3867–3872.

27 See for example, JC Ribot 1995 The causal structure of vulnerability: Its application to climate impact analysis. *GeoJournal,* 35(2), 119–122; B Smit and J Wandel 2006 Adaptation, adaptive capacity and vulnerability. *Global Environmental Change,* 16(3), 282–292.

28 USAID 2023 *Agriculture, Food and Water Security*. https://www.usaid.gov/kenya/agriculture-food-and-water-security. Accessed 13 October 2023.

29 See Government of Kenya 2013 *Ministry of Planning and Devolution 51: FAO 2019*. Agriculture. https://www.fao.org/kenya/en/.

30 Satistica 2020 Value added of the agriculture sector to the Gross Domestic Product (GDP) at current prices in Kenya from 2018 to 2022. https://www.statista.com/statistics/1167532/value-added-of-agriculture-to-the-gdp-at-current-prices-in-kenya/.

31 Kenya Climate Smart Agriculture Strategy-2017–2026. https://adaptation-undp.org/sites/default/files/resources/kenya_climate_smart_agriculture_strategy.pdf, 12.

32 The bulk of agricultural income in Kenya is derived from the livestock subsector. As a national census conducted in 2009 revealed that there are 17.5 million cattle, 3 million camels, 1.8 million donkeys, 17 million sheep, and 27.7 million goats in Kenya. See Government of Kenya 2010 Agriculture Sector Development Strategy 2010–2020. https://faolex.fao.org/docs/pdf/ken140935.pdf, 14; E Odoyo, B Bett and AS Murenga Mwimali 2019 Challenges of and responses to climate variability: A case study of peri-urban female farmers in Kenya. *World Journal of Agricultural Research*, 7(2), 61–68.

33 It has already been explained that most African countries including Kenya are susceptible to the severity of the impacts of climate change due to their geographic location and geologic variations, coupled with poor adaptive capacity. See R Mwebaza and LJ Kotzé 2009 Environmental governance and climate change in Africa: Legal perspectives. *Institute for Security Studies Monographs,* 167, 283; S Huq, H Reid, M Konate, A Rahman, Y Sokona and F Crick, F.l 2004 Mainstreaming adaptation to climate change in least developed countries (LDCs) *Climate Policy,* 4(1), 25–43.

34 Kenya Climate Smart Agriculture Strategy 2017–2026. https://adaptation-undp.org/sites/default/files/resources/kenya_climate_smart_agriculture_strategy.pdf, 12.

35 Apart from its impacts on crops and livestock, climate change could have also led to "alteration in fodder quality and quantity; change in host-pathogen interaction resulting in increased incidences of emerging diseases; and disease epidemics" in Kenya. See *Kenya climate smart agriculture strategy, 2017–2026.* https://adaptation-undp.org/sites/default/files/resources/kenya_climate_smart_agriculture_strategy.pdf. 17.

36 In addition to its impacts on Kenya's agriculture, climate change serves as vector for other calamities such as malaria and pests and is also a source of concern for energy resources in Kenya. In other words, climate change has attending impacts on not only the economic and physical infrastructure but also on social cohesion in Kenya. See Kenya climate smart agriculture strategy 2017–2026. https://adaptation-undp.org/sites/default/files/resources/kenya_climate_smart_agriculture_strategy.pdf. 25.

37 Since Kenya is classified as lower-middle income economy by 2014 World Bank indices, it is expected that the country will soon develop, and by implication, its GHG-emission profile will increase. As such, a "low carbon and climate resilient development pathway" as explained below has been recommended by the provisions of the National Climate Change Response Strategy 2010 and Kenya's blueprint for national development (Vision 2030). See World Bank 2014 Take-off Delayed? Kenya's economy facing headwinds in 2014 with a special focus on delivering primary health care services. https://documents1.worldbank.org/curated/ar/849741468273022634/pdf/892690WP0P14810ssue0100Final0June14.pdf.

38 E Odoyo, B Bett and AS Murenga Mwimali 2019 Challenges of and responses to climate variability: A case study of peri-urban female farmers in Kenya. *World Journal of Agricultural Research,* 7(2), 61–68.

39 S Erdoğan, DÇ Yıldırım and A Gedikli 2020 Natural resource abundance, financial development and economic growth: An investigation on next-11 countries. *Resources Policy,* 65, 101559; AK Raggl 2017 Natural resources, institutions, and economic growth: The case of Nigeria. *World Bank Policy Research Working Paper* no. 8153.

40 IK Maji, MS Habibullah and M Yusof-Saari 2016 Emissions from agricultural sector and financial development in Nigeria: An empirical study. *International Journal of Economics and Management,* 10(1).

41 Statista 2024 https://www.statista.com/topics/6914/oil-industry-in-nigeria/#dossierKeyfigures.

42 World Bank 2023 Agriculture, forestry, and fishing, value added (% of GDP) – Nigeria. https://data.worldbank.org/indicator/NV.AGR.TOTL.ZS?locations=NG

43 FAO 2021 Nigeria Agriculture at a Glance. https://www.fao.org/nigeria/fao-in-nigeria/nigeria-at-a-glance/en/. Accessed 25 October 2023.

44 See Worldometer 2016 World Emissions https://www.worldometers.info/co2-emissions/nigeria-co2-emissions/. In spite of Nigeria's position as a member of oil exporting countries (OPEC), the country's contribution to global GHG emissions is minimal compared to other countries such as China and Mexico. See UNFCCC https://di.unfccc.int/ghg_profiles/nonAnnexOne/NGA/NGA_ghg_profile.pdf.

45 More so, shifting cultivation as a major farming method (involving short fallow periods) adopted by peasant farmers in producing over 90% of total food production in Nigeria; when done in excess particularly in marginal lands, mountain areas or wetlands could lead to more GHG emission, resulting in habitat destruction and vegetation destruction for indigenous species of plant and animals within same ecological ranges. For example, about 351,000 hectares of land are destroyed annually in Nigeria because of shifting cultivation as well as overgrazing. See AI Kankara, GK Adamu, R Tukur and A Ibrahim 2013 Examining environmental policies and laws in Nigeria. *International Journal of Environmental Engineering and Management,* 4(3), 165–170.

46 DI Okorie and B Lin 2022 Emissions in agricultural-based developing economies: A case of Nigeria. *Journal of Cleaner Production,* 337,130570.

47 These variations in rainfall patterns in Nigeria could be attributed to "a tropical climate that is associated with a unimodal rainfall regime (that mostly occurs

between August and September) in the northern part, and a bimodal rainfall regime in the southern part of the country". See Federal Government of Nigeria 2018. National Drought Plan https://www.unccd.int/sites/default/files/country_profile_documents/1%2520FINAL_NDP_Nigeria.pdf. National Adaptation Strategy and Plan of Action for Climate Change in Nigeria 2011 https://csdevnet.org/wp-content/uploads/NATIONAL-ADAPTATION-STRATEGY-AND-PLAN-OF-ACTION.pdf.

48 According to the Federal Government of Nigeria, droughts occur in the northern part of the country mostly at the beginning of the rainy season in June and could at times last until August. And these droughts are attributed to farming-related activities especially bush burning which causes excessive build-up of heat on the earth's surface, causing meteorological changes which result in a reduction of rainfall, and reduced clouds. See Federal Government of Nigeria 2018 National Drought Plan, https://www.unccd.int/sites/default/files/country_profile_documents/1%2520FINAL_NDP_Nigeria.pdf, 19.
Historically, Nigeria witnessed the most devastating droughts in 1914, 1924, 1935, 1943, 1951–1954, 1972–1973, and 1991–1995. During these horrific years, droughts had led to enormous environmental degradation, massive crop failure, loss of livestock, human displacement migration, diseases, deaths, etc. By implication, poverty increased in Nigeria exponentially, and the ability of the country to achieve its sustainable development goals had been stagnated. See SJ Chang 2013 Solving the problem of carbon dioxide emissions. *Forest Policy and Economics*, 35, 92–97; Federal Government of Nigeria 2018 https://knowledge.unccd.int/sites/default/files/country_profile_documents/1%2520FINAL_NDP_Nigeria.pdf, 23–24.

49 SJ Chang 2013 Solving the problem of carbon dioxide emissions. *Forest Policy and Economics*, 35, 92–97.

50 C Ifejika Speranza 2010 Resilient adaptation to climate change in African Agriculture Studies, (No. 54). Studies, 54; T Stocker (Ed.) 2014 *Climate change 2013: The physical science basis: Working Group I contribution to the fifth assessment report of the Intergovernmental Panel on Climate Change*. Cambridge University Press, 1535.

51 First Approved Nigeria's Intended Nationally Determined Contribution. https://unfccc.int/sites/default/files/NDC/2022-06/Approved%20Nigeria%27s%20INDC_271115.pdf, 4.

52 Orie EG 2021 Climate change adaptation mechanism for sustainable development goal 1 in Nigeria: Legal Imperative. In *African handbook of climate change adaptation*. Springer International Publishing, 1–21, 1903.

53 A Calzadilla, T Zhu, K Rehdanz, RS Tol and C Ringler 2014 Climate change and agriculture: Impacts and adaptation options in South Africa. *Water Resources and Economics*, 5, 24–48.

54 During these periods, the annual precipitation rate in South Africa has decreased by 20 mm, and this is expected to further decrease by 40 mm in 2050, and sea level rose from 1.9 mm and 2.74 mm per annum, while temperature has been increasing by approximately 1°C. See F Giorgi 2019 Thirty years of regional climate modeling: Where are we and where are we going next? *Journal of Geophysical Research: Atmospheres*, 124(11), 5696–5723.

55 South Africa Department of Energy 2018. https://www.dmre.gov.za/Portals/0/Energy_Website/files/media/explained/2018-South-African-Energy-Sector-Report.pdf, 4–26.
Coal mining-related emissions in South Africa are among others, illustrated by the activities of Sasol at its Secunda plant in Mpumalanga Province, where coal is converted into diesel and other fuels. See Bank and Karsten 2020 Climate Change and South Africa: A Critical Analysis of the Earthlife Africa Johannesburg

and Another v Minister of Energy and Others 65662/16 (2017) Case and the Drive for Concrete Climate Practices. *Air, Soil and Water Research*, 13, 1–11. Sasol is an integrated energy and chemical company based in Sandton-Johannesburg, South Africa, and Secunda is an area which houses one of Sasol's power plants. The site is considered as one of the world's major points of CO_2 emission, considering the enormous amount of GHG that is emitted from dirty energy (coal) that is fired in order to provide fuel combustion for transport, power, communication networks, water, and sanitation, and other infrastructure services in South Africa. See https://www.sasol.com/about-sasol/regional-operating-hubs/southern-africa operations/secunda-chemicals-operations.

56 For example, coal mining which is an essential activity for energy generation and source of employment in South Africa accounts of 84% of GHG-induced climate change in the country. Department of Energy 2018 http://www.energy.gov.za/files/media/explained/2018-South-African-Energy-Sector-Report.pdf, 4–26; Carbon Brief 2018 https://www.carbonbrief.org/the-carbon-brief-profile-south-africa.

57 UNDP 2022 South Africa Climate Change Adaptation https://www.adaptation-undp.org/explore/africa/south-africa.

58 Department of Water Affairs and Forestry, Republic of South Africa (DWAF) 2004 https://www.gov.za/sites/default/files/gcis_document/201409/dwafanrep0304.pdf; E Mwendera and Y Atyosi 2018 A review of water storage for socio-economic development in South Africa. *Journal of Water Resource and Protection*, 10(3), 266.

59 CM Botai, JO Botai and AM Adeola 2018 Spatial distribution of temporal precipitation contrasts in South Africa. *South African Journal of Science*, 114(7–8), 70–78.

60 TG Lumsden, RE Schulze and BC Hewitson 2009 Evaluation of potential changes in hydrologically relevant statistics of rainfall in Southern Africa under conditions of climate change. *Water SA*, 35(5), 649–656.

61 Bank and Karsten 2020 Climate Change and South Africa: A Critical Analysis of the Earthlife Africa Johannesburg and Another v Minister of Energy and Others 65662/16 (2017) Case and the Drive for Concrete Climate Practices *Air, Soil and Water Research,* 13, 1–11.

62 E Mwendera and Y Atyosi 2018 A review of water storage for socio-economic development in South Africa. *Journal of Water Resource and Protection*, 10(03), 266–286.

63 See International Trade Administration 2021 South Africa - Country Commercial Guide. https://www.trade.gov/country-commercial-guides/south-africa-agricultural-sector. See also A Calzadilla, T Zhu, K Rehdanz, RS Tol and C Ringler 2014 Climate change and agriculture: Impacts and adaptation options in South Africa. *Water Resources and Economics*, 5, 24–48.

64 JK Benhin 2006 Climate change and South African agriculture: Impacts and adaptation options Centre for Environmental Economics and Policy in Africa CEEPA Discussion Paper, 21.

65 See JK Musango and C Peter 2007 A Bayesian approach towards facilitating climate change adaptation research on the South African agricultural sector. *Agrekon*, 46(2), 245–259. Examples of agricultural products that are exported from South Africa include corn, wool, sugar, peanuts, tobacco, etc. Ministry of Agriculture and Land Affairs 1998 Agricultural policy in South Africa. See https://web.archive.org/web/20170715212908/; http://www.gov.za:80/documents/agricultural-policy-south-africa-discussion-document.

66 Satistica 2020 South Africa: Distribution of gross domestic product (GDP) across economic sectors from 2012 to 2022, https://www.statista.com/statistics/371233/south-africa-gdp-distribution-across-economic-sectors/.

67 The Borgen Project 2019 Understanding Hunger in South Africa. https://borgenproject.org/understanding-hunger-in-south-africa/.
68 EM Zwane 2019 Impact of climate change on primary agriculture, water sources and food security in Western Cape, South Africa. *Jàmbá: Journal of Disaster Risk Studies*, 11(1), 1–7.
69 S Eriksen and K Brown 2011 Sustainable adaptation to climate change. *Climate and Development*, 3 (1), 3–6.
70 Oxford learner's dictionary 2014 http://www.oxfordlearnersdictionaries.com/definition/english/adaptation.
71 B Smit 1993 *Adaptation to climatic variability and change*. Report of the task force on climate adaptation (No. GU/G-CE04610). Guelph University, 1–28. The IPCC definition of adaptation as "the ability of a system to adjust to climate change (including climate variability and extremes), to moderate potential damages, to take advantages of opportunities, or to cope with consequences", was also motivated by a technical and non-technical definition of adaptation. See RA Pielke, JO Adegoke, TN Chase, CH Marshall, T Matsui and D Niyogi 2007. A new paradigm for assessing the role of agriculture in the climate system and in climate change. *Agricultural and Forest Meteorology*, 142(2–4), 234–254.
72 Human systems can be explained as those activities attributed to industrialization through the burning of fossil fuels, high population-growth rates, increasing reliance on fossil fuel-driven technologies, and land use effects (particularly urbanisation, agriculture, and deforestation) that increases the emission of GHG. See N Stern and NH Stern 2007 *The economics of climate change: The Stern review*. Cambridge University Press, 165.
73 Natural systems include variations in solar sunspot activities, volcanic eruption, ocean atmosphere interaction and integration, and continental drift.
74 LC Stringer, JC Dyer, MS Reed, AJ Dougill, C Twyman and D Mkwambisi 2009 Adaptations to climate change, drought and desertification: Local insights to enhance policy in southern Africa. *Environmental Science and Policy*, 12(7), 748–765; DR Nelson, WN Adger, and K Brown 2007 Adaptation to environmental change: Contributions of a resilience framework. *Annual Review of Environment and Resources*, 32(1), 395–419.
75 B Smit, I Burton, RJ Klein and R Street 1999 The science of adaptation: A framework for assessment. *Mitigation and Adaptation Strategies for Global Change*, 4, 199–213; Barnett and O'Neil 2010 Global Environmental Change, 211–213; KL O'Brien and J Wolf 2010 A values-based approach to vulnerability and adaptation to climate change. *Wiley Interdisciplinary Reviews: Climate Change*, 232–242.
76 The United States for instance pulled out of the Kyoto Protocol in 2001 due to political reasons. See Lenntech 2001 The United States and the Kyoto Protocol.https://www.lenntech.com/greenhouse-effect/kyoto-united-states.htm#:~:text=Nevertheless%2C%20in%202001%20the%20US,involved%20in%20the%20Kyoto%20Protocol. Likewise, as discussed in relevant chapters of this book, party politics and changes in governments were responsible for the protracted struggle and delay in enacting the *Nigerian Climate Change Act* and *the South African Climate Change Act*.
77 SH Eriksen, AJ Nightingale and H Eakin 2015 Reframing adaptation: The political nature of climate change adaptation. *Global Environmental Change*, (35), 523–533.
78 SH Eriksen, AJ Nightingale and H Eakin 2015 Reframing adaptation: The political nature of climate change adaptation. *Global Environmental Change*, (35), 523–533.
79 M Pelling 2010 *Adaptation to climate change: From resilience to transformation*. Routledge, 81.

80 Given that the primary aim of this book is to draw on the legal aspects of adaptation in the context of agriculture, subsequent discussions on the political aspects of adaptation would be minimal.
81 B Smit, I Burton, RJ Klein and R Street 1999 The science of adaptation: A framework for assessment. *Mitigation and Adaptation Strategies for Global Change,* 4, 199–213; Barnett and O'Neil 2010 Maladaptation. *Global Environmental Change,* 20, 211–213.
82 UNFCCC 2014 https://unfccc.int/news/adapatation-and-the-convention.
83 The need to localise adaptation to climate change has been emphasised in academic literature. See for example, JC Ribot 1995 The causal structure of vulnerability: Its application to climate impact analysis. *GeoJournal,* 35(2), 119–122; P Blaikie, T Cannon, I Davis, and B Wisner 2004. *At risk: Natural hazards, people's vulnerability and disasters.* Routledge, 21.
84 B Smit, I Burton, RJ Klein, and J Wandel 2000 An anatomy of adaptation to climate change and variability. In *Societal adaptation to climate variability and change.* Springer, 223–251.
85 L Berrang-Ford, JD Ford and J Paterson 2011 Are we adapting to climate change? *Global Environmental Change,* 21(1), 25–33.
86 B Smit, I Burton, RJ Klein, and J Wandel 2000 An anatomy of adaptation to climate change and variability. In *Societal adaptation to climate variability and change.* Springer, 223–251.
87 MParry et al., 2007 Climate Change 2007: Impacts, Adaptation and Vulnerability. Contribution of Working Group II to the Fourth Assessment Report of the Intergovernmental Panel on Climate Change. (4), Cambridge University Press; R Shaw, JM Pulhin and JJ Pereira 2010 Climate change adaptation and disaster risk reduction: An Asian perspective. In *Climate change adaptation and disaster risk reduction: An Asian Perspective. Emerald Group Publishing Limited,* 5, 1–18.
88 B Smit, I Burton, RJ Klein, and J Wandel 2000 An anatomy of adaptation to climate change and variability. *Climatic Change,* 1(45), 223–251.
89 R Mendelsohn 2012 Development and climate adaptation. In *Climate change and common sense. Essays in honour of Tom Schelling,* 245–259.
90 IPCC 2007 IPCC Fourth Assessment Report: Climate Change 2007. https://archive.ipcc.ch/publications_and_data/ar4/wg2/en/annexessglossary-a-d.html.
91 A Ignaciuk and D Mason-D'Croz 2014 Modelling adaptation to climate change in agriculture. *OECD Food Agriculture and Fisheries Papers,* 58.
92 J Ardo and AE Bashir 2007 Agroforestry-an adaptation to climate change in the Sahel. A Research Plan, 6, 1–12.
93 Planned adaptation can also be applied in other sectors of the economy such as the transport, business and urban planning sectors. JD Ford, L Berrang-Ford, A Bunce, McKay C, M Irwin and T Pearce 2015 The status of climate change adaptation in Africa and Asia. *Regional Environmental Change,* 15(5), 801–814.
94 JB Smith, T Dickinson, JD Donahue, I Burton, E Haites, RJ Klein and A Patwardhan 2011 Development and climate change adaptation funding: Coordination and integration. *Climate Policy,* 11(3), 987–1000; WN Adger, NW Arnell and EL Tompkins 2005 Adapting to climate change: Perspectives across scales. *Global Environmental Change,* 15(2), 77–86.
95 L Lebel, L Li, C Krittasudthacheewa, M Juntopas, T Vijitpan, T Uchiyama and D Krawanchid 2012 Mainstreaming climate change adaptation into development planning. *Stockholm Environment Institute,* 8, 32.
96 L Berrang-Ford, JD Ford and J Paterson 2011 Are we adapting to climate change? *Global Environmental Change,* 21(1), 25–33.
97 M Murthy, R Tiwari and NH Ravindranath 2011 Climate change and forests in India: adaptation opportunities and challenges. *Mitigation and Adaptation Strategies for Global Change,* 2(16), 161–175.

98 Global Humanitarian Forum 2009 Human Impact of Climate Change. http://www.gci.org.uk/Documents/GHF_2009_.pdf, 69.
99 B Lim, I Burton, E Spanger-Siegfried, EL Malone and S Huq 2005 *Adaptation policy frameworks for climate change*. Cambridge University Press, 15.
100 C Pettengell 2010 Climate change adaptation: Enabling people living in poverty to adapt. Oxford*:* Oxfam GB, 1–48.
101 B Smit, I Burton, RJ Klein, and J Wandel 2000 An anatomy of adaptation to climate change and variability. *Climatic Change*, 1(45), 223–251.
102 A Cliquet, C Backes, J Harris and P Howsam 2009 Adaptation to climate change legal challenges for protected areas. *Utrecht Law Review,* 5, 158–175.
103 For example, African countries as climate change hotspots are also vulnerable to the impacts of climate change because of a deterioration in the ecological base, poverty, high dependence on natural resources, and poor adaptive capacity, which is seen as the inability to make long-term adjustments in a change climatic condition. See R Mwebaza and LJ Kotzé 2009 Environmental governance and climate change in Africa: Legal perspectives. *Institute for Security Studies Monographs*, 167, 283.
104 WN Adger, NW Arnell and EL Tompkins 2005 Adapting to climate change: Perspectives across scales. *Global Environmental Change*, 15(2), 77–86.
105 Forest acts as carbon sinks that help in absorbing CO2, one of the most emitted GHGs responsible for climate change. See B Mayer 2018 *The international law on climate change*. Cambridge University Press, 238; FAO 2013 Climate change guidelines for forest managers. http://www.fao.org/3/i3383e/i3383e.pdf.
106 R Verheyen 2002 Adaptation to the impacts of anthropogenic climate change-the international legal framework. *Review of European Comparative and International Environmental Law*, 11(2), 129–143.
107 C Pettengell 2010 Climate Change Adaptation: Enabling People Living in Poverty to Adapt. Oxford: Oxfam GB, 1–48.
108 CB Field and VR Barros (Eds.) *2014 Climate change 2014–Impacts, adaptation and vulnerability: Regional aspects*. Cambridge University Press, 1758.
109 B Lim, I Burton, E Spanger-Siegfried, EL Malone and S Huq 2005 *Adaptation policy frameworks for climate change*. Cambridge University Press, 15.
110 C Pettengell 2010 Climate Change Adaptation: Enabling People Living in Poverty to Adapt. *Oxfam GB, Oxford,* 1–48.
111 S Davoudi, K Shaw, LJ Haider, AE Quinlan, GD Peterson, C Wilkinson and S Davoudi 2012 Resilience: A bridging concept or a dead end? *Planning Theory and Practice*, 13(2), 299–333. See also S Oberthür and HE Ott 1999 *The Kyoto protocol: International climate policy for the 21st century*. Springer Science and Business Media.
112 RJ Klein, RJ Nicholls and F Thomalla 2003 Resilience to natural hazards: How useful is this concept? *Global Environmental Change Part B: Environmental Hazards*, 5(1), 35–45.
113 FE Opiyo, OV Wasonga and MM Nyangito 2014 Measuring household vulnerability to climate-induced stresses in pastoral rangelands of Kenya: Implications for resilience programming. *Pastoralism*, 1(4), 1–15. Using drought as an example, certain pastoral and agro-pastoral areas and households are usually vulnerable to food insecurity because they lack the resilience to resist an external shock.
114 EL Tompkins and WN Adger 2004 Does adaptive management of natural resources enhance resilience to climate change? *Ecology and Society*, (9), 2.
115 M Pelling 2010 *Adaptation to climate change: From resilience to transformation*. Routledge, 81.
116 W Anita, M Dominic and A Neil 2010 *Climate change and agriculture impacts, adaptation and mitigation: Impacts, adaptation and mitigation*. OECD Publishing, 52.

117 G Jobbins, D Conway, S Fankhauser, B Gueye, E Liwenga, E Ludi and A Suleri 2016 Resilience, equity and growth in semi-arid economies a research agenda, Pathways to Resilience in Semi-arid Economies (PRISE) Working Paper, 1–36.
118 S Pavanello 2009 *Pastoralists' vulnerability in the Horn of Africa: Exploring political marginalisation, donors' policies and cross-border issues–Literature review*. London, UK: Humanitarian Policy Group (HPG) Overseas Development Institute, 1–47.
119 SH Eriksen and K O'Brien 2007 Vulnerability, poverty and the need for sustainable adaptation measures. *Climate Policy*, 7(4), 337–352.
120 C Walthall, J Hatfield, E Marshall, L Lengnick, P Backlund, S Adkins and J Bunce 2013 Climate change and agriculture in the United States: Effects and adaptation. USDA Technical Bulletin, 1953(1), 1–186.
121 B Lim, I Burton, E Spanger-Siegfried, EL Malone and S Huq 2005 *Adaptation policy frameworks for climate change*. Cambridge University Press, 49–60.
122 M Mehar, S Mittal and N Prasad 2016 Farmers coping strategies for climate shock: Is it differentiated by gender? *Journal of Rural Studies*, 100(44), 123–131.
123 W Anita, M Dominic and A Neil 2010 *Climate change and agriculture impacts, adaptation and mitigation: Impacts, adaptation and mitigation*. OECD Publishing, 62.
124 W Anita, M Dominic and A Neil 2010 *Climate change and agriculture impacts, adaptation and mitigation: Impacts, adaptation and mitigation*. OECD Publishing, 60.
125 BK Sovacool 2011 Hard and soft paths for climate change adaptation. *Climate Policy*, 11(4), 1177–1183; M Kumamoto and A Mills 2012 What African countries perceive to be adaptation priorities: Results from 20 countries in the Africa adaptation programme. *Climate and Development*, 4(4), 265–274.
126 NE Korres, JK Norsworthy, NR Burgos and DM Oosterhuis 2017 Temperature and drought impacts on rice production: An agronomic perspective regarding short-and long-term adaptation measures. *Water Resources and Rural Development*, 100(9), 12–27.
127 In this regard, migration could serve as a complementary strategy to adaptation in reducing vulnerability against the impacts of climate change in Africa. For example, the heavy flooding recorded along the Zambezi River in central Mozambique in 2001, 2007, and in 2008 caused migration. M Stal 2011 Flooding and relocation: The Zambezi River valley in Mozambique. *International Migration*, 49, e125–e145. This is also a basis for the contrary opinion that since the effects of climate change could spill over borders leading to the displacement of peoples, adaptation in the context of migration should be considered a public good that needs the cooperation of the international community. See for example G Gioli, T Khan, S Bisht and J Scheffran 2014 Migration as an adaptation strategy and its gendered implications: A case study from the Upper Indus Basin. *Mountain Research and Development* 34(3), 255–265; AK Magnan and T Ribera 2016 Global adaptation after Paris. *Science*, 352(6291), 1280–1282.
128 B Orlove 2009 The past, the present and some possible futures of adaptation, In WN Adger, I Lorenzoni, and K O'Brien (Eds.), *Adaptation to Climate Change: Thresholds, Values, Governance*, Cambridge, UK: Cambridge University Press, 131–163.
129 M Stal 2011 Flooding and relocation: The Zambezi River valley in Mozambique. *International Migration*, 49, e125–e145. This is also a basis for the contrary opinion that since the effects of climate change could spill over borders leading to the displacement of peoples, adaptation in the context of migration should be considered as a public good that needs the cooperation of the international community. See for example G Gioli, T Khan, S Bisht and J Scheffran 2014 Migration as an adaptation strategy and its gendered implications: A case study from the

Upper Indus Basin. *Mountain Research and Development* 34(3), 255–265; AK Magnan and T Ribera 2016 Global adaptation after Paris. *Science*, 352(6291), 1280–1282.

130 K Hampshire 2002 Fulani on the move: Seasonal economic migration in the Sahel as a social process. *Journal of Development Studies*, 38(5), 15–36; M Leighton 2006 *Desertification and migration. Governing global desertification*. Ashgate, 43–58.

131 UNFCCC 2009 Copenhagen Climate Change Conference - 7 Dec - 18 Dec 2009. https://unfccc.int/conference/copenhagen-climate-change-conference-december-2009.

132 B Lim, I Burton, E Spanger-Siegfried, EL Malone and S Huq 2005 *Adaptation policy frameworks for climate change*. Cambridge University Press, 49–60.

133 A Persson and RJ Klein 2008 *Mainstreaming adaptation to climate change into official development assistance: Integration of long-term climate concerns and short-term development needs*. Stockholm Environment Institute, Stockholm, 16.

134 See also SH Eriksen, K Brown and PM Kelly 2005 The dynamics of vulnerability: Locating coping strategies in Kenya and Tanzania. *Geographical Journal*, 171(4), 287–305.

135 The Committee on Approaches to Climate Change Adaptation 2010 https://www.env.go.jp/en/earth/cc/adapt_guide/pdf/approaches_to_adaptation_en.pdf, 17.

136 For example, a risk assessment exercise is required in considering existing financial and human resource constraints based on priority levels. In other words, risk assessment exercise enables policy makers to use high projected value for impacts that are likely to have a substantial risk in the target area. See for example, MR Anwar, DL Liu, I Macadam and G Kelly 2013 Adapting agriculture to climate change: A review. *Theoretical and Applied Climatology*, 113 (1–2), 225–245; R Zougmoré, S Partey, M Ouédraogo, B Omitoyin, T Thomas, A Ayantunde and A Jalloh 2016 Toward climate-smart agriculture in West Africa: A review of climate change impacts, adaptation strategies and policy developments for the livestock, fishery and crop production sectors. *Agriculture and Food Security*, 1(5), 26.

137 B Lim, I Burton, E Spanger-Siegfried, EL Malone and S Huq 2005 *Adaptation policy frameworks for climate change: Developing strategies, policies and measures*, Cambridge University Press, 185.

138 S Hallegatte 2009 Strategies to adapt to an uncertain climate change. *Global Environmental Change*, 19(2), 240–247.

139 WE Walker, M Haasnoot and JH Kwakkel 2013 Adapt or perish: A review of planning approaches for adaptation under deep uncertainty. *Sustainability*, 5(3), 955–979.

140 NE Korres, JK Norsworthy, NR Burgos and DM Oosterhuis 2017 Temperature and drought impacts on rice production: An agronomic perspective regarding short-and long-term adaptation measures. *Water Resources and Rural Development*, 100(9), 12–27; B Smit and MW Skinner 2002 Adaptation options in agriculture to climate change: A typology. *Mitigation and Adaptation Strategies for Global Change*, 7(1), 85–114.

141 LC Stringer, JC Dyer, MS Reed, AJ Dougill, C Twyman and D Mkwambisi 2009 Adaptations to climate change, drought and desertification: Local insights to enhance policy in southern Africa. *Environmental Science and Policy*, 12(7), 748–765.

142 P Mickwitz, F Aix, S Beck, D Carss, N Ferrand, C Görg and S van Bommel 2009 *Climate policy integration, Coherence and governance*, Partnership for European Environmental research (PEER) Report No. 2, 1–96.

143 P Mickwitz, F Aix, S Beck, D Carss, N Ferrand, C Görg and S van Bommel 2009 *Climate Policy Integration, Coherence and Governance*, Partnership for European Environmental Research (PEER) Report No. 2, 1–96.

144 P Mickwitz, F Aix, S Beck, D Carss, N Ferrand, C Görg and S van Bommel 2009 *Climate Policy Integration, Coherence and Governance*, Partnership for European Environmental Research (PEER) Report No. 2, 15.

145 IPCC 2007 https://www.ipcc.ch/site/assets/uploads/2018/02/ar4_syr_full_report.pdf, 732, 814, 832.

146 A Persson and RJ Klein 2008 *Mainstreaming adaptation to climate change into official development assistance: Integration of long-term climate concerns and short-term development needs*. Stockholm Environment Institute, Stockholm, 3.

147 W Lafferty and E Hovden 2003 Environmental policy integration: Towards an analytical framework. *Environmental Politics*, 12(3), 8–9.

148 RD Lasco, FB Pulhin, PA Jaranilla-Sanchez, RJP Delfino, R Gerpacio and K Garcia 2009 Mainstreaming adaptation in developing countries: The case of the Philippines. *Climate and Development*, 1(2), 130–146.

149 P Mickwitz, F Aix, S Beck, D Carss, N Ferrand, C Görg and S van Bommel 2009 Climate Policy Integration, Coherence and Governance, Partnership for European Environmental Research (PEER) Report No. 2, 15.

150 RK Craig 2010 Stationarity is dead-long live transformation: Five principles for climate change adaptation law. *Harvard Environmental Law Review*, 34(1), 9.

151 MO Oulu 2011 Mainstreaming climate adaptation in Kenya. *Climate Law*, 2(3), 375–394.

152 MO Oulu 2011 Mainstreaming climate adaptation in Kenya. *Climate Law*, 2(3), 375–394.

153 W Lafferty and E Hovden 2003 Environmental policy integration: Towards an analytical framework. *Environmental Politics*, 12(3), 1–22.

154 H Hettiarachchi and R Ardakanian 2016 Managing Water, Soil, and Waste in the Context of Global Change. In H Hettiarachchi and R Ardakanian (Eds.), *Environmental resource management and the Nexus approach,* Cham: Springer, 1–7.

155 L Berrang-Ford, JD Ford and J Paterson 2011 Are we adapting to climate change? *Global Environmental Change,* 21(1), 25–33. Material factors, institutional factors, cognitive factors, and behavioural factors also considered as potential barriers to adaptation. See A Aaheim and M Aasen 2008 What do we know about the economics of adaptation? *CEPS Policy Brief*, 150.

156 S Hallegatte, M Fay and EB Barbier 2018 Poverty and climate change: Introduction. *Environment and Development Economics*, 23(3), 217–233; A Bauer, J Feichtinger and R Steurer 2012 The governance of climate change adaptation in 10 OECD countries: Challenges and approaches. *Journal of Environmental Policy and Planning*, 14(3), 279–304.

157 R Jones 2010 A risk management approach to climate change adaptation. In R A C Nottage, D S Wratt, J F Bornman, K Jones (Eds.) *Climate change adaptation in New Zealand: Future scenarios and some sectoral perspectives*, Wellington, 10–25.

158 H Hettiarachchi and R Ardakanian 2016 Managing Water, Soil, and Waste in the Context of Global Change. In H Hettiarachchi and R Ardakanian (Eds.), *Environmental resource management and the Nexus approach*, Cham: Springer, 1–7.

159 SC Moser and JA Ekstrom 2010 A framework to diagnose barriers to climate change adaptation. *Proceedings of the National Academy of Sciences,* 107(51), 22026–22031.

160 M Mandryk, P Reidsma, K Kartikasari, M van Ittersum and B Arts 2015 Institutional constraints for adaptive capacity to climate change in Flevoland's agriculture. *Environmental Science and Policy*, 48, 147–162; E Glaas, A Jonsson, M Hjerpe and Y Andersson-Sköld 2010 Managing climate change vulnerabilities: Formal institutions and knowledge use as determinants of adaptive capacity at the local level in Sweden. *Local Environment*, 15(6), 525–539; P Huntjens, L Lebel, C Pahl-Wostl, J Camkin, R Schulze and N Kranz 2012 Institutional design propositions for the governance of adaptation to climate change in the water sector. *Global Environmental Change,* 22(1), 67–81.

161 I Baker, A Peterson, G Brown and C McAlpine 2012 Local government response to the impacts of climate change: An evaluation of local climate adaptation plans. *Landscape and Urban Planning*, 107(2), 127–136; W Adger, S Dessai, M Goulden, M Hulme, I Lorenzoni, RD Nelson and A Wreford 2009 Are there social limits to adaptation to climate change? *Climatic Change*, 3(93), 335–354; P Mukheibir, N Kuruppu, A Gero and J Herriman 2013 Overcoming cross-scale challenges to climate change adaptation for local government: A focus on Australia. *Climatic Change*, 2(121), 271–283. See also MT Islam and M Nursey-Bray 2017 Adaptation to climate change in agriculture in Bangladesh: The role of formal institutions. *Journal of Environmental Management*, 200, 347–358; S Burch 2010 Transforming barriers into enablers of action on climate change: Insights from three municipal case studies in British Columbia, Canada. *Global Environmental Change*, 20(2), 287–297; SC Moser and JA Ekstrom 2010 A framework to diagnose barriers to climate change adaptation. *Proceedings of the National Academy of Sciences*, 107(51), 22026–22031; E Glaas, A Jonsson, M Hjerpe and Y Andersson-Sköld 2010 Managing climate change vulnerabilities: Formal institutions and knowledge use as determinants of adaptive capacity at the local level in Sweden. *Local Environment*, 15(6), 525–539; RM Wise, I Fazey, MS Smith, SE Park, HC Eakin, EA Van Garderen and B Campbell 2014 Reconceptualising adaptation to climate change as part of pathways of change and response. *Global Environmental Change*, 28, 325–336.

162 L Gomez-Echeverri 2010 *The reformed financial mechanism of the UNFCCC promoting transparency and accountability*. Oxford Institute for Energy Studies, 40.

163 H Brown, JN Nkem, DJ Sonwa and Y Bele 2010 Institutional adaptive capacity and climate change response in the Congo Basin forests of Cameroon. *Mitigation and Adaptation Strategies for Global Change,* 15(3), 263–282.

164 L Jones 2010 *Overcoming social barriers to adaptation*. Overseas Development Institute, Background Note 2.

165 RJ Klein, GF Midgley, BL Preston, M Alam, FG Berkhout, K Dow and A Thomas Adaptation Opportunities, Constraints, and Limits. In *Climate change 2014: Impacts, adaptation, and vulnerability. Part A: Global and sectoral aspects. Contribution of working group II to the fifth assessment report of the intergovernmental panel on climate change*, Cambridge: Cambridge University Press, 899.

166 J. JR Porter, L Xie, AJ Challinor, K Cochrane, SM Howden, MM Iqbal and MI Travasso 2014 Food security and food production systems. *Climate change 2014: Impacts, adaptation, and vulnerability*, (Part A), 485–533.

167 M Kandlinkar and J Risbey 2000 *Agricultural Impacts of Climate Change. If Adaptation Is the Answer, What Is the Question? Climatic Change*, 45, 529–539.

168 DR Nelson, WN Adger, and K Brown 2007 Adaptation to environmental change: Contributions of a resilience framework. *Annual Review of Environment and Resources*, 32(1), 395–419.

169 RLS Castellano and J Moroney 2018 Farming adaptations in the face of climate change. *Renewable Agriculture and Food Systems,* 33(3), 206–211.

170 J Blesh and S A Wolf 2014 Transitions to agroecological farming systems in the Mississippi River Basin: Toward an integrated socioecological analysis. *Agriculture and Human Values*, 31(4), 621–635.

171 S Li, L Juhász-Horváth, PA Harrison, L Pintér and MD Rounsevell 2017 Relating farmer's perceptions of climate change risk to adaptation behaviour in Hungary. *Journal of Environmental Management*, 185, 21–30.

172 C Walthall, J Hatfield, E Marshall, L Lengnick, P Backlund, S Adkins and J Bunce 2013 *Climate change and agriculture in the United States: Effects and adaptation*. USDA Technical Bulletin 1953, 2.

173 QP Chiotti and T Johnston 1995 Extending the boundaries of climate change research: A discussion on agriculture. *Journal of Rural Studies*, 11 (3), 335–350; RS Tol, S Fankhauser and JB Smith 1998 The scope for adaptation to climate change: What can we learn from the impact literature? *Global Environmental Change*, 8 (2), 109–123; B Smit, I Burton, RJ Klein and R Street 1999 The science of adaptation: A framework for assessment. *Mitigation and Adaptation Strategies for Global Change*, 4 (3–4), 199–213; CR Bryant, B Smit, M Brklacich, TR Johnston, J Smithers, Q Chiotti and B Singh 2000 Adaptation in Canadian agriculture to climatic variability and change. In *Societal adaptation to climate variability and change. Climatic Change*, 45(1), 181–201.

174 I Burton, KW Robert and WF Gilbert 1993 *The environment as hazard*. Guilford Press, 92; TCarter, MParry, HHarasawa and SNishioka 1994 IPCC technical guidelines for assessing climate change impacts and adaptations. In *Part of the IPCC Special Report to the First Session of the Conference of the Parties to the UN Framework Convention on Climate Change,* https://www.ipcc.ch/site/assets/uploads/2018/03/ipcc-technical-guidelines-1994n-1.pdf; and J Smithers and B Smit 1997 Human adaptation to climatic variability and change. *Global Environmental Change*, 7(2), 129–146.

175 See NV Fedoroff, DS Battisti, RN Beachy, PJ Cooper, DA Fischhoff, CN Hodges … and JK Zhu 2010 Radically rethinking agriculture for the 21st century. *Science*, 327(5967), 833-834; N Ahouissoussi, JE Neumann and JP Srivastava 2014 Building Resilience to Climate Change in South Caucasus Agriculture. *Directions in Development Agriculture and Rural Development*. World Bank Publications. http://hdl.handle.net/10986/18033.

176 S Dong, L Wen, S Liu, X Zhang, J Lassoie, S Yi and Y Li 2011 Vulnerability of worldwide pastoralism to global changes and interdisciplinary strategies for sustainable pastoralism. *Ecology and society*, 16(2), 1-23.

177 As a corollary, there are also off-farm adaptation strategies which refer to the ways by which "households prepare for and cope with a reduction in food and income through managing their consumption and capital assets". See CR Bryant, B Smit, M Brklacich, TR Johnston, J Smithers, Q Chiotti and B Singh 2000 Adaptation in Canadian agriculture to climatic variability and change. In *Societal adaptation to climate variability and change. Climatic Change*, 45(1), 181–201.

178 AN Abdulai and A Abdulai 2016 Allocative and scale efficiency among maize farmers in Zambia: A zero-efficiency stochastic frontier approach. *Applied Economics*, 48(55), 5364–5378.

179 A Abdulai 2018 Simon Brand Memorial address-the challenges and adaptation to climate change by farmers in sub-Saharan Africa. *Agrekon*, 57(1), 28–39; GW Kassie 2017 The Nexus between livelihood diversification and farmland management strategies in rural Ethiopia. *Cogent Economics and Finance*, 5(1), 1275087.

180 Land topography could be changed through land contouring and terrace farming, or through the construction of diversions and water-storage systems. These methods have the advantage of reducing rainwater runoff and soil erosion as well as retention of soil moisture and nutrients. See B Smit 1993 Adaptation to climatic variability and change. Report of the Task Force on Climate Adaptation No. GU/G--CE04610. Guelph University, 1–28; WE Easterling 1996 Adapting North American agriculture to climate change in review. *Agricultural and Forest Meteorology*, 80(1), 1–53; R De Loë, R Kreutzwiser and L Moraru 2001 Adaptation options for the near term: Climate change and the Canadian water sector. *Global Environmental Change*, 11(3), 231–245.

181 Alternative fallow is a farming practice where arable lands are without sowing for a certain period to allow the land to replenish its nutrients to further store organic matter. Fallowing while has been noted for its moisture- and nutrient-conservation

potentials in the event of frequent and reoccurring droughts. See Q Chiotti, T Johnston, B Smit and B Ebel 1997 Agricultural response to climatic change: A preliminary investigation of farm-level adaptation in Southern Alberta. UK: Cab International, 201–218; A Hucq, Kowalski J, L Gutek and R Gray 2000 Achieving understanding in agricultural GHG emission reduction, In D Scott, B Jones, J Audrey, R Gibson, P Key, L Mortsch and K Warriner (Eds.), *Climate Change Communication: Proceedings of an international conference, Kitchener-Waterloo, Canada, Environment Canada* 3–17; G Delcourt and Van GC Kooten 1995 How resilient is grain production to climatic change? Sustainable agriculture in a dryland cropping region of western Canada. *Journal of Sustainable Agriculture*, 5(3), 37–57.

182 S Klassen and J Gilpin 1999 Alberta irrigation in the old and new millennium. *Canadian Water Resources Journal*, 24(1), 61–70. See also B Smit 1993 Adaptation to climatic variability and change. *Report of the task force on climate adaptation (No. GU/G--CE04610)*. Guelph University, 1–28: WE Easterling 1996 Adapting North American agriculture to climate change in review. *Agricultural and Forest Meteorology*, 80(1), 53.

183 Irrigation has been noted to "improve and enable diversification of crops and livestock productions in light of climate-related changes". See M Brklacich, D McNabb, C Bryant and J Dumanski 1997 Adaptability of agricultural systems to global climate change: A Renfrew County, Ontario. *Canada Pilot Study*, 185–200.

184 GW Kassie 2017 The Nexus between livelihood diversification and farmland management strategies in rural Ethiopia. *Cogent Economics and Finance*, 5(1), 1275087.

185 Spraying of pesticides on farms can reduce the risks of climate change-induced pest on crops and livestock. See M Brklacich, D McNabb, C Bryant and J Dumanski 1997 Adaptability of agricultural systems to global climate change: A Renfrew County, Ontario. *Canada Pilot Study*, 185–200; A Hucq, J Kowalski, L Gutek and R Gray 2000 Achieving understanding in agricultural GHG emission reduction. In D Scott, B Jones, J Audrey, R Gibson, P Key, L Mortsch and K Warriner (Eds.), *Climate Change Communication: Proceedings of an international conference, Kitchener-Waterloo, Canada, Environment Canada* 3–17.

186 QP Chiotti and T Johnston 1995 Extending the boundaries of climate change research: A discussion on agriculture. *Journal of Rural Studies*, 11(3), 335–350.

187 Q Chiotti, T Johnston, B Smit and B Ebel 1997 Agricultural response to climatic change: A preliminary investigation of farm-level adaptation in Southern Alberta. In B Ilbery, Q Chiotti and TRickard (Eds.), *Agricultural restructuring and sustainability: a geographical perspective*, Wallingford (United Kingdom): CAB INTERNATIONAL, 201–218.

188 JA Delgado, PM Groffman, MA Nearing, T Goddard, D Reicosky, R Lal and P Salon 2011 Conservation practices to mitigate and adapt to climate change. *Journal of Soil and Water Conservation*, 66(4), 118A–129A.

189 H Blanco and R Lal 2010 *Principles of soil conservation and management*. Springer, 68.

190 H Blanco and R Lal 2009 Crop residue removal impacts on soil productivity and environmental quality. *Critical Reviews in Plant Science*, 28(3), 139–163; A Bot and J Bonites 2001 *Conservation Agriculture: Case Studies for Latin America and Africa*. FAO Soils Bulletin No. 78. Food and Agriculture Organization, 41.

191 JP Kaye and M Quemada 2017 Using cover crops to mitigate and adapt to climate change. A review. *Agronomy for Sustainable Development*, 37(1), 1–17.

192 AM Palese, N Vignozzi, G Celano, AE Agnelli, M Pagliai and C Xiloyannis 2014 Influence of soil management on soil physical characteristics and water storage in a mature rainfed olive orchard. *Soil and Tillage Research*, 144, 96–109.

193 RT Conant 2009 Challenges and opportunities for carbon sequestration in grassland systems: A Technical report on grassland management and climate change

mitigation Prepared for the Plant Production and Protection Division, Food and Agriculture Organization of the United Nations (FAO). *Integrated Crop Management*, 9, 1–67. See also GA Peterson and DG Westfall 2004 Managing precipitation use in sustainable dryland agroecosystems. *Annals of Applied Biology*, 144(2), 127–138.

194 FAO 2008 The State of Food And Agriculture. *Biofuels: prospects, risks and opportunities* http://www.fao.org/3/i0100e/i0100e00.htm.

195 H Blanco, and R Lal 2008 *Principles of soil conservation and management* Vol. 167169. Springer; RT Conant, M Easter, K Paustian, A Swan and S Williams 2007 Impacts of periodic tillage on soil C stocks: A synthesis. *Soil and Tillage Research*, 95(1–2), 1–10.

196 See for example, R Lal and WC Moldenhauer 1987 Effects of soil erosion on crop productivity. *Critical Reviews in Plant Sciences*, 5(4), 303–367.

197 SJ Capon, LE Chambers, R Mac Nally, RJ Naiman, P Davies, N Marshall and SE Williams 2013 Riparian ecosystems in the 21st century: Hotspots for climate change adaptation? *Ecosystems*, 16(3), 359–381.

198 S Hallegatte 2009 Strategies to adapt to an uncertain climate change. *Global Environmental Change*, 19(2), 240–247.

199 J Dumanski, DR Coote, G Luciuk and C Lok 1986 Soil conservation in Canada. *Journal of Soil and Water Conservation*, 41(4), 204–210; WE Easterling 1996 Adapting North American agriculture to climate change in review. *Agricultural and Forest Meteorology*, 80(1), 1–53.

200 J Smithers and A Blay-Palmer 2001 Technology innovation as a strategy for climate adaptation in agriculture. *Applied Geography*, 21(2), 175–197.

201 X Zhu, R Clements, A Quezada, J Torres and J Haggar 2011 Technologies for climae change adaptation-agriculture sector, 40; M Albaji, M Golabi, AR Hooshmand and M Ahmadee 2016 Investigation of surface, sprinkler and drip irrigation methods using GIS. *Jordan Journal of Agricultural Sciences*, 12(1), 211–222; DS Wilks 1992 Adapting stochastic weather generation algorithms for climate change studies. *Climatic Change*, 22(1), 67–84.

202 M Brklacich, D McNabb, C Bryant and J Dumanski 1997 Adaptability of agricultural systems to global climate change: A Renfrew County, Ontario, *Canada Pilot Study*, 185–200.

203 B Smit and MW Skinner 2002 Adaptation options in agriculture to climate change: A typology. *Mitigation and Adaptation Strategies for Global Change*, 7(1), 85–114.

204 WE Easterling 1996 Adapting North American agriculture to climate change in review. *Agricultural and Forest Meteorology*, 80(1), 1–53; R De Loë, R Kreutzwiser and L Moraru 2001 Adaptation options for the near term: Climate change and the Canadian water sector. *Global Environmental Change*, 11(3), 231–245.

205 K Warner, N Ranger, S Surminski, M Arnold, J Linnerooth-Bayer, E Michel-Kerjan and C Herweijer 2009 Adaptation to climate change: Linking disaster risk reduction and insurance. *United Nations International Strategy for Disaster Reduction*, Geneva 3.

206 CR Bryant, B Smit, M Brklacich, TR Johnston, J Smithers, Q Chiotti and B Singh 2000 Adaptation in Canadian agriculture to climatic variability and change. In *Societal adaptation to climate variability and change. Climatic Change*, 45(1), 181–201.

207 B Smit and MW Skinner 2002 Adaptation options in agriculture to climate change: A typology. *Mitigation and Adaptation Strategies for Global Change*, 7(1), 85–114.

208 Xia Y, Song Y and Pu M 2010 Analysis on The mutual aid fund system based on farmers cooperatives [J]. *Issues in Agricultural Economy*, 4, 29–33.

An index-based insurance is a form of insurance whereby a premium holder is compensated for a loss with reference to an index, for instance, poor weather or low rainfall. An indemnity-based insurance for agriculture is a form of insurance scheme where the farmer is paid an indemnity by the insurer for agricultural products damaged by drought or pest (also known as "multirisk" insurance). N Ntukamazina, RN Onwonga, R Sommer, JC Rubyogo, CM Mukankusi, J Mburu and R Kariuki 2017 Index-based agricultural insurance products: Challenges, opportunities and prospects for uptake in sub-Sahara Africa. *Journal of Agriculture and Rural Development in the Tropics and Subtropics,* 118(2), 171–185.

An indemnity-based insurance is usually paid following a costly damage assessment. As such, the insurance is largely subsidised and more than half of the premiums are paid by the government. For example, PROPAGRO in Brazil, INS in Costa Rica, CCIS in India, ANAGSA and the FONDEN programme in Mexico, PCIC in the Philippines, Agroseguro in Spain, and FCIC in the USA, for which every respective government pays for more than half of the premiums. See V Molini, MKeyzer, B van den Boom, W Zant, and N Nsowah-Nuamah 2010 Safety Nets and index-based insurance: Historical assessment and Semiparametric Simulation for Northern Ghana. *Economic Development and Cultural Change*, 58(4), 671–712.

Both forms of agricultural insurance schemes have been widely used in countries such as Ethiopia, Kenya, Malawi, and Tanzania to reduce the risk of droughts on crops and livestock. See PB Hazell 1992 The appropriate role of agricultural insurance in developing countries. *Journal of International Development,* 4(6), 567–581.

209 Herdt RW 2010 Development aid and agriculture. In *Handbook of agricultural economics,* (4), 3253–3304.

210 B Smit and MW Skinner 2002 Adaptation options in agriculture to climate change: A typology. *Mitigation and Adaptation Strategies for Global Change,* 7(1), 85–114. Invasive alien species are classified as either animals, plants, or other organisms that breed in places other than their natural habitats. They include Green Crab or the European crab (which is usually carried by ships in ballast water, eventually sold as fish bait to fishers). Other examples of invasive alien species include the Killer Algae, Sea Walnut, Veined Rapa Whelk, and the Zebra Mussel. Invasive alien species and noxious weeds are examples of the most common pests and diseases that affect crops and livestock. Climate change serves as vector for invasive alien species, as a result of which biodiversity is lost, while ecosystem services and human settlements become less resilient. See for example, P Pyšek, M Vilà, PE Hulme, D Simberloff, S Bacher, TM Blackburn and JT Carlton 2020 Scientists' warning on invasive alien species. *Biological Reviews,* 95(6), 1511–1534; Szyniszewska 2007 *Invasive species and climate change.* https://climate.org/archive/topics/ecosystems/invasivespecies.html; IUCN 2021 Invasive Alien Species: A major and growing threat to biodiversity and sustainable development. https://iucn.org/our-work/topic/invasive-alien-species.

211 J Ssozi, S Asongu and VH Amavilah 2019 The effectiveness of development aid for agriculture in Sub-Saharan Africa. *Journal of Economic Studies*, 46(2), 1–22.

212 Government of South Africa 2022 Guidelines for Presidential Employment Stimulus (PESI) Beneficiaries. https://www.dalrrd.gov.za/images/PESI/Guidelines/PESIguidelineENGLISH.pdf

213 The Anchor Borrowers' programme started operation in 2015 as a single-digit Loan. The programme is intended to boost the agricultural sector in Nigeria by creating the required economic linkage between small-scale subsistence farmers and agro-processors. Among others, it is expected that the programme would promote access to quality agricultural inputs by small-scale subsistence farmers.

See NIRSAL Microfinance Bank 2022 Anchor Borrowers Program (ABP) https://nmfb.com.ng/anchor-borrowers-program.

214 See K Hansen *Washington Post,* March 29, 2021.

215 See the International Institute for Sustainable Development 2010. http://sdg.iisd.org/events/the-hague-conference-on-agriculture-food-security-and-climate-change/.

216 FAO, 2015 Climate change and food security: risks and responses. https://www.fao.org/3/i5188e/I5188E.pdf 72.

217 MP Senyolo, TB Long, V Blok and O Omta 2018 How the characteristics of innovations impact their adoption: An exploration of climate-smart agricultural innovations in South Africa. *Journal of Cleaner Production*, 172, 3825–3840.

218 F Matteoli, J Schnetzer and H Jacobs 2020 Climate-smart agriculture (CSA): An integrated approach for climate change management in the agriculture sector. In *Handbook of climate change management: Research, leadership, transformation*, Cham: Springer International Publishing, 1–29.

219 FAO, 2010 Success Stories on Climate-Smart Agriculture. https://www.fao.org/3/i3817e/i3817e.pdf, 15.

220 FAO, 2010 Success Stories on Climate-Smart Agriculture. https://www.fao.org/3/i3817e/i3817e.pdf, 2.

221 Detailed discussion on the CSA strategies of Kenya, Nigeria, and South Africa are discussed further under the respective chapters of the three countries.

222 MA Altieri and CI Nicholls 2009 Agroecology, small farms, and food sovereignty. *Monthly Review,* 61(3), 102–113.

223 A sprinkler is an irrigation method, where through a network of pipes, pumps, and valves, water is applied on agricultural product similar to rainfall, but in a controlled way. IP Kruzhilin, AS Ovchinnikov, NV Kuznetsova, OV Kozinskaya, SD Fomin, VS Bocharnikov and ES Vorontsova 2018 Water pressure monitoring in irrigation piping as quality management tools of sprinkler irrigation. *ARPN Journal of Engineering and Applied Sciences*, 13(3), 4181–4184.

224 J Ensor and R Berger 2009 Community-based adaptation and culture in theory and practice. In *Adapting to climate change: Thresholds, values, governance*, Cambridge University Press, 227–239.

225 See Section 1.4.3 above.

226 C Rademacher-Schulz and ES Mahama 2012 *Rainfall, food security, and human mobility: Case study Ghana*. United Nations University Institute for Environment and Human Security.

227 P Newborne and NR Gansaonré 2017 Agriculture, Water, Climate and Migration in semi-arid lands in Burkina Faso. 'Pathways to Resilience in Semi-arid Economies' (PRISE) Working Paper. https://idl-bnc-idrc.dspacedirect.org/server/api/core/bitstreams/57ee78c3-6daf-4e03-a3b0-901ae572f639/content.

228 S Pavanello 2009 Pastoralists' vulnerability in the Horn of Africa: Exploring political marginalisation, donors' policies and cross-border issues–Literature review. Humanitarian Policy Group (HPG) Overseas Development Institute, London, 1–47.

229 M Desta 2008 *Epidemiology of child psychiatric disorders in Addis Ababa*, Ethiopia CARE International in Ethiopia; J Markakis 2004 *Pastoralism on the margin*. Minority Rights Group International, 2–37.

230 The assets of pastoralists have been classified as human assets (education, health, and nutrition), natural assets (grazing land and water sources), financial assets (livestock and credit), social assets (livestock and intra-community social support), and physical assets (livestock herd and infrastructure). S Pavanello 2009 *Pastoralists' vulnerability in the Horn of Africa: Exploring political marginalisation, donors' policies and cross-border issues–Literature review*. London: Humanitarian Policy Group (HPG) Overseas Development Institute, 1–47.

Livestock deserves special mention because it is considered as the main asset for pastoralists, which serve as a source of food, income, and reserve of wealth, at the same time used for gifts, exchanges, fines, etc., as a means of solidifying social relationships. For pastoralists, livestock is their bank and supermarkets. See A Taruvinga, V Muchenje and A Mushunje 2013 Climate change impacts and adaptations on small-scale livestock production. *International Journal of Development and Sustainability,* 2(2), 664–685.

231 K Rybarczyk 2021 The herder-farmer conflict in Nigeria – is there hope for the future? Institute of Development Studies. https://www.ids.ac.uk/opinions/the-herder-farmer-conflict-in-nigeria-is-there-hope-for-the-future/.

232 An example of this migration in Kenya occurred between 2004 and 2005, when farmers migrated to regions noted for poor soil quality to other regions with high soil quality. CL Gray 2011 Soil quality and human migration in Kenya and Uganda. *Global Environmental Change*, 21(2), 421–430.

233 N Sobania 2019 Pastoralist migration and colonial policy: A case study from northern Kenya. In *The Ecology of Survival*, Routledge, 219–239.

234 See International Organisation for Migration 2022. 15 Million People Face Humanitarian Crisis Due to Drought in the Horn of Africa. https://www.iom.int/news/15-million-people-face-humanitarian-crisis-due-drought-horn-africa.

235 E Bryan, C Ringler, B Okoba, C Roncoli, S Silvestri and M Herrero 2013 Adapting agriculture to climate change in Kenya: Household strategies and determinants. *Journal of Environmental Management*, 114, 26–35.

236 See D Dodman and D Mitlin 2013 Challenges for community-based adaptation: Discovering the potential for transformation. *Journal of International Development*, 25(5), 640–659.

237 H Wright, S Vermeulen, G Laganda, M Olupot, E Ampaire and ML Jat 2014 Mainstreaming adaptation in developing countries: The case of the Philippines. *Climate and Development*, 1(2), 318–328.

238 See H Reid, M Alam, R Berger, T Cannon, S Huq and A Milligan 2009 Community-based adaptation to climate change: An overview. *Participatory Learning and Action*, 60(1), 11–33.

239 See for example, J Ayers and T Forsyth 2009 Community-based adaptation to climate change. *Environment: Science and Policy for Sustainable Development*, 51(4), 22–31.

240 See for instance, L Bizikova, E Crawford, M Nijnik and R Swart 2014 Climate change adaptation planning in agriculture: Processes, experiences and lessons learned from early adapters. *Mitigation and Adaptation Strategies for Global Change*, 19(4), 411–430.

241 MW Skinner, RG Kuhn and AE Joseph 2001 Agricultural land protection in China: A case study of local governance in Zhejiang Province. *Land Use Policy*, 18(4), 329–340; J Smithers and B Smit 1997 Human adaptation to climatic variability and change. *Global Environmental Change*, 7(2), 129–146. In addition, agro-forestry techniques can be a source for animal fodder and fruits for human consumption. A Ajayi 2006 'An assessment of farmers' willingness to pay for extension services using the contingent valuation method (CVM): The case of Oyo State, Nigeria. *Journal of Agricultural Education and Extension,* 12(2), 97–108. Above all, agro-forestry has the advantage of "improving water management and crop output stability". DE Mercer 2004 Adoption of agroforestry innovations in the tropics: A review. *Agroforestry Systems,* 61(1–3), 311–328; S Franzel, GL Denning, JPB Lillesø and AR Mercado 2004 Scaling up the impact of agroforestry: Lessons from three sites in Africa and Asia. *Agroforestry Systems,* 61, 329–344.

242 Research institutions are saddled with the task of translating scientific information and indigenous knowledge into action. KPC Rao, LV Verchot and J Laarman 2007

Adaptation to climate change through sustainable management and development of agroforestry systems. *Journal of SAT Agricultural Research*, 4(1), 1–30.
243 NJ Noble, T Iorliam and CJ Arene 2014 Economic effects of climate change on Sorghum in North Central Nigeria. *International Journal of Development and Sustainability*, 3(7), 1517–1525.
244 Y Kim, JB Smith, C Mack, J Cook, J Furlow, JL Njinga and M Cote 2017 A perspective on climate-resilient development and national adaptation planning based on USAID's experience. *Climate and Development*, 9(2), 141–151.
245 However, NAPs must not be confused with National Adaptation Programme of Action (NAPA) as these are designed to operate for a short-term, while a NAP is designed for medium- and long-term operation and is usually integrated with national sustainable development plans. See G Kissinger and T Namgyel 2014 NAPAs and NAPs in least developed countries, *Climate and Development Knowledge Network (CDKN). South Africa.* https://cdkn.org/sites/default/files/files/LDC-briefing_NAPs.pdf, 1–4.
246 See for example Government of Canada 2011 Federal Adaptation Policy Framework for climate change https://www.canada.ca/en/environment-climate-change/services/climate-change/federal-adaptation-policy-framework.html#_5.
247 DS Lemmen, FJ Warren, and J Lacroix 2007 *From impacts to adaptation: Canada in a changing climate 2007: Synthesis*, Canada.
248 L Bizikova, E Crawford, M Nijnik and R Swart 2014 Climate change adaptation planning in agriculture: Processes, experiences and lessons learned from early adapters. *Mitigation and Adaptation Strategies for Global Change*, 19(4), 411–430.
249 P Debaeke, S Pellerin and E Scopel 2017 Climate-smart cropping systems for temperate and tropical agriculture: Mitigation, adaptation and trade-offs. *Cahiers Agricultures*, 26(3), 34002.
250 In this instance, there are heat- and drought-resistant maize varieties developed by the international Maize and Wheat Improvement Centre (CIMMYT) and their partners, and used by farmers in Malawi, South Africa, Tanzania, and Zimbabwe. See Salinger, Sivakumar and Motha 2005 Increasing climate variability and change. *Climatic Change*, 70, 341–362.
251 The Committee on Approaches to Climate Change Adaptation 2010 https://www.env.go.jp/en/earth/cc/adapt_guide/pdf/approaches_to_adaptation_en.pdf.
252 R Jones 2010 A risk management approach to climate change adaptation *In* RA Nottage, DS Wratt, J F Bornman and K Jones 2010 Climate change adaptation in New Zealand. Future scenarios and some sectoral perspectives, *New Zealand Climate Change Centre, NIWA, Wellington*. 10–25; LC Stringer, JC Dyer, MS Reed, AJ Dougill, C Twyman and D Mkwambisi 2009 Adaptations to climate change, drought and desertification: Local insights to enhance policy in southern Africa. *Environmental Science and Policy,* 12(7), 748–765.
253 C Arcidiacono 2018 Engineered solutions for animal heat stress abatement in livestock buildings. *Agricultural Engineering International CIGR Journal*, 20.
254 WE Walker, M Haasnoot and JH Kwakkel 2013 Adapt or perish: A review of planning approaches for adaptation under deep uncertainty. *Sustainability*, 5(3), 955–979.
255 E Ajani, R Mgbenka and M Okeke 2013 Use of indigenous knowledge as a strategy for climate change adaptation among farmers in sub-Saharan Africa: Implications for policy. *Asian Journal of Agricultural Extension, Economics and Sociology,* 2(1), 23-40.
256 E Ajani, R Mgbenka and M Okeke 2013 Use of indigenous knowledge as a strategy for climate change adaptation among farmers in sub-Saharan Africa: Implications for policy. *Asian Journal of Agricultural Extension, Economics and Sociology*, 2(1), 23–40.

257 SM Howden, JF Soussana, FN Tubiello, N Chhetri, M Dunlop and H Meinke 2007 Adapting agriculture to climate change. *Proceedings of the National Academy of* Sciences, 104(50), 19691–19696.

258 B Lim, I Burton, E Spanger-Siegfried, EL Malone and S Huq 2005 *Adaptation policy frameworks for climate change*. Cambridge University Press, 49–60; Climate-ADAPT2017. Adaptation options.https://climate-adapt.eea.europa.eu/knowledge/adaptation-information/adaptation-measures.

259 WN Adger et al. 2009 WN Adger,SDessai, M Goulden, M Hulme, I Lorenzoni, DR Nelson, and A Wreford Are there social limits to adaptation to climate change? *Climatic Change*, 93, 335–354.

260 M Murthy, R Tiwari and NH Ravindranath 2011 Climate change and forests in India: Adaptation opportunities and challenges. *Mitigation and Adaptation Strategies for Global Change*, 16, 161–175.

261 M Murthy, R Tiwari and NH Ravindranath 2011 Climate change and forests in India: Adaptation opportunities and challenges. *Mitigation and Adaptation Strategies for Global Change*, 16, 161–175.

262 M Tafere, J Olivier and M Jordaan 2013 Major human and Natural Disruptions that facilitated Vegetation cover removal in the Ethiopia Highlands. *Journal of Environment and Earth Science*, 3(9), 52–67.

263 M Chambwera, G Heal, C Dubeux, S Hallegatte, L Leclerc, A Markandya and JE Neumann 2014 Economics of adaptation.In CB Field, VR Barros, DJ Dokken, KJ Mach, MD Mastrandrea, TE Bilir, M Chatterjee, KL Ebi, YO Estrada, RC Genova, B Girma, ES Kissel, AN Levy, S MacCracken, PR Mastrandrea, and LL White (Eds.), *Climate Change 2014: Impacts, Adaptation, and Vulnerability. Part A: Global and Sectoral Aspects. Contribution of Working Group II to the Fifth Assessment Report of the Intergovernmental Panel on Climate Change*. Cambridge, United Kingdom and New York, NY, USA: Cambridge University Press, 1–33.

264 M Keshavarz, H Maleksaeidi and E Karami 2017 Livelihood vulnerability to drought: A case of rural Iran. *International Journal of Disaster Risk Reduction*, 21, 223–230.

265 Transaction costs that affect adaptation could also include lack of access to markets and "weak extension and credit institutions". See TT Deressa, RM Hassa, C Ringler, T Alemu and M Yesuf 2009 Determinants of farmers' choice of adaptation methods to climate change in the Nile Basin of Ethiopia. *Global Environmental Change*, 19(2), 248–255.

266 AE Piggott-McKellar, KE McNamara, PD Nunn and JE Watson 2019 What are the barriers to successful community-based climate change adaptation? A review of grey literature. *Local Environment*, 2(4), 374–390; S De Jalón García, S Silvestri, A Granados and A Iglesias 2015 Behavioural barriers in response to climate change in agricultural communities: An example from Kenya. *Regional Environmental Change*, 15, 851–865.

267 Information asymmetries consist of the required information and the mode of technology to support information dissemination. A Carlsson-Kanyama, H Carlsen, and KH Dreborg 2013 Barriers in municipal climate change adaptation: Results from case studies using backcasting. *Futures*, 49, 9–21.

268 NL Engle 2011 Adaptive capacity and its assessment. *Global Environmental Change*, 21(2), 647–656; P Mukheibir, N Kuruppu, A Gero and J Herriman 2013 Overcoming cross-scale challenges to climate change adaptation for local government: A focus on Australia. *Climatic Change*, 2(121), 271–283.

269 M Chambwera, G Heal, C Dubeux, S Hallegatte, L Leclerc, A Markandya and JE Neumann 2014 Economics of adaptation. In CB Field, VR Barros, DJ Dokken, KJ Mach, MD Mastrandrea, TE Bilir, M Chatterjee, KL Ebi, YO Estrada, RC Genova, B Girma, ES Kissel, AN Levy, S MacCracken, PR Mastrandrea, and LL White (Eds.), *Climate Change 2014: Impacts, Adaptation, and Vulnerability. Part A: Global*

and Sectoral Aspects. Contribution of Working Group II to the Fifth Assessment Report of the Intergovernmental Panel on Climate Change. Cambridge, United Kingdom and New York, NY, USA: Cambridge University Press, 1–33.

270 R Mwebaza and LJ Kotzé 2009 Environmental governance and climate change in Africa: Legal perspectives. *Institute for Security Studies Monographs,* (167), 283.

271 L Lipper, P Thornton, BM Campbell, T Baedeker, A Braimoh, M Bwalya and EF Torquebiau 2014 Climate-smart agriculture for food security. *Nature Climate Change,* 4(12), 1068–1072.

References

Aaheim A and Aasen M 2008 What do we know about the economics of adaptation? *CEPS Policy Brief,* 150.

Abdulai A 2018 *Simon Brand Memorial address-the challenges and adaptation to climate change by farmers in sub-Saharan Africa.* Agrekon, 57(1), 28–39.

Abdulai AN and Abdulai A 2016 Allocative and scale efficiency among maize farmers in Zambia: A zero-efficiency stochastic frontier approach. *Applied Economics,* 48(55), 5364–5378.

Adger WN, Arnell, NW and Tompkins, EL 2005 Adapting to climate change: Perspectives across scales. *Global Environmental Change,* 15(2), 75–76.

Adger WN, Nelson DR and Brown K 2007 Adaptation to environmental change: Contributions of a resilience framework. *Annual Review of Environment and Resources,* 32(1), 395–419.

Ajani E, Mgbenka R and Okeke M 2013 Use of indigenous knowledge as a strategy for climate change adaptation among farmers in sub-Saharan Africa: Implications for policy. *Asian Journal of Agricultural Extension, Economics & Sociology,* 2(1), 24–40.

Ajayi A 2006 'An assessment of farmers' willingness to pay for extension services using the contingent valuation method (CVM): The case of Oyo State, Nigeria. *Journal of Agricultural Education and Extension,* 12(2), 97–108.

Albaji M, Golabi M, Hooshmand AR and Ahmadee M 2016 Investigation of surface, sprinkler and drip irrigation methods using GIS. *Jordan Journal of Agricultural Sciences,* 12(1), 211–222.

Altieri MA 2009 Agroecology, small farms, and food sovereignty. *Monthly Review,* 61(3), 102–113.

Angelo MJ 2009 Corn, carbon, and conservation, rethinking US agricultural policy in a changing global environment. *George Mason Law Review,* 17, 593.

Angelo MJ and Du Plessis A (Eds.) 2017 *Research handbook on climate change and agricultural law.* Edward Elgar Publishing, 20.

Anita W, Dominic M and Neil A 2010 *Climate change and agriculture impacts, adaptation and mitigation: Impacts, adaptation and mitigation.* OECD Publishing, 1–62.

Antle JM, Capalbo SM, Elliott ET and Paustian KH 2004 Adaptation, spatial heterogeneity, and the vulnerability of agricultural systems to climate change and CO2 fertilization: An integrated assessment approach. Climatic Change, 64(3), 289–315.

Antwi-Agyei P, Dougill AJ, Agyekum TP and Stringer LC 2018 Alignment between nationally determined contributions and the sustainable development goals for West Africa. *Climate Policy,* 18(10), 1296–1312.

Antwi-Agyei P, Dougill AJ, Stringer LC and Codjo SN 2018 Adaptation opportunities and maladaptive outcomes in climate vulnerability hotspots of northern Ghana. *Climate Risk Management,* 19, 83–93.

Anwar MR, Liu DL, Macadam I and Kelly G 2013 Adapting agriculture to climate change: A review. *Theoretical and Applied Climatology*, 113(1–2), 225–245.

Arcidiacono C 2018 Engineered solutions for animal heat stress abatement in livestock buildings. *Agricultural Engineering International: CIGR Journal* 20.

Ardo J and Bashir AE 2007 Agroforestry-an adaptation to climate change in the Sahel. *A Research Plan*, 6, 1–12.

Arnell NW 2004 Climate change and global water resources: SRES emissions and socio-economic scenarios. *Global Environmental Change*, 14(1), 31–52.

Awange JL, Ogalo L, Bae KH, Were P, Omondi P, Omute P and Omullo M 2008 Falling Lake Victoria water levels: Is climate a contributing factor? *Climatic Change*, 89, 281–297.

Ayers J, and Forsyth T 2009. Community-based adaptation to climate change. *Environment: Science and Policy for Sustainable Development*, 51(4), 22–31.

Baker I, Peterson A, Brown G and McAlpine C 2012 Local government response to the impacts of climate change: An evaluation of local climate adaptation plans. *Landscape and Urban Planning*, 107(2), 127–136.

Bauer A, Feichtinger J and Steurer R 2012 The governance of climate change adaptation in 10 OECD countries: Challenges and approaches. *Journal of Environmental Policy and Planning*, 14(3), 279–304.

Benhin JK 2006 Climate change and South African agriculture: Impacts and adaptation options Centre for Environmental Economics and Policy in Africa CEEPA Discussion Paper, 21.

Berrang-Ford L, Ford JD and Paterson J 2011 Are we adapting to climate change? *Global Environmental Change*, 21(1), 25–33.

Bizikova L, Crawford E, Nijnik M and Swart R 2014 Climate change adaptation planning in agriculture: Processes, experiences and lessons learned from early adapters. *Mitigation and Adaptation Strategies for Global Change*, 19(4), 411–430.

Blaikie P, Cannon T, Davis I, and Wisner B 2004. *At risk: Natural hazards, people's vulnerability and disasters*. Routledge.

Blanco H and Lal R 2008 *Principles of soil conservation and management*. Springer 167169.

Blanco E and Razzaque J 2011 *Globalisation and natural resources law: Challenges, key issues and perspectives.* Edward Elgar Publishing, 203.

Blesh J and Wolf S A 2014 Transitions to agroecological farming systems in the Mississippi River Basin: Toward an integrated socioecological analysis. *Agriculture and Human Values*, 31(4), 621–635.

Bot A and Bonites J 2001 *Conservation Agriculture: Case Studies for Latin America and Africa*. FAO Soils Bulletin No.78, Food and Agriculture Organization, 41.

Botai CM, Botai J O and Adeola AM 2018 Spatial distribution of temporal precipitation contrasts in South Africa. *South African Journal of Science*, 114(7–8), 70–78.

Brklacich M, McNabb D, Bryant C and Dumanski J 1997 Adaptability of agricultural systems to global climate change: A renfrew County, Ontario, Canada Pilot Study, 185–200.

Brown H, Nke JN, Sonwa DJ and Bele Y 2010 Institutional adaptive capacity and climate change response in the Congo Basin forests of Cameroon. *Mitigation and Adaptation Strategies for Global Change*, 15(3), 263–282.

Bryan E, Deressa TT, Gbetibouo GA and Ringler C 2009 Adaptation to climate change in Ethiopia and South Africa: Options and constraints. *Environmental Science and Policy*, 12(4), 413–426.

Bryan E, Ringler C, Okoba B, Roncoli C, Silvestri S and Herrero M 2013 Adapting agriculture to climate change in Kenya: Household strategies and determinants. *Journal of Environmental Management*, 114, 26–35.
Bryant CR, Smit B, Brklacich M, Johnston TR, Smithers J, Chiotti Q and Singh B 2000 Adaptation in Canadian agriculture to climatic variability and change. In *Societal adaptation to climate variability and change*. Springer, 45(1), 181–201.
Burch S 2010 Transforming barriers into enablers of action on climate change: Insights from three municipal case studies in British Columbia, Canada. *Global Environmental Change*, 20(2), 287–297.
Burton I 1993 *The environment as hazard*. Guilford Press, 92.
Calzadilla A, Zhu T, Rehdanz K, Tol RS and Ringler C 2014 Climate change and agriculture: Impacts and adaptation options in South Africa. *Water Resources and Economics*, 5, 24–48.
Capon SJ, Chambers LE, Mac Nally R, Naiman RJ, Davies P, Marshall N and Williams SE 2013 Riparian ecosystems in the 21st century: Hotspots for climate change adaptation? *Ecosystems*, 16(3), 359–381.
Carlsson J, Köhlin G and Ekbom A 2016 *The political economy of evaluation: International aid agencies and the effectiveness of aid*. Springer, 109.
Carter et al. 1994 Report of Working Group II of the IPCC. https://www.ipcc-wg3.de/special-reports/.files-images/ipcc-technical-guidelines-1994n.pdf.
Castellano RLS and Moroney J 2018 Farming adaptations in the face of climate change. *Renewable Agriculture and Food Systems*, 33(3), 206–211.
Chambwera M, Heal G, Dubeux C, Hallegatte S, Leclerc L, Markandya A and Neumann JE 2014 Economics of adaptation.In CB Field, VR Barros, DJ Dokken, KJ Mach, MD Mastrandrea, TE Bilir, M Chatterjee, KL Ebi, YO Estrada, RC Genova, B Girma, ES Kissel, AN Levy, S MacCracken, PR Mastrandrea, and LL White (Eds.), *Climate Change 2014: Impacts, Adaptation, and Vulnerability. Part A: Global and Sectoral Aspects. Contribution of Working Group II to the Fifth Assessment Report of the Intergovernmental Panel on Climate Change*. Cambridge, United Kingdom and New York, NY, USA: Cambridge University Press, 1–33.
Chang SJ 2013 Solving the problem of carbon dioxide emissions. *Forest Policy and Economics*, 35, 92–97.
Chiotti QP and Johnston T 1995 Extending the boundaries of climate change research: A discussion on agriculture. *Journal of Rural Studies*, 11(3), 335–350.
Chiotti Q, Johnston T, Smit B and Ebel B 1997 *Agricultural response to climatic change: A preliminary investigation of farm-level adaptation in Southern Alberta*. UK: Cab International, 201–218.
Climate Policy Integration, Coherence and Governance, *Partnership for European Environmental Research* 2009 PEER Report No. 2, 15.
Cliquet A, Backes C, Harris J and Howsam P 2009 Adaptation to climate change Legal challenges for protected areas. *Utrecht Law Review*, 5, 158–175.
Conant RT 2009 Rebuilding resilience: Sustainable land management for climate mitigation and adaptation. *Resources Management and Environment Department*, 6252.
Conant RT, Easter M, Paustian K, Swan A and Williams S 2007 Impacts of periodic tillage on soil C stocks: A synthesis. *Soil and Tillage Research*, 95(1-2), 1–10.
Craig RK 2010 Stationarity is dead-long live transformation: Five principles for climate change adaptation law. *Harvard Environmental Law Review*, 34(1), 9.
Davoudi S, Shaw K, Haider LJ, Quinlan AE, Peterson GD, Wilkinson C and Davoudi, S 2012 Resilience: A bridging concept or a dead end? *Planning Theory and Practice*, 13(12), 299–333.

De Jalón García S, Silvestri S, Granados A and Iglesias A 2015 Behavioural barriers in response to climate change in agricultural communities: An example from Kenya. *Regional Environmental Change*, 15, 851–865.

De Loë R, Kreutzwiser R and Moraru L 2001 Adaptation options for the near term: Climate change and the Canadian water sector. *Global Environmental Change*, 11(3), 231–245.

Debaeke P, Pellerin S and Scopel E 2017 Climate-smart cropping systems for temperate and tropical agriculture: Mitigation, adaptation and trade-offs. *Cahiers Agricultures*, 26(3), 34002.

Delcourt G and Van Kooten GC 1995 How resilient is grain production to climatic change? Sustainable agriculture in a dryland cropping region of western Canada. *Journal of Sustainable Agriculture*, 5(3), 37–57.

Delgado JA, Groffman PM, Nearing MA, Goddard T, Reicosky D, Lal R and Salon, P 2011 Conservation practices to mitigate and adapt to climate change. *Journal of Soil and Water Conservation*, 66(4), 118A–129A.

Deressa TT, Hassa RM, Ringler C, Alemu T and Yesuf M 2009 Determinants of farmers' choice of adaptation methods to climate change in the Nile Basin of Ethiopia. *Global Environmental Change*, 19(2), 248–255.

Desta M 2008 *Epidemiology of child psychiatric disorders in Addis Ababa, Ethiopia*. Doctoral dissertation, Barn-och ungdomspsykiatri.

Dodman D and Mitlin D 2013 Challenges for community-based adaptation: Discovering the potential for transformation. *Journal of International Development*, 25(5), 640–659.

Dong G, Wang L, Cui Y, Elston R and Chen F 2013 The spatiotemporal pattern of the Majiayao cultural evolution and its relation to climate change and variety of subsistence strategy during late Neolithic period in Gansu and Qinghai Provinces, northwest China. *Quaternary International*, 316, 155–161.

Dumanski J, Coote DR, Luciuk G and Lok C 1986 Soil conservation in Canada. *Journal of Soil and Water Conservation*, 41(4), 204–210.

Easterling WE 1996 Adapting North American agriculture to climate change in review. *Agricultural and Forest Meteorology*, 80(1), 1–53.

Ensor J and Berger R 2009 Community-based adaptation and culture in theory and practice. In *Adapting to climate change: Thresholds, values, governance*, Cambridge University Press, 227–239.

Erdoğan S, Yıldırım DÇ and Gedikli A 2020 Natural resource abundance, financial development and economic growth: An investigation on Next-11 countries. *Resources Policy*, 65, 101559.

Eriksen SH and O'Brien K 2007 Vulnerability, poverty and the need for sustainable adaptation measures. *Climate Policy*, 7(4), 337–352.

Eriksen S and Brown K 2011 Sustainable adaptation to climate change. *Climate and Development*, 3(1), 3–6.

Eriksen SH, Brown K and Kelly PM 2005 The dynamics of vulnerability: Locating coping strategies in Kenya and Tanzania. *Geographical Journal*, 171(4), 287–305.

Eriksen SH, Nightingale A J and Eakin H 2015 Reframing adaptation: The political nature of climate change adaptation. *Global Environmental Change*, 35, 523–533.

Falkenmark M 1997 Meeting water requirements of an expanding world population. *Philosophical Transactions of the Royal Society of London. Series B: Biological Sciences*, 352(1356), 929–936.

Falkenmark M and Rockström J 2006 The new blue and green water paradigm: Breaking new ground for water resources planning and management. *Journal of Water Resources Planning and Management*, 132(3), 129–132.

Fedoroff NV, Battisti DS, Beachy RN, Cooper PJ, Fischhoff DA, Hodges CN ... and Zhu JK 2010 Radically rethinking agriculture for the 21st century. *Science*, 327(5967), 833–834.

Field CB, Barros V, Stocker TF and Dahe Q (Eds.) 2012 *Managing the risks of extreme events and disasters to advance climate change adaptation: Special report of the Intergovernmental Panel on Climate Change*. Cambridge University Press, 231–436.

Field CB and Barros VR (Eds.) 2014 *Climate change 2014–Impacts, adaptation and vulnerability: Regional aspects*. Cambridge University Press, 1758.

Foden WB and Young BE (Eds.) 2016 *IUCN SSC guidelines for assessing species' vulnerability to climate change. Version 1.0. Occasional Paper of the IUCN Species Survival Commission No.* 59. IUCN Species Survival Commission, 114.

Ford JD, Berrang-Ford L, Bunce A, McKay C, Irwin M and Pearce T 2015 The status of climate change adaptation in Africa and Asia. *Regional Environmental Change*, 15(5), 801–814.

Franzel S, Denning GL, Lillesø JPB and Mercado AR 2004 Scaling up the impact of agroforestry: Lessons from three sites in Africa and Asia. *Agroforestry Systems*, 61, 329–344.

Gioli G, Khan T, Bisht S and Scheffran J 2014 Migration as an adaptation strategy and its gendered implications: A case study from the Upper Indus Basin. *Mountain Research and Development*, 34(3), 255–265.

Giorgi F 2019 Thirty years of regional climate modeling: Where are we and where are we going next? *Journal of Geophysical Research: Atmospheres*, 124(11), 5696–5723.

Gitz V, Meybeck A, Lipper L, Young CD and Braatz S 2016 Climate change and food security: Risks and responses. *Food and Agriculture Organization of the United Nations (FAO) Report*, 2–4.

Glaas E, Jonsson A, Hjerpe M and Andersson-Sköld Y 2010 Managing climate change vulnerabilities: Formal institutions and knowledge use as determinants of adaptive capacity at the local level in Sweden. *Local Environment*, 15(6), 525–539.

Glatzle A 2014 Questioning key conclusions of FAO publications 'Livestock's Long Shadow' (2006). *Appearing Again in 'Tackling Climate Change Through Livestock' Pastoralism*, 4(1), 1–6.

Godde CM, Boone RB, Ash AJ, Waha K, Sloat LL, Thornton PK and Herrero M 2020 Global rangeland production systems and livelihoods at threat under climate change and variability. *Environmental Research Letters*, 15(4), 044021.

Gomez-Echeverri L 2010 *The reformed financial mechanism of the UNFCCC promoting transparency and accountability*. Oxford Institute for Energy Studies, 40.

Grossman L 1977 Man-environment relationships in anthropology and geography. *Annals of the Association of American Geographers*, 67(1), 126–144.

Guerrero AB and Muñoz E 2018 Life cycle assessment of second-generation ethanol derived from banana agricultural waste: Environmental impacts and energy balance. *Journal of Cleaner Production*, 174, 710–717.

Hallegatte S 2009 Strategies to adapt to an uncertain climate change. *Global Environmental Change*, 19(2), 240–247.

Hallegatte S, Fay M and Barbier EB 2018 Poverty and climate change: Introduction. *Environment and Development Economics*, 23(3), 217–233.

Hampshire K 2002 Fulani on the move: Seasonal economic migration in the Sahel as a social process. *Journal of Development Studies*, 38(5), 15–36.

Hazell PB 1992 The appropriate role of agricultural insurance in developing countries. *Journal of International Development*, 4(6), 567–581.

Herdt RW 2010 Development aid and agriculture. In *Handbook of agricultural economics*, (4), 3253–3304.

Hettiarachchi H and Ardakanian R 2016 Managing Water, Soil, and Waste in the Context of Global Change. In H Hettiarachchi and R Ardakanian (Eds.), *Environmental resource management and the Nexus approach*, Cham: Springer, 1–7.

Houghton RA 2005 Tropical deforestation as a source of greenhouse gas emissions. In *Tropical deforestation and climate change*, Jul 1, 13.

Howden SM, Soussana JF, Tubiello FN, Chhetri N, Dunlop M and Meinke H 2007 Adapting agriculture to climate change. *Proceedings of the National Academy of Sciences*, 104(50), 19691–19696.

Hucq A, Kowalski J, Gutek L and Gray R 2000 *Achieving understanding in agricultural GHG emission reduction*. In D Scott, B Jones, J Audrey, R Gibson, P Key, L Mortsch and K Warriner (Eds.), *Climate Change Communication: Proceedings of an international conference, Kitchener-Waterloo, Canada, Environment Canada*, 3–17.

Huho JM and Kosonei RC 2014 Understanding extreme climatic events for economic development in Kenya. *IOSR Journal of Environmental Science, Toxicology and Food Technology*, 8(2), 14–24.

Hulme M, Doherty R, Ngara T, New M and Lister D 2001 African climate change: 1900–2100. *Climate Research*, 17(2), 145–168.

Huntjens P, Lebel L, Pahl-Wostl C, Camkin J, Schulze R and Kranz N 2012 Institutional design propositions for the governance of adaptation to climate change in the water sector. *Global Environmental Change*, 22(1), 67–81.

Ifejika Speranza C 2010 Resilient adaptation to climate change in African Agriculture Studies 54, 3.

Ignaciuk A and Mason-D'Croz D 2014 Modelling adaptation to climate change in agriculture. *OECD Food* Agriculture and Fisheries Papers, 58.

Islam MT and Nursey-Bray M 2017 Adaptation to climate change in agriculture in Bangladesh: The role of formal institutions. *Journal of Environmental Management*, 200, 347–358.

Jobbins G, Conway D, Fankhauser S, Gueye B, Liwenga E, Ludi E and Suleri A 2016 *Resilience, equity and growth in semi-arid economies a research agenda*, 1–36.

Jones R 2010 A risk management approach to climate change adaptation. In R A C Nottage, D S Wratt, J F Bornman, K Jones (Eds.), *Climate change adaptation in New Zealand: Future scenarios and some sectoral perspectives*, 10–25.

Jones L 2010 Overcoming social barriers to adaptation. Overseas Development Institute, Background Note, 1–7.

Kankara AI, Adamu GK, Tukur R and Ibrahim A 2013 Examining environmental policies and laws in Nigeria. *International Journal of Environmental Engineering and Management*, 4(3), 165–170.

Kassie GW 2017 The Nexus between livelihood diversification and farmland management strategies in rural Ethiopia. *Cogent Economics and Finance*, 5(1), 1275087.

Kaye JP and Quemada M 2017 Using cover crops to mitigate and adapt to climate change. A review. *Agronomy for Sustainable Development*, 37(1), 1–17.

Keshavarz M, Maleksaeidi H and Karami E 2017 Livelihood vulnerability to drought: A case of rural Iran. *International Journal of Disaster Risk Reduction*, 21, 223–230.

Kim Y, Smith JB, Mack C, Cook J, Furlow J, Njinga JL and Cote M 2017 A perspective on climate-resilient development and national adaptation planning based on USAID's experience. *Climate and Development*, 9(2), 141–151.

Kissinger G and Namgyel T 2014 *NAPAs and NAPs in least developed countries, Climate and Development Knowledge Network (CDKN). South Africa*. https://cdkn.org/sites/default/files/files/LDC-briefing_NAPs.pdf, 1–4.

Klassen S and Gilpin J 1999 Alberta irrigation in the old and new millennium. *Canadian Water Resources Journal*, 24(1), 61–70.

Klein RJ, Nicholls RJ and Thomalla F 2003 Resilience to natural hazards: How useful is this concept? *Global Environmental Change Part B: Environmental Hazards*, 5(1), 35–45.

Korres NE, Norsworthy JK, Burgos NR and Oosterhuis DM 2017 Temperature and drought impacts on rice production: An agronomic perspective regarding short-and long-term adaptation measures. *Water Resources and Rural Development*, 100(9), 12–27.

Kruzhilin IP, Ovchinnikov AS, Kuznetsova NV, Kozinskaya OV, Fomin SD, Bocharnikov VS and Vorontsova ES 2018 Water pressure monitoring in irrigation piping as quality management tools of sprinkler irrigation. *ARPN Journal of Engineering and Applied Sciences*, 13(3), 4181–4184.

Kumamoto M and Mills A 2012 What African countries perceive to be adaptation priorities: Results from 20 countries in the Africa adaptation programme. *Climate and Development*, 4(4), 265–274.

Lafferty W and Hovden E 2003 Environmental policy integration: Towards an analytical framework. *Environmental Politics*, 12(3), 1–22.

Lal R and Moldenhauer WC 1987 Effects of soil erosion on crop productivity. *Critical Reviews in Plant Sciences*, 5(4), 303–367.

Lasco RD, Pulhin FB, Jaranilla-Sanchez PA, Delfino RJP, Gerpacio R and Garcia K 2009 Mainstreaming adaptation in developing countries: The case of the Philippines. *Climate and Development*, 1(2), 130–146.

Lebel L, Li L, Krittasudthacheewa C, Juntopas M, Vijitpan T, Uchiyama T and Krawanchid D 2012 *Mainstreaming climate change adaptation into development planning*. Stockholm Environment Institute, 8.

Leighton M 2006 *Desertification and migration. Governing global desertification*. Ashgate, 43–58.

Li S, Juhász-Horváth L, Harrison PA, Pintér L and Rounsevell MD 2017 Relating farmer's perceptions of climate change risk to adaptation behaviour in Hungary. *Journal of Environmental Management*, 185, 21–30.

Lim B, Burton I, Spanger-Siegfried E, Malone EL and Huq S 2005 *Adaptation policy frameworks for climate change*. Cambridge University Press, 49–60.

Lipper L, Thornton P, Campbell BM, Baedeker T, Braimoh A, Bwalya M and Torquebiau EF 2014 Climate-smart agriculture for food security. *Nature Climate Change*, 4(12), 1068–1072.

Lumsden TG, Schulze RE and Hewitson BC 2009 Evaluation of potential changes in hydrologically relevant statistics of rainfall in Southern Africa under conditions of climate change. *Water SA*, 35(5), 649–656.

Magnan AK and Ribera T 2016 Global adaptation after Paris. *Science*, 352(6291), 1280–1282.

Maji IK, Habibullah MS and Yusof-Saari M 2016 Emissions from agricultural sector and financial development in Nigeria: An Empirical study. *International Journal of Economics and Management*, 10(1), 1–15.

Malak DA, McGlade K, Pascual D and Pla E 2017 *Adapting to climate change: An assessment of vulnerability and risks to human security in the Western Mediterranean Basin*. Springer, 13–61.

Mandryk M, Reidsma P, Kartikasari K, van Ittersum M and Arts B 2015 Institutional constraints for adaptive capacity to climate change in Flevoland's agriculture. *Environmental Science and Policy*, 48, 147–162.

Markakis J 2004 *Pastoralism on the margin*. Minority Rights Group International, 2–37.

Matteoli F, Schnetzer J and Jacobs H 2020 Climate-smart agriculture (CSA): An integrated approach for climate change management in the agriculture sector. In *Handbook of climate change management: Research, leadership, transformation*. Cham: Springer International Publishing, 409–437.

Mayer B 2018 *The international law on climate change*. Cambridge University Press, 35–53.

McGuire JM, Morton LW, Arbuckle Jr, JG, and Cast AD 2015 Farmer identities and responses to the social–biophysical environment. *Journal of Rural Studies*, 39, 145–155.

Mehar M, Mittal S and Prasad N 2016 Farmers coping strategies for climate shock: Is it differentiated by gender? *Journal of Rural Studies,* 100(44), 123–131.

Mendelsohn R 2012 Development and climate adaptation. In *Climate change and common sense. Essays in Honour of Tom Schelling,* 245–259.

Mendelsohn R, Dinar A and Dalfelt A 2000 *Climate change impacts on African agriculture*. Preliminary analysis prepared for the World Bank, 25.

Mercer DE 2004 Adoption of agroforestry innovations in the tropics: A review. *Agroforestry Systems,* 61(1–3), 311–328.

Mickwitz P, Aix F, Beck S, Carss D, Ferrand N, Görg C and van Bommel S 2009 *Climate policy integration, Coherence and governance,* Partnership for European Environmental research (PEER) Report No. 2, 1–96.

Molini et al. 2010 Safety Nets and index-based insurance: Historical assessment and semiparametric simulation for Northern Ghana. *Economic Development and Cultural Change,* 58(4), 671–712.

Moser SC and Ekstrom JA 2010 A framework to diagnose barriers to climate change adaptation. *Proceedings of the National Academy of Sciences,* 107(51), 22026–22031.

Mukheibir P, Kuruppu N, Gero A and Herriman J 2013 Overcoming cross-scale challenges to climate change adaptation for local government: A focus on Australia. *Climatic Change,* 2(121), 271–283.

Murthy M, Tiwari R and Ravindranath NH 2011 Climate change and forests in India: Adaptation opportunities and challenges. *Mitigation and Adaptation Strategies for Global Change,* 16, 161–175.

Musango JK and Peter C 2007 A Bayesian approach towards facilitating climate change adaptation research on the South African agricultural sector. *Agrekon,* 46(2), 245–259.

Mwebaza R and Kotzé LJ 2009 Environmental governance and climate change in Africa: Legal perspectives. *Institute for Security Studies Monographs,* 167, 283.

Mwendera E and Atyosi Y 2018 A review of water storage for socio-economic development in South Africa. *Journal of Water Resource and Protection,* 10(3), 266.

Nelson DR, Adger WN, and Brown K 2007 Adaptation to environmental change: Contributions of a resilience framework. *Annual Review of Environment and Resources,* 32(1), 395–419.

Noble NJ, Iorliam T and Arene CJ 2014 Economic effects of climate change on Sorghum in North Central Nigeria. *International Journal of Development and Sustainability*, 3(7), 1517–1525.
Ntukamazina N, Onwonga RN, Sommer R, Rubyogo JC, Mukankusi CM, Mburu J and Kariuki R 2017 Index-based agricultural insurance products: Challenges, opportunities and prospects for uptake in sub-Sahara Africa. *Journal of Agriculture and Rural Development in the Tropics and Subtropics (JARTS)*, 118(2), 171–185.
Oberthür S and Ott HE 1999 *The Kyoto protocol: International climate policy for the 21st century*. Springer Science & Business Media.
O'Brien KL and Wolf J 2010 A values-based approach to vulnerability and adaptation to climate change. *Wiley Interdisciplinary Reviews: Climate Change*, 1(2), 232–242.
Odoyo E, Bett B and Murenga Mwimali AS 2019 Challenges of and responses to climate variability: A case study of peri-urban female farmers in Kenya. *World Journal of Agricultural Research*, 7(2), 61–68.
Okorie DI and Lin B 2022 Emissions in agricultural-based developing economies: A case of Nigeria. *Journal of Cleaner Production*, 337, 130570.
Opiyo FE, Wasonga OV and Nyangito MM 2014 Measuring household vulnerability to climate-induced stresses in pastoral rangelands of Kenya: Implications for resilience programming. *Pastoralism,* 1(4), 1–15.
Orie EG 2021 Climate change adaptation mechanism for sustainable development goal 1 in Nigeria: Legal Imperative. In *African handbook of climate change adaptation*. Springer International Publishing, 1–21, 1903.
Orlove B 2009 The past, the present and some possible futures of adaptation, In WN Adger, I Lorenzoni, and K O'Brien (Eds.), *Adaptation to Climate Change: Thresholds, Values, Governance,* Cambridge, UK: Cambridge University Press, 131–163.
Oulu MO 2011 Mainstreaming climate adaptation in Kenya. *Climate Law,* 2(3), 375–394.
Pachauri RK, Allen MR, Barros VR, Broome J, Cramer W, Christ R and van Ypserle JP 2014 Climate change 2014: Synthesis report. In *Contribution of Working Groups I, II and III to the fifth assessment report of the Intergovernmental Panel on Climate Change*, 151.
Palese AM, Vignozzi N, Celano G, Agnelli AE, Pagliai M and Xiloyannis C 2014 Influence of soil management on soil physical characteristics and water storage in a mature rainfed olive orchard. *Soil and Tillage Research*, (144), 96–109.
Parry et al. 2007 *Climate Change 2007: Impacts, Adaptation and Vulnerability*. Contribution of Working Group II to the Fourth Assessment Report of the Intergovernmental Panel on Climate Change. (4), Cambridge University Press.
Pavanello S 2009 Pastoralists' vulnerability in the Horn of Africa: Exploring political marginalisation, donors' policies and cross-border issues–Literature review. Humanitarian Policy Group (HPG) Overseas Development Institute London, UK, 1–47.
Pelling M 2010 *Adaptation to climate change: From resilience to transformation*. Routledge, 81.
Persson A and Klein RJ 2008 *Mainstreaming adaptation to climate change into official development assistance: Integration of long-term climate concerns and short-term development needs*. Stockholm Environment Institute, 16.
Peterson GA and Westfall DG 2004 Managing precipitation use in sustainable dryland agroecosystems. *Annals of Applied Biology,* 144(2), 127–138.

Pettengell C 2010 Climate change adaptation: Enabling people living in poverty to adapt. Oxford*:* Oxfam GB*,* 7, 1–48.

Pielke R, Prins G, Rayner S and Sarewitz D 2007 Lifting the taboo on adaptation. *Nature*, 445(7128), 597–598.

Piggott-McKellar AE, McNamara KE, Nunn PD and Watson JE 2019 What are the barriers to successful community-based climate change adaptation? A review of grey literature. *Local Environment,* 2(4), 374–390.

Porter JR and Semenov MA 2005 Crop responses to climatic variation. *Philosophical Transactions of the Royal Society B: Biological Sciences*, 360(1463), 2021–2035.

Porter JR, Howden M and Smith P 2015 Considering agriculture in IPCC assessments. *Nature Climate Change*, 7(10), 680–683.

Pyšek P, Vilà M, Hulme PE, Simberloff D, Bacher S, Blackburn TM and Carlton JT 2020 Scientists' warning on invasive alien species. *Biological Reviews*, 95(6), 1511–1534.

Rademacher-Schulz C and Mahama ES 2012 Rainfall, food security, and human mobility: Case study Ghana. United Nations University Institute for Environment and Human Security 132Raggl 2017 World Bank Policy Research Working Paper, 8153.

Rao KPC, Verchot LV and Laarman J 2007 Adaptation to climate change through sustainable management and development of agroforestry systems. *Journal of SAT Agricultural Research,* 4(1), 1–30.

Reid H, Alam M, Berger R, Cannon T, Huq S and Milligan A 2009 Community-based adaptation to climate change: An overview. *Participatory Learning and Action*, 60(1), 11–33.

Ribera LA, Hons FM and Richardson JW 2004 An economic comparison between conventional and no-tillage farming systems in Burleson County, Texas. *Agronomy Journal*, 96(2), 415–424.

Ribot JC 1995 The causal structure of vulnerability: Its application to climate impact analysis. *GeoJournal,* 35(2), 119–122.

Ringersma J, Batjes NH and Dent D 2003 Green water: Definitions and data for assessment (No. 2003/2). ISRIC-World Soil Information, 1–50.

Ringler C, Zhu T, Cai X, Koo J and Wang D 2010 Climate change impacts on food security in sub-Saharan Africa. *Insights from Comprehensive Climate Change Scenarios*, 114, 1–28.

Salinger MJ, Sivakumar MVK and Motha R 2005 Reducing vulnerability of agriculture and forestry to climate variability and change: Workshop summary and recommendations. In *Increasing climate variability and change*. Springer Dordrecht, 70, 341–362.

Senyolo MP, Long TB, Blok V and Omta O 2018 How the characteristics of innovations impact their adoption: An exploration of climate-smart agricultural innovations in South Africa. *Journal of Cleaner Production*, 172, 3825–3840.

Skinner MW, Kuhn RG and Joseph AE 2001 Agricultural land protection in China: A case study of local governance in Zhejiang Province. *Land Use Policy*, 18(4), 329–340.

Smit B 1993 Adaptation to climatic variability and change. Report of the Task Force on Climate Adaptation No. GU/G-CE04610. Guelph University, 1–28.

Smit B and Pilifosova O 2003 From adaptation to adaptive capacity and vulnerability reduction. In *Climate change, adaptive capacity and development,* 8(9), 1–28.

Smit B and Skinner MW 2002 Adaptation options in agriculture to climate change: A typology. In *Mitigation and adaptation strategies for global change*, 7(1), 85–114.

Smit B and Wandel J 2006 Adaptation, adaptive capacity and vulnerability. *Global Environmental Change*, 16(3), 282–292.

Smit B, Burton I, Klein RJ and Street R 1999 The science of adaptation: A framework for assessment. In *Mitigation and adaptation strategies for global change*, 4, 199–213.

Smit B, Burton I, Klein RJ, and Wandel J 2000 An anatomy of adaptation to climate change and variability. In *Societal adaptation to climate variability and change*. Springer, *Climatic Change* 1(45), 223–251.

Smith JB, Dickinson T, Donahue JD, Burton I, Haites E, Klein RJ and Patwardhan A 2011 Development and climate change adaptation funding: Coordination and integration. *Climate Policy*, 11(3), 987–1000.

Smithers J and Smit B 1997 Human adaptation to climatic variability and change. *Global Environmental Change*, 7(2), 129–146.

Smithers J and Blay-Palmer A 2001 Technology innovation as a strategy for climate adaptation in agriculture. *Applied Geography*, 21(12), 175–197.

Sobania 2019 Pastoralist migration and colonial policy: A case study from northern Kenya. In *The Ecology of Survival*, Routledge, 219–239.

Sovacool BK 2011 Hard and soft paths for climate change adaptation. *Climate Policy*, 11(4), 1177–1183.

Spires, Shackleton and Cundill 2014 *Climate and Development*, 6(3), 277–287.

Ssozi J, Asongu S and Amavilah VH 2019 The effectiveness of development aid for agriculture in Sub-Saharan Africa. *Journal of Economic Studies*, 46(2), 1–22.

Stal M 2011 Flooding and relocation: The Zambezi River valley in Mozambique. *International Migration*, 49, e125–e145.

Stern N and Stern NH 2007 *The economics of climate change: The Stern review*. Cambridge University Press, 165.

Stocker T (Ed.) 2014 *Climate change 2013: The physical science basis: Working Group I contribution to the fifth assessment report of the Intergovernmental Panel on Climate Change*. Cambridge University Press, 1535.

Stringer LC, Dyer JC, Reed MS, Dougill AJ, Twyman C and Mkwambisi D 2009 Adaptations to climate change, drought and desertification: Local insights to enhance policy in southern Africa. *Environmental Science and Policy*, 12(7), 748–765.

Tafere M, Olivier J and Jordaan M 2013 Major human and Natural Disruptions that facilitated Vegetation cover removal in the Ethiopia Highlands. *Journal of Environment and Earth Science*, 3(9), 52–67.

Taruvinga A, Muchenje V and Mushunje A 2013 Climate change impacts and adaptations on small-scale livestock production. *International Journal of Development and Sustainability*, 2(2), 664–685.

Thornton PK and Herrero M 2014 Climate change adaptation in mixed crop–livestock systems in developing countries. *Global Food Security*, 3(2), 99–107.

Thorpe A 2009 Enteric fermentation and ruminant eructation: The role (and control?) of methane in the climate change debate. *Climatic Change*, 93(3), 407–431.

Tol RS, Fankhauser S and Smith JB 1998 The scope for adaptation to climate change: What can we learn from the impact literature? *Global Environmental Change*, 8(2), 109–123.

Tompkins EL and Adger WN 2004 Does adaptive management of natural resources enhance resilience to climate change? *Ecology and Society*, (9), 2.

Verheyen R 2002 Adaptation to the impacts of anthropogenic climate change-the international legal framework. *Review of European Comparative and International Environmental Law*, 11(2), 129–143.

Vermeulen SJ, Campbell BM and Ingram JS 2012 Climate change and food systems. *Annual Review of Environment and Resources*, 37(1), 195–222.

Walker WE, Haasnoot M and Kwakkel JH 2013 Adapt or perish: A review of planning approaches for adaptation under deep uncertainty. *Sustainability*, 5(3), 955–979.

Walthall C, Hatfield J, Marshall E, Lengnick L, Backlund P, Adkins S and Bunce J 2013 *Climate change and agriculture: Effects and adaptation*. USDA Technical Bulletin, 1935.

Warner K, Ranger N, Surminski S, Arnold, M, Linnerooth-Bayer J, Michel-Kerjan E and Herweijer C 2009 *Adaptation to climate change: Linking disaster risk reduction and insurance*. United Nations International Strategy for Disaster Reduction, Geneva 1–30.

Warner K, Hamza M, Oliver-Smith A, Renaud F and Julca A 2010 Climate change, environmental degradation and migration. *Natural Hazards*, (55), 689–715.

Warner K, Zommers Z, Wreford A, Hurlbert M, Viner D, Scantlan J and Tamang C 2019 Characteristics of transformational adaptation in climate-land-society interactions. *Sustainability*, 11(2), 356.

Wilks DS 1992 Adapting stochastic weather generation algorithms for climate change studies. *Climatic Change*, 22(1), 67–84.

Wise RM, Fazey I, Smith MS, Park SE, Eakin HC, Van Garderen EA and Campbell B 2014 Reconceptualising adaptation to climate change as part of pathways of change and response. *Global Environmental Change*, 28, 325–336.

Wright H, Vermeulen S, Laganda G, Olupot M, Ampaire E and Jat ML 2014 Farmers, food and climate change: Ensuring community-based adaptation is mainstreamed into agricultural programmes. *Climate and Development*, 1(2), 318–328.

Xia Y, Song Y and Pu M 2010 Analysis on the mutual aid fund system based on farmers' cooperatives. *Issues in Agricultural Economy*, 4, 29–33.

Zhu X, Clements R, Quezada A, Torres J and Haggar J 2011 Technologies for climate change adaptation. *Agriculture Sector*, 40.

Zougmoré R, Partey S, Ouédraogo M, Omitoyin B, Thomas T, Ayantunde A and Jalloh A 2016 Toward climate-smart agriculture in West Africa: A review of climate change impacts, adaptation strategies and policy developments for the livestock, fishery and crop production sectors. *Agriculture and Food Security*, 1(5), 26.

Zwane EM 2019 Impact of climate change on primary agriculture, water sources and food security in Western Cape, South Africa. *Jàmbá: Journal of Disaster Risk Studies*, 11(1), 1–7.

2 Climate Change Adaptation, Agriculture, and the Law

2.1 Introduction

The previous chapter broadly explained adaptation as a means of coping with the impacts of agriculture-induced climate change on important resources such as soil and water, as well as a strategy for enabling agricultural products especially crops and livestock to adjust to the impacts of climate change. The chapter indicated that adaptation strategies and measures, whether in a broad context or in the context of agriculture, cannot be effective, except if guided by well-developed and effective laws. Law is thus relevant for practices and projects that can either reduce climate-related risks or realise opportunities associated with climate change. This emphasis means that the law could facilitate anticipatory adaptation to the required extent by providing a set of minimum standards and obligations.

Hence, this chapter will highlight both climate change-related laws and agriculture-based laws at the international and regional levels, and where necessary, references to some national laws will be made as examples. The aim of highlighting the regulatory framework is to lay the foundation for subsequently investigating whether the laws of Kenya, Nigeria, and South Africa also support adaptation, particularly in the context of agriculture. The analysis is guided by the following question: what is the role of the law in relation to adaptation, and to what extent can the law facilitate adaptation in the context of agriculture?

Providing answers to the question above will require a brief discussion of the meaning of law in relation to adaptation and why the law is important to adaptation. In this regard, the first part of the chapter will explain how the law supports the aims of adaptation that were previously discussed.[1] The second part of the chapter will discuss the role of law in promoting all previously discussed forms of adaptation in the context of agriculture.[2] For this reason, laws that control agricultural production (fertiliser application and chemical use laws) and laws that promote agricultural production (water management, land use and soil conservation laws) will be discussed.

DOI: 10.4324/9781003536437-2

2.2 The Law in the Context of Climate Change Adaptation

According to Oxford Advanced Learner's Dictionary,[3] "law is a whole system of rules that everyone in a country or society must obey". More aptly described:

> [l]aw refers to an organic mechanism whereby certain claims of right are socially established as collectively enforced norms and other claims of right are denied such status. These norms might be formally established and enforced through legislatures, courts, and executives, or informally established and enforced through custom and informal, but often highly effective, collective action.[4]

In addition, law is portrayed as a tool for coordinating institutional and governance action by which targets, obligations, specific roles, and responsibilities for parties are defined.[5] Law as an instrument for regulating human behaviour can manifest either as "hard laws" or "soft laws". Hard laws include constitutions, rules, directives, and regulations,[6] in addition to legislation and technical policies, and they are legally binding.[7]

Soft laws are not legally binding, but their compliance is based on recognition, "peer pressure and competition".[8] Soft laws are also called quasi-legal instruments, and they are so described because their legally binding force is virtually absent or somehow weak compared to hard laws.[9] Soft laws sometimes consist of pre-legislation policies, best practices, industry codes, standards,[10] environmental stewardship[11] and reporting schemes, among others. [12] Amid these examples of soft laws, policies (especially pre-legislation policies) deserve special attention. In most instances, pre-legislation policies cannot be separated from the law in any governance setting because; pre-legislation policies are mostly translated into hard laws, while hard laws serve as tools for policy implementation and operationalisation.[13] The climate change governance landscape for instance as presented in subsequent chapters of this book, dealing with Kenya, Nigeria, and South Africa, is heavily dominated by pre-legislation policies.

Hence, both hard laws and soft laws (simply described as "the law"), in the context of adaptation, can serve as a tool or mechanism that is applied in order to "translate the commitments made by a country under Multilateral Environmental Agreements (MEAs) into nationally enforceable targets and plans for government…"[14] In more specific terms, law in the context of adaptation could mean any set of directive rules, statutes, or treaties and policies, which prescribes:

> [l]ow carbon and climate resilient development, by mandating, guiding and coordinating relevant institutional and governance action; regulating adverse behaviour; and symbolically elevating climate change concerns to the status of national legislation.[15]

Given that the goal of adaptation is to generally foster low-carbon and climate-resilient development, law is required (as an instrument of command and control) in order to achieve this goal.[16] Thus, climate change adaptation like any other concept can only function effectively within a framework, which also serves as an embodiment of action plans and policies.[17]

The role of the law in facilitating a low-carbon climate-resilient development through adaptation can be discerned from its agenda, which is basically to guide, coordinate, and regulate the behaviour of actors and institutions involved with adaptation governance. [18] In a strict sense, law is an important element in adaptation governance, for it provides standards, mechanisms, and/or a framework for the implementation of adaptation strategies. In this regard, Mwebaza and Kotzé[19] observe that the law functions as a regulatory tool and a standard-setting mechanism for actors (state and non-state) and entities (responsible for promoting adaptation at the global, regional, and national levels).

In other words, the law can be used to provide directives and standards to actors, as well as establish institutions, for tackling high-level uncertainty related to the irreversible impacts of climate change on the environment and socio-economic activities of vulnerable societies.[20] For example, the Intergovernmental Panel on Climate Change (IPCC) as an institution saddled with the responsibility of providing scientific guidance to all sectors of the economy and stakeholders involved with climate change adaptation is established and regulated by the law.[21] Rumble[22] adds that administrative structures (mandate, functions, and organisational procedure) of climate change institutions such as the IPCC are entrenched and formalised by the law.[23]

Hence, the law enables intergovernmental institutions such as the United Nations Environmental Program (UNEP) and the Food and Agricultural Organization (FAO), and (specifically) countries to implement long-term adaptation measures and facilitate quick short-term responses to the impacts of climate change.[24] In the same manner, participation in adaptation planning, access to information or decision-making, and implementation by private entities such as non-governmental organisations (NGOs) and corporations are rights given by the law.[25]

Law is important in actualising adaptation goals because it "create[s] rights and obligations that identify duty bearers and rights holders".[26] Similarly, adaptation strategies require the backing of the law in order to actualise their set of objectives – previously discussed as reducing vulnerability, increasing resilience, and boosting the adaptive capacity of human systems and natural systems at the national level, regional level, and international level. By way of illustration, adaptation strategies such as the development of early warning systems – in the form of observation and monitoring technologies that forecast the occurrence of severe weather events – and the construction of flood-defence structures such as the placing of hydrosacks and sandbags, building of dikes, and the use of water-extraction machines can only be legitimately and effectively executed when they are supported by the law.[27]

More specifically, law clarifies the apportionment of losses and benefits accruing from adaptation measures in any given society.[28] In this regard, the law can serve as an instrument which supports economic and financial incentives such as insurance, which has been previously explained as a form of adaptation.[29] Similarly, in the context of adaptation, the law is an important instrument that could be used to:

> [r]ecognize and give effect to the critical differences between "no regrets" measures that should be undertaken immediately, such as information gathering and reductions in pollution, and longer-term adaptation plans and strategies, which should be based on greater understanding of the actual climate change impacts to particular socio-ecological systems than we currently possess.[30]

In other words, the law could serve as a tool for mainstreaming adaptation actions at the regional and national levels.[31] Meaning that the law could foster flexible, collaborative, and voluntary adaptation governance initiatives in a multi-level and decentralised setting.[32] Further, it has been rightly submitted that:

> [t]he law plays a crucial role in structuring the concept of mainstreaming as it ushers in the optimal exercise of (government) functions, whereby the agriculture sector factors climate change considerations into the planning, design, budgeting and implementation of national policies for the good of the entire nation.[33]

By implication, the law is suitable in supporting the objectives of adaptation, when guided by certain principles,[34] which are mainly provided by the international legal order.

2.3 Principles of Climate Change Adaptation Laws

Given that the objectives of adaptation are to enhance adaptive capacity, boost resilience, and reduce vulnerability, any law that is enacted to facilitate adaptation must also be tailored to support those objectives. In specific terms, adaptation laws are expected to foster: the role of the government in protecting key infrastructure, plans for substantive adaptation measures that address vulnerability relative to underlying demographic differences such as income disparity and cultural sensitivity, clearly defined participatory process, as well as a comprehensive agenda that considers the trade-offs between adaptation and mitigation.[35] Therefore, the basic principles which must be found in an effective adaptation law among others consist of: the principles of monitoring and studying circumstances, promotion of resilience, long-term planning and coordination, flexible regulatory goals, and the recognition of uncertain nature of climate change.[36]

Kulovesi, Hollo, and Mehling[37] are of the opinion that the existence of the aforementioned principles in key legal instruments such as the UNFCCC, the Kyoto Protocol, and the Paris Agreements has aided the formulation and implementation of adaptation policies and measures at the regional level and national level.[38] For example, the UNFCCC (as a legal framework) provides some basic guiding principles, which among others include the principles of equity,[39] solidarity,[40] precaution,[41] sustainability,[42] and good neighbourliness,[43] which inform effective adaptation laws at the regional level.[44] These guiding principles are also expected to give impetus to existing national adaptation laws in a broad and generic context, and they could even inform the development of new, specific agriculture-focused adaptation laws national. Moreover, the principles of the UNFCCC are intended to guide adaptation planning by institutions such as disaster management agencies, housing agencies, public health organisations, and local governments.[45] These key legal instruments which support the principles of adaptation are now discussed in more details.

2.3.1 The United Nations Framework Convention on Climate Change

The UNFCCC has been introduced as the principal international legal framework governing climate change.[46] Contiades and Fotiadou[47] further describe the UNFCCC as a legal instrument which provides an essential forum for dialogue and decision-making that enables countries to adapt to the impacts of climate change. In other words, the UNFCCC serves as a focal point for most adaptation regulations and decision-making at the international level, regional level, and national level.[48] According to Bodansky, Brunnée, and Rajamani,[49] the guiding principles provided by the UNFCCC in article 3 oblige States to foster cooperation related to adaptation, while also considering national peculiarities in terms of vulnerability and development needs.[50]

The principles of law which are relevant to adaptation under the UNFCCC are communicated through its annual Conferences of the Parties (COPs) which serves as basis for developing National Adaptation Plan of Actions (NAPAs).[51] The importance of NAPAs has been further recognised at the COPs of the UNFCCC as critical in enabling developing countries to assess their vulnerabilities, mainstream climate change risks, and address their adaptation needs.[52] The Cancun Agreements that were adopted in 2010 is an example of a COP decision.[53] The Cancun Agreements placed adaptation as a high priority (and at the same level with mitigation) on the international climate change governance agenda.[54] The Cancun Agreements established the Cancun Adaptation Framework (CAF) with the intent to, among others things, plan, prioritise, and implement adaptation actions through international cooperation.[55] In addition, the Cancun Agreements are generally intended to assess adaptation actions, including economic, social, and environmental evaluation of adaptation options as well as strengthening institutional capacities for adaptation.[56] In other words, the social and economic dimension of adaptation has already been firmly established by the UNFCCC. The preamble to the UNFCCC partly states that:

> responses to climate change should be coordinated with social and economic development in an integrated manner with a view to avoiding adverse impacts on the latter, taking into full account the legitimate priority needs of developing countries for the achievement of sustained economic growth and the eradication of poverty.[57]

In addition to containing certain guiding principles that are relevant to adaptation, the UNFCCC also promotes the objectives of adaptation (reducing vulnerability, enhancing adaptive capacity, and boosting resilience) by obliging member states to formulate, implement, publish, and regularly update national and regional programmes containing measures to facilitate comprehensive adaptation to climate change.[58] For example, article 4(1) (e) of the UNFCCC requires member states to make specific adaptation plans against the impacts of climate change in vulnerable sectors such as water resources and agriculture.[59]

However, it has been argued that the provisions of the UNFCCC, especially article 4(1)b, are not entirely clear about what adaptation in the context of agriculture is, and how it should be achieved. As a legal instrument and a framework for adaptation, the UNFCCC is expected to adequately provide the required legal incentive for addressing the impacts of climate change on vulnerable sectors especially agriculture and water resources in the context of African and developing countries.[60] Likewise, the general provisions of the UNFCCC are not clear on the rate of greenhouse gas (GHG)-emission reduction targets required by developed countries, and it entirely excluded developing countries from the reduction commitments.[61] In addition, the role of the private sector in advancing the objectives of adaptation has not been clearly outlined by the provisions of article 4(1) (b) of the UNFCCC,[62] even though private sectors (particularly NGOs and transnational corporations) have been noted to play important roles in facilitating adaptation in societies that are vulnerable to the impacts of climate change, due to their financial strength and technological prowess.[63] Another concern raised about the UNFCCC is that, while developed countries are under an obligation to generally provide financial support for adaptation in rural areas of developing countries through the Special Climate Change Fund (SCCF),[64] the Least Developed Countries Fund (LDCF),[65] and the Green Climate Fund (GCF),[66] the type of adaptive measure, time-frame, and institutional infrastructure for channelling such funding assistance to developing countries have not been specifically outlined.[67] More specifically, Nanda[68] believes that it is not lack of funding that prevents adaptation from reaching its potentials under the UNFCCC, especially in African and developing countries, but it is lack of clear legal provisions.

In order to partly improve the provisions of the UNFCCC in supporting the objectives of adaptation, the Kyoto Protocol was negotiated as an additional protocol to the UNFCCC.[69]

2.3.2 The Kyoto Protocol

The Kyoto Protocol elaborated upon some of the obscure provisions of the UNFCCC. One such elaboration is exemplified by the guidance for developed countries on how to meet their emission-reduction targets. Through specific guidelines, the Kyoto Protocol has tasked annex I parties (developed countries) to "minimize the adverse impacts of climate change on non-annex I parties (developing countries)".[70] This, as a response measure, is intended to also help developed countries reduce their emission targets.[71]

Again, unlike the UNFCCC which places stringent obligation only on developed countries to provide financial support for adaptation in developing countries, the Kyoto Protocol places a somewhat less onerous obligation on developing country parties in a bid to promote adaptation. In a better and clearer wording than the provisions of article 3 of the UNFCCC, article 10(b)(i) of the Kyoto Protocol requires developing countries to formulate, implement, publish, and regularly update programmes and measures that facilitate adequate adaptation for various sectors at the national and regional levels.[72] Equally, the Kyoto Protocol promotes adaptation by urging all parties (developed and developing countries) to:

> [c]ooperate in the promotion of effective modalities for the development, application and diffusion of environmentally sound technologies, know-how, practices and processes pertinent to climate change in particular to developing countries.[73]

More specifically, the Kyoto Protocol obliges parties to "integrate climate change considerations into their social, economic and environmental policies and actions".[74] This kind of integration has been previously explained as mainstreaming.[75] However, in terms of procedure, the provisions of the Kyoto Protocol do not specifically provide information on how national adaptation policies should be integrated into the communication that must be submitted to the secretariat of the UNFCCC.[76] This procedural concern emanating from the Kyoto Protocol, among others, necessitated the need for the negotiation of an additional legal instrument known as the Paris Agreement, to cater for a new "global goal for adaptation".[77] In other words, the Paris Agreement replaces the Kyoto Protocol.[78] Therefore, any subsequent mention of the Kyoto Protocol in this book is merely made in order to provide the context for the development of adaptation law.

2.3.3 The Paris Agreement

The Paris Agreement through its "global goal for adaptation"[79] identifies adaptation as a means of addressing the challenges of climate change through local, subnational, national, regional, and international efforts and collaborations.[80] In order to facilitate adaptation, parts of the objectives of the Paris Agreement, as stated in article 2, are to "increase the ability to adapt... and foster climate resilience ... in a manner that does not threaten food production".[81] For present purposes, these objectives appear to be broader than that of the UNFCCC, and they also directly concern agricultural productions.[82]

Yet again, unlike the UNFCCC and Kyoto Protocol which place more obligations on developed nations, and less on developing nations in addressing the impacts of climate change, the Paris Agreement creates obligations for both developed and developing nations (based on their respective capacities) to address the impacts of climate change through adaptation.[83] This is technically described as the principle of Common-but differentiated responsibility and respective capabilities (CBDR and RC). and it is a basic requirement for all parties to the Paris Agreement in fulfilling the requirements of global goals for adaptation. In other words, Parties to the Paris Agreement are mandated to create effective national (and even regional) adaptation plans and policies that could also boost the resilience of socio-economic systems.[84] More specifically, article 7 of the Paris Agreement requires Parties to prepare, submit, and update their national adaptation communications at periodic intervals.[85] Apart from boosting the visibility of a country's climate action (both mitigation and adaptation) profile, the benefits of submitting a national adaptation communications also include the reinforcement of international support for developing country parties, as well as the enhancement of learning and understanding of differing adaptation needs (for individual countries) in line with the global goal for adaptation.[86]

Given that adaptation is a strategy for addressing the impacts of climate change on important sectors of the economy such as agriculture, energy, transport, forestry, and mining, it is noteworthy that the laws which promote adaptation are equally enacted in light of this interconnection.[87] For this reason, the next section of this chapter will discuss the role of the law in facilitating adaptation in the context of agriculture.

2.4 Agricultural Adaptation Laws

Agriculture is a sector of the economy that partly acts as a driver to climate change; at the same time, it is a sector of the economy that is seriously affected by the impacts of climate change.[88] Hence, the role of the law in promoting adaptation in the context of agriculture is twofold. First, the law is required to regulate certain practices that support crop and livestock production – such as the application of fertiliser and chemical use that were previously identified as drivers of climate change.[89] Second, the law can be applied to promote and actualise some short-term, medium-term, and long-term measures that among others include irrigation, technology transfer, and insurance as adaptation strategies for addressing the impacts of climate change on agricultural products, especially crops and animal fodder and other supportive resources such as soil and water.[90]

However, given the intricate nature of agriculture which connects with other aspects of the environment, agriculture-focused adaptation laws are somehow also sourced from different but related sources such as environmental law, land tenure law, soil conservation laws, fertiliser and chemical use laws, and water management laws, among others.[91] For example, crop production cannot be adequately regulated without food and safety law or land tenure laws.[92] Also, the regulation of livestock production cannot be discussed without referring to cattle-grazing laws or animal-health laws.[93] Similarly, the laws that regulate irrigation as an adaptation strategy are generally derived from water management laws.[94]

The purpose of these divergent but related agriculture-focused laws is not only to safeguard the environment from agriculture-induced climate change but also to enable agriculture to adapt to the impacts of climate change. By way of illustration, agriculture-focused adaptation laws drawn from land use laws and soil conservation laws are required in prohibiting certain land use activities such as farming with unacceptable methods of ploughing, burning of residues on the ground, or the use of sludge on agricultural land, and other practices which could lead to climate change.[95]

As a requirement, agriculture-focused adaptation laws must be amenable to alteration in the event of innovative circumstances while still serving their purpose of "maintaining life and ecological balance in an unseeable future".[96] These requirements are by extension also provided by key adaptation laws such as the UNFCCC, Kyoto Protocol, and the Paris Agreement.[97] In relation to agriculture, the UNFCCC generally requires all Parties to:

> [p]romote and cooperate in the development, application and diffusion, including transfer, of technologies, practices and processes that control, reduce or prevent anthropogenic emissions of greenhouse gases in all relevant sectors, including the agriculture sector[s].[98]

Furthermore and in the context of agriculture, a provision of the UNFCCC states that:

> The ultimate objective of this Convention and any related legal instruments that the Conference of the Parties may adopt is to achieve, in accordance with the relevant provisions of the Convention, stabilization of greenhouse gas concentrations in the atmosphere at a level that would prevent dangerous anthropogenic interference with the climate system. Such a level should be achieved within a time frame sufficient to allow ecosystems to adapt naturally to climate change, to ensure that food production is not threatened and to enable economic development to proceed in a sustainable manner.[99]

In particular, the Kyoto Protocol requires parties to prepare and implement national and regional adaptation measures that can preserve the soil and improve carbon sinks.[100] The objective of the Paris Agreement on the need for agriculture-focused adaptation is clearly stated in article 2, and is already explained above.[101]

In view of the fact that agricultural adaptation strategies are implemented based on the vulnerability of the impacts of climate change and the socio-economic dimension of a location, agriculture-focused adaptation laws are equally applied relatively and contextually.[102] As such, agriculture-focused adaptation laws are explained in the following sections based on what Härtel[103] described as laws for controlling agricultural production and laws for promoting agricultural production.

2.4.1 Laws for controlling agricultural production

It was explained in Chapter 1 that agricultural production methods such as the application of fertilisers and chemical uses, as well as animal-rearing techniques that are responsible for climate change, can be adjusted through adaptation.[104] These agricultural adaptation strategies, which are generally implemented at the farm level through technological and financial interventions, are broadly supported by laws at different levels (international, regional, etc.) and in different jurisdictions. Prominent among these laws is the Paris Agreement, which provides a set of norms that facilitate some technological adaptation strategies at the farm level through an initiative known as the Koronivia Joint Work on Agriculture (KJWA).[105] The norms provided by the KJWA are intended to enhance "soil health and fertility, water management, manure management and livestock management systems".[106]

In addition, the Paris Agreement requires all member parties to strengthen efforts in technological use and its transfer as part of adaptation measures against the impacts of climate change on agriculture.[107] Under the Paris Agreement, developed countries are obliged to provide to developing countries the technical know-how and skills for environmentally sound technologies.[108] Likewise, the UNFCCC also supports agricultural adaptation strategies. By virtue of the CAF, all member parties are mandated to "develop, and diffuse technologies, practices and processes for adaptation".[109]

As previously explained, where technological interventions have failed or are insufficient in promoting adaptation, farmers usually resort to insurance in order to acquire the vital resources needed to recoup farm produce lost or damaged by the impacts of climate change such as droughts and other extreme weather events.[110] In most instances, an effective insurance system usually thrives on a well-developed legal mechanism. In the context of agricultural adaptation, the UNFCCC and the Kyoto Protocol provide the legal stimulus for insurance, by requesting parties to consider insurance as part of their adaptation actions.[111] As an international legal framework for adaptation, the Paris Agreement also recognises insurance as an option for adaptation in the context of agriculture.[112] Consequently, and in a more specific context, the role of the law in supporting the adjustment of agricultural production methods that act as drivers of climate change is further illustrated in fertiliser and chemical use regulation as well as livestock-production laws.[113]

2.4.1.1 Fertiliser application and chemical-use Laws

There are several MEAs which regulate the application of fertilisers and chemicals on agricultural lands. Notable among them are the *Basel Convention on the Control of Transboundary Movement of Hazardous Wastes and their Disposal* of 1989[114]; the *Rotterdam Convention on the Prior Informed Consent Procedure for Certain Hazardous Chemicals and Pesticides in International Trade* of 1998[115]; and the *Stockholm Convention on Persistent Organic Pollutants* of 2001.[116] These MEAs are collectively knowns as the Basel, Rotterdam and Stockholm (BRS) Conventions.[117] The role of the BRS Conventions in preventing or minimising the effects of fertilisers and chemicals on agricultural lands is expressed in their main objective, which is to "protect human health and the environment from hazardous chemicals (including pesticides) and wastes".[118]

The BRS Conventions are complemented by an *International Code of Conduct for the Effective Use and Management of Fertilizers* that was issued by the FAO in 2019 (simply described as FAO code of conduct on fertiliser use).[119] The key purpose of the FAO code of conduct on fertiliser use is to:

> [p]romote practices that can improve soil health, including nutrient recycling and agronomic and land management, by providing regulations that relates to the sale, distribution and labelling of fertilizer products wherever appropriate.[120]

Furthermore, the FAO code of conduct on fertiliser use is aimed at promoting best practice in fertiliser use and management at the farm level and national level, through educational programmes and capacity development for all stakeholders.[121] To achieve its key purpose, the FAO code of conduct on fertiliser use provides specific norms that guide national regulatory agencies for pesticides in enacting legislation that regulates pesticides according to local needs.[122]

Before the *International Code of Conduct for the Effective Use and Management of Fertilizers*, the FAO had developed an *International Code of Conduct on Pesticide Management* (ICCPM) in 2014.[123] The ICCPM exists to provide guidance on the use and distribution of pesticides and other chemicals that could constitute a challenge on the environment.[124] The type of environmental challenge envisaged in this regard could be chemical- and pesticide-induced GHG emissions.

Not only fertiliser application and chemical use ought to be adjusted through agriculture-focused adaptation laws, livestock production also requires adjustment through the same laws. This is necessary to avoid excessive emission of methane which has been previously identified as one of the major sources of climate change, soil erosion, and the destruction of valuable plant species.[125] The role of the law in supporting the adjustment of livestock production could be much clearer when discussed under livestock-production laws.

2.4.1.2 Livestock production Laws

Livestock production consists of many components such as "animal management process, land allocation system and land consolidation procedure".[126] These components require regulation because if left unchecked or when poorly managed, they could lead to overgrazing and by extension land degradation, soil erosion and the loss of root system of native plants.[127] In a similar manner, livestock production laws are necessary in order to promote sustainable livestock production practices and at the same time protect the environment from GHG emissions.[128]

The laws that regulate livestock production are mostly drawn from many sources that are found at the international level and national level. For example, the *Convention on Biological Diversity* (CBD) of 1992[129] has an objective that is relevant to the protection of the environment from GHG-intensive livestock production practices.[130] Also, Agenda 21 of the UNEP, although non-binding, provides certain guidelines in the context of international environmental law that can assist in sustainable livestock production practices.[131] For example, a section of Agenda 21 provides:

> Governments at the appropriate level, with the support of the relevant international and regional organizations, should ... promote in-country programmes and related physical infrastructure for animal livestock conservation and breed development, as well as for strengthening national capacities to take preventive action when breeds are endangered.[132]

The Ugandan *Cattle Grazing Act*[133] could serve as an example of a law that regulates cattle grazing at the national level. The law partly provides that:

> A veterinary officer, with respect to land within his or her area or a district administration, with respect to land within the area of its jurisdiction, may make orders—(b) prescribing the maximum number of cattle that any person may be permitted to graze on any particular area of land.[134]

The provision of the Ugandan *Cattle Grazing Act* above could be relevant in promoting adaptation, especially since overgrazing was previously identified as a cause of soil erosion and climate change.[135] Given that agriculture and supporting resources such as water and soil are also affected by climate change, laws are required in enabling these resources to adapt.

2.4.2 Laws for promoting agricultural production

As an important sector of the economy, agriculture is generally regulated by the supply chain production law, marketing and sales law, and other laws that strengthen the sector such as insurance laws.[136] It has also been observed that agriculture law intersects with other intricate and dynamic sets of related laws, such as environmental law, food law, health and safety law, and is implemented by both public and private actors.[137] Examples of these laws in a generic sense are presented below.

2.4.2.1 Water-management Laws

Lack of water supply, or limited water supply, or excess water supply constitutes a significant threat to the environment and agricultural practices, especially crop and livestock production.[138] Therefore, laws are required to boost the adaptive capacity of water systems (soil water and blue water) against the impacts of climate change.[139] For example, the law can be applied to boost the adaptive capacity of water systems when enacted to protect water-storage facilities (aquifers) against the impacts of climate change such as drought or flooding.[140] Specifically, water laws play critical roles in:

> [p]roviding the underlying mandate for improving water quality, generating and sharing data, prioritizing and allocating quantities for use, and regulating demand through incentives and pricing mechanisms, while maintaining or improving ecosystem health.[141]

Most importantly, Adler[142] emphasises that water law provides rules that define the rights and interests of competing users – in the light of scarcity caused by droughts, for instance. Iza and Stein[143] further add that water law provides guiding principles on how water can be used in a sustainable manner, and how it can be less harmful to the environment.[144] In short, law sets some conditions through guiding principles and goals that could facilitate the adaptive capacity of water. The most important guiding principles of water law in this regard are the principle of equitable and reasonable utilisation and the principle of prevention of significant harm, while a notable goal of water law is that of conservation and restoration of water systems.[145]

The principle of equitable and reasonable utilisation and the principle of prevention of significant harm are supported by the provisions of international law and regional laws such as the *United Nations Economic Commission for Europe Convention on the Protection and Use of Transboundary Watercourses and International Lakes* (UNECE *Water Convention*) of 1992,[146] the *Revised African Convention on Conservation of Nature and Natural Resources (RACNNR)* of 2003,[147] the South African Development Commission *(SADC) Protocol on Shared Water System* of 1995,[148] and the European Union (EU) *Water Framework Directive* of 2000.[149] The principle of reasonable and equitable use has also been elaborated in other water laws such as the *Berlin Rules on Water Resources* of 2004, the 1992 *Convention on the Protection and Use of Transboundary Watercourses and International Lakes* and the 1997 *UN Convention on the Law of the Non- Navigational Uses of International Watercourses.*[150]

In specific terms, the principles of equitable and reasonable utilisation and that of prevention of significant harm require State parties:

> [n]ot to use water in such a manner as to prevent or otherwise limit other riparian [S]tates from making full use of their equitable and reasonable entitlements in relation to that shared water.[151]

The principle of equitable and reasonable utilisation and the principle of prevention of significant harm of water law are applied while taking certain factors into consideration such as the geographical context, hydrology, and climate as well as the economic and social needs of a State.[152] The conservation and restoration of water systems are goals of water law, and they are incorporated in most water policies or strategies[153] in order to ensure the sustainable use of water and its availability at all times in spite of supervening challenges such as droughts.[154]

In order to achieve the goals of conservation and restoration, parties to the UNECE *Water Convention* are expected to prevent and manage the negative impact of climate change on water at the international level; by developing and implementing strategies that incorporate the principle of the "ecosystems approach"[155] or other strategies that will promote sustainable water use.[156] In addition to the conservation and restoration of water systems, the goal of the UNECE is:

> [t]o, prevent, control and reduce transboundary impact of water, and to use transboundary waters in a reasonable and equitable way, and ensure their sustainable management.[157]

These goals of the UNECE on adaptive water management are similar to the goals of the African regional legal framework on adaptive water management, which in this regard is represented by the *RACNNR* as well as the *SADC Protocol on Shared Water System.*

As a regional law, the *RACNNR* requires parties to promote the principle of equitable and reasonable utilisation and prevention of significant harm through an established commission known as Inter-State Commission for the management and equitable utilisation and development of natural resources.[158] Water is included as constituent natural resources by the *RACNNR.*[159] Similarly, the *RACNNR* also promotes the conservation and restoration of water systems as part of its objectives, though in different wordings. Accordingly, State parties to the *RACNNR* are required to design policies for the utilisation, preservation, and development of soil water and blue water within their jurisdiction, and in conjunction with other party members in case of shared watercourses.[160] In accordance with the goals of conservation and restoration of water systems, parties to the *RACNNR* are mandated to implement adaptive water-management strategies that could improve the "qualitative and quantitative levels of soil water and blue water, while also preventing pollutants and water borne diseases".[161] In addition to the *RACNNR*, the *SADC Protocol on Shared Water System* (as a legal instrument that regulates the watercourses of the Congo and Zambezi Rivers) also incorporates the notable principle of equitable and reasonable utilisation and prevention of significant harm and the goals of conservation and restoration of water systems.[162]

The *Water Framework Directive* of the EU is another legal framework that combines the principles of equitable and reasonable utilisation and prevention of significant harm as well as the objective of conservation and restoration of the water system at the regional level.[163] As a law with a regional scope, the *Water Framework Directive*:

> [p]rovides a holistic legal framework for all water issues, including water quality and water quantity issues, both of surface and ground waters, as well as coastal waters in the European Union.[164]

As well, the *Water Framework Directive* establishes the required institutional framework for the management of river basin districts, plans, and programmes in EU Member States.[165]

2.4.2.2 Land-use and soil-conservation Laws

According to Roosevelt,[166] "a nation that destroys its soils, destroys itself". Based on this assertion, agricultural lands serve as basic national wealth and shall only be used for agricultural purposes in such a way so as not to damage their fertility and health. Farmers resort to some agricultural practices such as conservation agriculture, organic agriculture, and agro ecology in order to achieve optimum agricultural yield.[167] Aside from their potentials in mitigating GHG emission, these agricultural practices if left unchecked also have some negative repercussions on agricultural lands which among others include contamination, degradation, and the general destruction of agricultural land.[168]

Consequently, land-use and soil-conservation laws are required in order to regulate how individuals and communities make use of agricultural lands.[169] In this context, the *Voluntary Guidelines for Responsible Governance of Tenure (VGGT) of Land, Fisheries and Forestry in the Context of National Food Security* of 2012,[170] the *Voluntary Guidelines for Sustainable Soil Management (VGSSM)*,[171] the revised *World Soil Charter* 2015,[172] and the *RACNNR* of 2003[173] serve as examples of land-use laws and soil-conservation laws.

As a legal instrument for regulating the use of agricultural lands, the *VGGT* has two main objectives. First, the *VGGT* aims to protect agricultural lands from practices that could endanger their fertility, and to control the conditions of land use that could lead to climate change.[174] Second, the *VGGT* seeks to improve the fertility of good agricultural lands, and to promote the restoration of damaged agricultural lands.[175] To achieve this second objective, the *VGGT* provides legal and policy guidelines to both communities and individuals on how to avoid agricultural practices that could affect agricultural lands and possibly lead to climate change.[176] A section of the *VGGT* legal and policy guidelines reads as follows:

> [s]tates should ensure that legitimate tenure rights to land, fisheries and forests of all individuals, communities or peoples likely to be affected, with an emphasis on farmers, small-scale food producers and vulnerable and marginalized people, are respected and protected by laws, policies, strategies and actions related to climate change.[177]

As for the *VGSSM*, it provides some guiding principles to stakeholders on sustainable crop-production practices with less carbon-retention potential.[178] The purpose of these guiding principles is partly highlighted in the main objectives of the *VGSSM*, which are to boost soil fertility and to facilitate carbon sequestration through sustainable soil management.[179]

Similar to the *VGSSM*, the revised World Soil Charter 2015 also provides some legal and policy principles that are aimed at both promoting sustainable soil management and preventing soil degradation. The principles provided by the *World Soil Charter* are also intended to assist in controlling carbon concentration and other activities that could lead to climate change.[180] Notable principles of sustainable soil management provided by the *World Soil Charter* include among others:

> [t]he incorporation of the principles and practices of sustainable soil management into policy guidance and legislation at all levels of government, ideally leading to the development of a national soil policy; and explicit consideration of the role of soil management practices in planning for adaptation to, and mitigation of, climate change and maintaining biodiversity[181]

Apart from the *World Soil Charter*, there are other legal instruments that were enacted to preserve natural resources, though at the regional level. For example, the *RACNNR* of 2003 mandates parties to adopt and implement strategies that will prevent land degradation and misuse of soil, and promote sustainable land use in Africa.[182] The types of strategies recommended by the *RACNNR* include any adaptation strategy that can improve soil quality or combat soil erosion, as well as that which can prevent further deterioration of the physical, chemical, and biological properties of the soil.[183]

Apart from the regional laws discussed above, there are also national laws such as the *Kyrgyz Law on Soil Fertility and Protection of Agricultural Lands*[184] and the Kenyan *Land Act*[185] that were enacted to ensure sustainable soil management at the national level.[186] These national laws (especially the *Kenyan Land Act* and other similar national laws), which seek to protect the soil from degradation (and to rehabilitate soil degraded) as a result of droughts and floods, will be analysed in subsequent chapters of this book.

2.4.2.3 Agricultural insurance Law

Crops and livestock need to be insured against risk and losses, often caused by thefts and accidents. Climate change simply adds to the list of risks and losses that could affect agriculture. The regulation of insurance especially for (agricultural) losses emanating from climate change impacts started gaining prominence almost the same time as the UNFCCC came into force. The UNFCCC urged developing countries to consider insurance services as part of adaptation efforts for critical sectors of the economy vulnerable to climate change, such as agriculture.[187] The Kyoto Protocol and the Paris Agreement also contain provisions on insurance similar to the provisions of the UNFCCC.[188]

Based on the provisions of the UNFCCC, Kyoto Protocol, and the Paris Agreement, the concept of agricultural insurance has been recognised and integrated into adaptation strategies for agriculture against climate change-induced losses (such as drought, invasive and alien species, and floods) both at the international and the national levels.[189] Parties to the UNFCCC, Kyoto Protocol, and the Paris Agreement are also mandated to make provision for agriculture insurance in their respective countries according to their specific needs and vulnerabilities.[190] In other words, since adaptation actions are in most cases locally implemented, local laws also play vital roles in providing an enabling environment for insurance to achieve its potential in promoting adaptation in the context of agriculture. For example, in 2014, the Republic of Korea re-enacted a legislation for agricultural insurance known as the *Agricultural and Fishery Disaster Insurance Act (AFDIA)*.[191] The aim of the law is to promote stable agricultural practices (through insurance) that would avoid or moderate loss of agricultural products.[192] Another example of a country with national legislation for agricultural insurance is Azerbaijan. The Azerbaijan law known as the *Agrarian Insurance Law*[193] sets rules for insuring against farming risks and provides the economic and organisational mechanism for agricultural insurance. Types of agricultural products covered by the *Agrarian Insurance Law*, among others, include perennial plants and vegetable products, livestock, and aquaculture. The *Agrarian Insurance Law* mainly seeks to regulate the protection of the aforementioned agricultural products from risks and damages occurring from natural disasters such as climate change, wild fires, infectious diseases, and pests, wild animal attacks as well as third-party acts.[194]

Hence on the basis of the discussions above, law facilitates adaptation in the context of agriculture at the international level, regional level, and the national level.

2.5 Conclusion

This chapter argued that law is important and necessary in facilitating adaptation in a broad context and in the context of agriculture. The laws that regulate adaptation in a broad context exist in key instruments such as the UNFCCC, Kyoto Protocol, and the Paris Agreement, in addition to several other related laws and policies. In the context of agriculture, this chapter observed that adaptation is instead regulated by several but related laws such as water laws, soil conservation laws, fertiliser, and chemical-use laws. The chapter highlighted that each of the laws is required in adjusting agricultural production methods that serve as potential sources of climate change, and in promoting CSA production methods.

Given that the impacts of climate change on agriculture are global, but mostly felt at the local level, it is essential to also adapt to the impacts of climate change on agriculture at the local level with the support of national adaptation laws. The next three chapters of this book will identify and analyse the policies and laws that are relevant to adaptation and agriculture in three African focus countries (Kenya, Nigeria, and South Africa), with the aim of ascertaining whether these policies and laws contain the type of provisions that could enable and promote adaptation in the context of agriculture.

Notes

1 The aims of adaptation include reducing vulnerability, enhancing adaptive capacity, and boosting resilience of human and natural systems against the impacts of climate change. See Chapter 1, Section 1.4.2.

2 These forms of adaptation in the context of agriculture were previously explained as on-farm adaptation strategies, technology-based adaptation strategies, and agricultural insurance.

3 AS Hornby, AP Cowie and JW Lewis 1994 *Oxford advanced learner's dictionary*. Oxford University Press, 841.

4 JW Dellapenna 2014 Patterns of water law. In *The global water system in the anthropocene: Challenges for Science and Governance*. Springer, 401–413.

5 For example, see O Rumble 2019 Facilitating African climate change adaptation through framework laws. *Carbon and Climate Law Review*, 13(4), 237–245; E Scotford, S Minas and A Macintosh 2017 Climate change and National Laws across Commonwealth. *Commonwealth Law Bulletin*, 43(3–4), 318–361; NW Adger, WN Arnell and EL Tompkins 2005 Successful adaptation to climate change across scales. *Global environmental change*, 15(2), 77–86.

6 It should however be noted that Regulations are made to cover specificities that are usually missing in generic laws. Similarly, Regulations are made (pursuant to oversight functions) by government agencies and departments which are tasked with enforcing a law. Hence, legislative approvals are not required to make Regulations. See Coble 2015 What's the Difference Between Laws and Regulations?

https://www.findlaw.com/legalblogs/law-and-life/whats-the-difference-between-laws-and-regulations/.

7 E Scotford, S Minas and A Macintosh 2017 Climate change and National Laws across Commonwealth Countries. *Commonwealth Law Bulletin*, 43(3–4), 318–361.

8 H Bulkeley, L Andonova, K Bäckstrand, M Betsill, D Compagnon, R Duffy and S VanDeveer 2012 Governing climate change transnationally: Assessing the evidence from a database of sixty initiatives. *Environment and Planning C: Government and Policy*, 30(4), 591–612.

9 BH Druzin 2017 Why does soft law have any power anyway? *Asian Journal of International Law*, 7(2), 361–378.

10 A prime example of codes and standards is the "ecolabels", which are labels on products that attest to their environmental attributes. Ecolabels are found in more than 450 products in about 25 industrial sectors spread across 197 countries. See ECOLABLE INDEX. http://www.ecolabelindex.com.

11 Environmental stewardship means the "responsible use and protection of the natural environment through conservation and sustainable practices," e.g., reduced waste of water, best landscape management, etc. P Rowland 2009 *National inventory of environmental management systems in Australian agriculture*. RIRDC. https://www.environmentyou.eu/images/2017/ems/EMS_in_Australian_Agriculture.pdf, 1–124; RJ Berry (Ed.) 2006 *Environmental stewardship: Critical perspectives, past and present*, (Vol. 56). A and C Black.

12 Reporting scheme allows both private and public organisations to improve their environmental performance through continuous self-assessment methods. Reporting scheme has been found effective in providing the basis for decision-making. S Dobbs and C Van Staden 2016 Motivations for corporate social and environmental reporting: New Zealand evidence. *Sustainability Accounting, Management and Policy Journal*, 7(3), 449–472.

13 See GeeksforGeeks 2019 Difference between Policy and Law https://www.geeksforgeeks.org/difference-between-policy-and-law/

14 FAO 2020 *Agriculture and climate change*, Law and governance in support of climate smart griculture and international climate change goals. https://openknowledge.fao.org/bitstreams/c1a366bc-0667-4cfa-aeeb-ead2fa93cbd6/download, 1–376.

15 O Rumble 2019 Facilitating African climate change adaptation through framework laws. *Carbon and Climate Law Review*, 13(4), 237–245.

16 RK Craig 2010 Stationarity is dead-long live transformation: Five principles for climate change adaptation law. *Harvard Environmental Law Review*, 34, 9; JB Ruhl 2010 Climate change adaptation and the structural transformation of environmental law. *Environmental Law*, 40, 363; J McDonald and C Megan, 2014 Legal strategies for adaptive management under climate change. *Journal of Environmental Law*, 26(1), 25–53; M Mehling 2015 The comparative law of climate change: A research agenda. *Review of European, Comparative and International Environmental Law*, 24(3), 341–352.

17 T Townshend, S Fankhauser, A Matthews, C Feger, J Liu and T Narciso 2011 Legislating climate change on a national level. *Environment: Science and Policy for Sustainable Development*, 53(5) 5–17; A Clare, S Fankhauser and C Gennaioli 2017 The national and international drivers of climate change legislation. *Trends in Climate Change Legislation*, 19–36.; S Fankhauser, C Gennaioli and M Collins 2014 Domestic dynamics and international influence: What explain the passage of climate change legislation? *Centre for Climate Change Economics and Policy Working Paper*, 156; M Nachmany and J Setzer 2018 Policy briefs global trends in climate change legislation and litigation: 2018 snapshot. In *Grantham*

Research Institute on climate change and the environment, https://www.lse.ac.uk/GranthamInstitute/wp-content/uploads/2018/04/Global-trends-in-climate-change-legislation-and-litigation-2018-snapshot-1.pdf, 1–8.
18 E Scotford, S Minas and A Macintosh 2017 Climate change and national laws across Commonwealth countries. *Commonwealth Law Bulletin*, 43(3), 318–361.
19 R Mwebaza and LJ Kotzé 2009 Environmental governance and climate change in Africa: Legal perspectives. *Institute for Security Studies Monographs,* 167, 283.
20 McDonald 2011 The role of law in adapting to climate change. *Wiley Interdisciplinary Reviews: Climate Change*, 2(2), 283–295.
21 See F Yamin and J Depledge 2004 *The international climate change regime*: A guide to rules, institutions and procedures. Cambridge University Press, 9.
22 O Rumble 2019 Facilitating African climate change adaptation through framework laws. *Carbon and Climate Law Review*, 13(4), 237–245.
23 See also JR Barnett and M Webber 2010 Accommodating migration to promote adaptation to climate change. *World Bank Policy Research Working Paper*, (5270).
24 WN Adger, S Agrawala, QM Mirza, C Conde, K O'Brien, J Pulhin and K Takahashi 2007 Assessment of adaptation practices, options, constraints and capacity. In ML Parry, OF Canziani, JP Palutikof, PJ van der Linden, CE Hanson (Eds.) Climate change 2007: *I*mpacts adaptation and vulnerability: *Contribution of Working Group II to the Fourth Assessment Report of the Intergovernmental Panel on Climate Change*, 729–730. For example, through the Climate Change Act of 2008, the United Kingdom and Northern Ireland were able to design a long-term comprehensive national adaptation programme.
25 R Few, K Brown and EL Tompkins 2007 Public participation and climate change adaptation: Avoiding the illusion of inclusion. *Climate Policy*, 7(1), 46–59; Peel 2007 *Environmental and Planning Law Journal,* 24(2), 90–105; FAO 2020 *Agriculture and climate change*, Law and governance in support of climate smart agriculture and international climate change goals. https://openknowledge.fao.org/bitstreams/c1a366bc-0667-4cfa-aeeb-ead2fa93cbd6/download, 1–376.
26 FAO 2020 *Agriculture and climate change*. https://www.fao.org/3/cb1593en/CB1593EN.pdf. FAO, 82.
27 BK Sovacool 2011 Hard and soft paths for climate change adaptation. *Climate Policy*, 11(4), 1177–1183; M Kumamoto and A Mills 2012 What African countries perceive to be adaptation priorities: Results from 20 countries in the Africa adaptation programme. Climate and Development, 4(4), 265–274; EL Tompkins and WN Adger 2004 Does adaptive management of natural resources enhance resilience to climate change? Ecology and Society, 9(2), 10.
28 T Bonyhady, A Macintosh and J McDonald 2010 Adaptation to climate change: Law and policy. Sydney: Federation Press. 1–304.
29 See Chapter 1, Section 1.5.1.3. In addition, see also T Bonyhady, A Macintosh and J McDonald 2010 Adaptation to climate change: Law and policy. Sydney: Federation Press. 1–304.
30 RK Craig 2010 Stationarity is dead-long live transformation: Five principles for climate change adaptation law. *Harvard Environmental Law Review*, 34, 9.
31 See also K Kulovesi, M Mehling and EJ Hollo 2013 Introduction: Climate change and the law. In: E Hollo, K Kulovesi, M Mehling (Eds.), *Climate Change and the Law. Ius Gentium: Comparative Perspectives on Law and Justice,* vol. 21. Dordrecht: Springer, 1–8.
32 JB Ruhl 2010 Climate change adaptation and the structural transformation of environmental law. *Environmental Law,* 40, 363; O Rumble 2019 Facilitating African climate change adaptation through framework laws. *Carbon and Climate Law Review*, 13(4), 237–245.

33 See R Kibugi 2017 Adaptation to climate change in smallholder agriculture in Kenya: The role of law. In *Research handbook on climate change and agricultural law.* Edward Elgar Publishing, 367–384.
34 RK Craig 2010 Stationarity is dead-long live transformation: Five principles for climate change adaptation law. *Harvard Environmental Law Review,* 34, 9; A Kaswan 2012 Seven principles for equitable adaptation. *Sustainable Development Law and Policy,* 13(1), 41.
35 See A Kaswan 2012 Seven principles for equitable adaptation. *Sustainable Development Law and Policy,* 13(1), 41.
36 JB Ruhl and J Salzman 2012 Climate change meets the law of the horse. *Duke Law Journal,* 62(5), 975–1027; VB Flatt 2012 Adapting laws for a changing world: A systemic approach to climate change adaptation. *Florida Law Review,* 64(1), 269; O Afinowi 2018 An outline and critical assessment of the role of planning laws in the regulatory framework of climate change adaptation in South Africa and Nigeria. PhD thesis University of Cape Town, 65.
37 K Kulovesi, M Mehling and EJ Hollo 2013 Introduction: Climate change and the law. In: E Hollo, K Kulovesi and M Mehling (Eds.), *Climate Change and the Law. Ius Gentium: Comparative Perspectives on Law and Justice,* vol. 21. Dordrecht: Springer, 1–8.
38 See also R Maguire 2013 Foundations of international climate law: Objectives, principles and methods. In *Climate change and the law.* Springer, 83–110.
39 Article 3(1) UNFCCC.
40 Article 3(2) UNFCCC.
41 Article 3(3) UNFCCC.
42 Article 3(4) UNFCCC.
43 Article 3(5) UNFCCC.
44 See also RK Craig 2010 Stationarity is dead-long live transformation: Five principles for climate change adaptation law. *Harvard Environmental Law Review,* 34, 9.
45 PP Driessen and HF van Rijswick 2011 Normative aspects of climate adaptation policies. *Climate Law,* 24, 559–581.
46 See Chapter 1, Section 1.3.3.
47 X Contiades and A Fotiadou 2012 Social rights in the age of proportionality: Global economic crisis and constitutional litigation. *International Journal of Constitutional Law,* 10(3), 660–686.
48 Art. 3(3) UNFCCC; E Sussman, DC Major, R Deming and PR Esterman 2010 Climate change adaptation: Fostering progress through law and regulation. *New York University Law School Environmental Law Journal,* 18, 60.
49 D Bodansky, J Brunnée and L Rajamani 2017 *International climate change law.* Oxford University Press, 118; FAO 2020 *Agriculture and climate change.* https://www.fao.org/3/cb1593en/CB1593EN.pdf. FAO, 20.
50 The type of corporation sanctioned by the UNFCCC calls on Parties to promote and facilitate "public participation in addressing climate change and its effects and developing adequate responses". See O Afinowi 2018 *An outline and critical assessment of the role of planning laws in the regulatory framework of climate change adaptation in South Africa and Nigeria*. PhD thesis, University of Cape Town.
51 The COP, as the highest decision-making body of the UNFCCC has previously highlighted in Chapter 1, Section 1.3.3. It is during the COP that Parties to the UNFCCC submit their national communications and emission inventories for assessment. See UNFCCC Conference of the Parties (COP). https://unfccc.int/process/bodies/supreme-bodies/conference-of-the-parties-cop.
52 UNFCCC 2007 Climate Change: Impacts, Vulnerabilities And Adaptation In Developing Countries. https://unfccc.int/resource/docs/publications/impacts.pdf, 13.

53 It should be noted that the CAF emerged out of the COP 16 meeting of the UNFCCC as an example of international cooperation for the facilitation of adaptation processes in developing countries. The objectives of the CAF, among others, are to strengthen the implementation of the UNFCCC and to ensure no gap between the first and second commitment period in the implementation of the Kyoto Protocol, and to facilitate adaptation action on water. See UNFCCC 2010Report of the Conference of the Parties on its sixteenth session, held in Cancun from 29 November to 10 December 2010. Part one: Proceedings. https://unfccc.int/documents/6525; K Kulovesi, M Mehling and EJ Hollo 2013 Introduction: Climate Change and the Law. In: E Hollo, K Kulovesi, M Mehling (Eds.), *Climate Change and the Law. Ius Gentium: Comparative Perspectives on Law and Justice*, vol. 21. Dordrecht: Springer, 1–8.

54 Decision 1/CP.16, FCCC/CP/2010/7/Add.1. https://unfccc.int/resource/docs/2010/cop16/eng/07a01.pdf. See also R Knutti, J Rogelj, J Sedláček and EM Fischer 2016 A scientific critique of the two-degree climate change target. *Nature Geoscience*, 9(1), 13.

55 Decision1/CP.16,FCCC/CP/2010/7/Add.1.https://unfccc.int/resource/docs/2010/cop16/eng/07a01.pdf; S Ndoko 2015 Introducing the NAP process Commission on Genetic Resources for Food and Agriculture, FAO - 8–9 April, Rome, Italy, UNFCCC Secretariat. https://www.fao.org/fileadmin/templates/nr/documents/CGRFA/NAP_process.pdf, 1–6. See also Liu 2011 The Cancun Agreements. *Environmental Law Review*, 13(1), 43–49.

56 L Rajamani 2011 The Cancun climate Agreements: Reading the text, subtext and tea leaves. *International and Comparative Law Quarterly*, 60(2), 499–519; UNFCCC 2013 The State of Adaptation under the United Nations Framework Convention on Climate Change 2013 Thematic Report. https://unfccc.int/files/adaptation/cancun_adaptation_framework/adaptation_committee/application/pdf/ac_2013_report_high_res.pdf.

57 See the text of the United Nations Framework Convention On Climate Change. https://unfccc.int/files/essential_background/background_publications_htmlpdf/application/pdf/conveng.pdf, 6.

58 Article 4(1) UNFCCC. Party members of the UNFCCC are also expected to collaborate with NGOs in promoting adaptation measures through education, training, and climate change awareness campaigns.

59 It should be noted that the type of adaptation plans encouraged by the UNFCCC must be done in a precautionary and strategic manner in order not to cause maladaptation. See R Verheyen 2002 Adaptation to the impacts of anthropogenic climate change-the international legal framework. *Review of European Community and International Environmental Law*, 11(2), 129–143; M Grasso 2010 An ethical approach to climate adaptation finance. *Global Environmental Change*, 20(1), 74–81; E Albrecht, M Schmidt, M Mißler-Behr and SP Spyra (Eds.) 2014 *Implementing adaptation strategies by legal, economic and planning instruments on climate change*. Springer, 4.

60 See M Grasso 2010 An ethical approach to climate adaptation finance. *Global Environmental Change*, 20(1), 74–81; L Dong, F Zhiming and Y Yanzhao 2010 Evaluation of Material-based Development of China Using GIS Methods. *China Journal of Population, Resources and Environment*, 8(3), 17–24; J Verschuuren 2017 Climate change and agriculture under the United Nations Framework Convention on Climate Change and related documents. In *Research handbook on climate change and agricultural law*, 21–46.

61 Rather, the provision of the UNFCCC gives member States a degree of flexibility or discretion on what to consider as adequate in terms of adaptation. See for example, R Verheyen 2002 Adaptation to the impacts of anthropogenic climate change-the international legal framework. *Review of European Community and International Environmental Law*, 11(2), 129–143.

62 This is because the UNFCCC, as an international treaty, only governs adaptation measures by state actors. See R Verheyan 2002 Adaptation to the impacts of anthropogenic climate change-the international legal framework. *Review of European Community and International Environmental Law*, 11(2), 129–143.
63 R Verheyan 2002 Adaptation to the impacts of anthropogenic climate change-the international legal framework. *Review of European Community and International Environmental Law*, 11(2), 129–143.
64 The SCCF was established in 2001 to cater to specific adaptation needs of developing countries that are parties to the UNFCCC. See *The Global Environment Facility* 2022 Special Climate Change Fund – SCCF. https://www.thegef.org/what-we-do/topics/special-climate-change-fund-sccf.
65 See *The Global Environment Facility* 2022 Least Developed Countries Fund – LDCF. https://www.thegef.org/what-we-do/topics/least-developed-countries-fund-ldcf.
66 See The Green Climate Fund. https://www.greenclimate.fund/. See also Art.12 Kyoto Protocol.
The Green Climate Fund also finances the Clean Development Mechanism – one of the three market-based mechanisms provided by the Kyoto Protocol. It was designed to achieve the twin purpose of assisting industrialized countries (Annex 1 countries) to comply with their GHG-emission reduction commitments in a cost-effective way, and to promote sustainable development (SD) in developing countries, through the transfer of low-carbon technologies from Annex 1 countries. See HS Usman and A Berti Suman 2018 The value of a human rights-based approach to the climate change mitigation policy. In JA Daniels (Ed.), *Advances in Environmental Research*, Volume 63. New York: Nova Science Publishers.
67 J Paavola and WN Adger 2006 Fair adaptation to climate change. *Ecological Economics*, 56(4), 594–609; A Lesnikowski, J Ford, R Biesbroek, L Berrang-Ford and SJ Heymann 2016 National-level progress on adaptation. *Nature Climate Change*, 6(3), 261–264.
68 P Nanda 2015 Global Environmental Governance and the Global South In S Alam, S Atapattu, CG Gonzalez and J Razzaque (Eds.), *International Environmental Law and the Global South*. Cambridge: Cambridge University Press. 130–151.
69 J Truby and M Deehring 2017 The Oxford handbook of International Climate Change Law. *Journal of Environmental Law*, 29(1), 197–201.
70 See articles 2(3) and 3 (14) of the Kyoto Protocol.
71 TG Puthucherril 2012 International law on climate change adaptation has the time come for a new protocol. *Macquarie Journal of International and Comparative Environmental Law*, 8(2), 42–74.
72 This has also been provided by the Bali Action Plan of 2007, where each country is encouraged to identify its own environmental priorities and needs consistent with regional and international strategies as defined by regional and MEAs. See UNEP 2005 *UNEP Governing Council/Global Ministerial Environment Forum*. https://www.un.org/press/en/2005/unep266.doc.htm.
73 Article 10(c) Kyoto Protocol. In the same context, Puthucherril believes that both mitigation and adaptation strategies could benefit from the deployment of "environmentally sound technologies". TG Puthucherril 2012 International law on climate change adaptation has the time come for a new protocol. *Macquarie Journal of International and Comparative Environmental Law*, 8(2), 42–74.
74 Article 4(1) (b) Kyoto Protocol; DM Pallangyo 2013 Climate change and Africa: the normative framework of the African Union (Doctoral dissertation). North West University, Potchefstroom, South Africa, 1–375.
75 See Chapter 1, Section 1.4.4.
76 Article 10(b) (ii) Kyoto Protocol.
77 N Hall and Å Persson 2017 Global climate adaptation governance: Why is it not legally binding? *European Journal of International Relations*, 24(3), 540–566.

78 Britannica 2022 Paris Agreement international treaty (2015) https://www.britannica.com/topic/Paris-Agreement-2015.
79 The global goal of adaptation is provided by article 7 of the Paris Agreement. The goal is about "enhancing adaptive capacity, strengthening resilience, reducing vulnerabilities, and ensuring adequate adaptation response". See X Ngwadla and S El-Bakri 2016 *The global goal for adaptation under the Paris Agreement: Putting ideas into action*, Climate and Development Knowledge Network (CDKN). South Africa, https://policycommons.net/artifacts/1445643/the-global-goal-for-adaptation-under-the-paris-agreement/2077406/, 41.
80 UNFCCC 2015 Paris Agreement. https://unfccc.int/sites/default/files/english_paris_agreement.pdf.; A Dzebo and J Stripple 2015 Transnational adaptation governance: An emerging fourth era of adaptation. *Global Environmental Change,* 100(35), 423–435.
81 Article 2 (1) (b) Paris Agreement.
82 The agricultural sector has not been specifically contemplated by the UNFCCC.
83 A Savaresi 2016 The Paris Agreement: A new beginning? *Journal of Energy and Natural Resources Law* 34(1), 16–26; M Mehling 2015 The comparative law of climate change: A research agenda. *Review of European, Comparative and International Environmental Law,* 24(3), 341–352; L Rajamani 2016 The 2015 Paris Agreement: Interplay between hard, soft and non-obligations. *Journal of Environmental Law,* 28(2), 337–358.
84 Article 7(9) Paris Agreement. These national adaptation plans are further discussed in the relevant chapters of this book.
85 Article 7, paragraphs 10 and 11. For the purposes of further emphasis, "National Adaptation Communications are prepared to provide information on a country's national circumstances concerning adaptation by communicating its plans and priorities; highlighting its implementation and support needs and displaying its achievements." See https://unfccc.int/topics/adaptation-and-resilience/workstreams/adaptation-communications.
86 X Ngwadla and S El-Bakri 2016 *The global goal for adaptation under the Paris agreement: Putting ideas into action*, Climate and Development Knowledge Network (CDKN). South Africa, https://policycommons.net/artifacts/1445643/the-global-goal-for-adaptation-under-the-paris-agreement/2077406/, 41.
87 M Nachmany and J Setzer 2018 Policy briefs Global trends in climate change legislation and litigation: 2018 snapshot. In *Grantham Research Institute on Climate Change and the Environment,* https://www.lse.ac.uk/GranthamInstitute/wp-content/uploads/2018/04/Global-trends-in-climate-change-legislation-and-litigation-2018-snapshot-1.pdf, 1–8.
88 See generally, Chapter 1, Section 1.2.
89 See Chapter 1, Section 1.2.
90 J Ensor and R Berger 2009 Community-based adaptation and culture in theory and practice. In *Adapting to climate change: Thresholds, values, governance*. Cambridge University Press. 227–239.
91 ND Hamilton 1990 The Study of agriculture law in the United States: Education, organization and practice. *Arkansas Law Review,* 43, 503–522.
92 Land tenure is described as "the bundle of rights and responsibilities under which land is held, used, transferred, and succeeded". See SJ La Croix 2002 *Land tenure*: An introduction. East-West Center, 4. Examples of rights that can be held in respect of land include access to the land, control of products from the land, succession, etc., while the responsibilities connected to land tenure are all the obligations required in maintaining the land, in addition to all the conditions for occupying the land and the duration for holding the land. In most African and developing countries, land tenure rights can either be formally acquired (when

granted by the State) or informally acquired (when granted by customary or traditional institutions). See DP Kehoe 2006 *Landlords and tenants. A companion to the Roman Empire,* 298–311. A detailed explanation of how land tenure rights promote adaptation is given below.

93 FAO 2020 *Agriculture and climate change.* https://www.fao.org/3/cb1593en/CB1593EN.pdf, 106.

94 FAO 2020 *Agriculture and climate change.* https://www.fao.org/3/cb1593en/CB1593EN.pdf, 158.

95 FAO 2020 *Agriculture and climate change.* https://www.fao.org/3/cb1593en/CB1593EN.pdf, 158.

96 A Kiss and D Shelton 2007 *Guide to international environmental law.* Brill, 45.

97 This has been previously explained above. See article 4, UNFCCC; JA Leggett 2020 *The United Nations Framework Convention on Climate Change, the Kyoto Protocol, and the Paris Agreement: A summary,* UNFCCC: New York, NY, USA. 1–15.

98 Article 4(1) (c) UNFCCC. See also Art 10(c) KP for a similar provision.

99 See article 2 of the UNFCCCC.

100 Article 10(b)(i) of the Kyoto Protocol; A Telesetsky 1999 The Kyoto Protocol. *Ecology Law Quarterly,* 26(4), 797–813.

101 See Section 2.3.3 above.

102 QP Chiotti and T Johnston 1995 Extending the boundaries of climate change research: A discussion on agriculture. *Journal of Rural Studies,* 11(3), 335–350; RS Tol, S Fankhauser and JB Smith 1998 The scope for adaptation to climate change: what can we learn from the impact literature? *Global Environmental Change,* 8(2), 109–123; B Smit, I Burton, RJ Klein and R Street 1999 The science of adaptation: A framework for assessment. In *Mitigation and adaptation strategies for global change,* 4, 199–213; CR Bryant, B Smit, M Brklacich, TR Johnston, J Smithers, Q Chiotti and B Singh 2000 Adaptation in Canadian agriculture to climatic variability and change. In *Societal adaptation to climate variability and change.* Springer, 45, 181–201.

103 I Härtel 2018 *Handbook of agri-food law in China Germany, European Union: Food Security, food safety, sustainable use of resources in agriculture.* Springer., 496.

104 See generally, Chapter 1, Section 1.5.

105 UNFCC 2017 Decision/CP.23. Koronivia joint work on agriculture. https://unfccc.int/files/meetings/bonn_nov_2017/application/pdf/cp23_auv_agri.pdf. The KJWA was part of the outcome of the COP 23 held in November 2018 at Koronivia, Fiji. The KJWA acknowledges the significance of the agriculture sectors in facilitating climate change adaptation and mitigation. See FAO 2018 Koronivia Joint Work on Agriculture: Analysis of Submissions https://openknowledge.fao.org/server/api/core/bitstreams/6bdc7ab0-b1a4-411c-ae0b-018866282cb0/content.

106 FAO 2020 *Agriculture and climate change.* https://www.fao.org/3/cb1593en/CB1593EN.pdf, 13–14.

107 Articles 10 and 13 Paris Agreement.

108 The concept of technology transfer as an adaptation measure under the Paris Agreement also extends to capacity development and the provision of enabling environments for such transfers to take place. See article 10 (6) of the Paris Agreement.

109 UNFCCC 2011 Pre-event on early warning: Third Global Platform for Disaster Risk Reduction International Conference Centre Geneva. https://www.preventionweb.net/files/globalplatform/entry_presentation~microsoftpowerpointunfcccpreeventew.pdf, 14.

110 See Chapter 1, Section 1.5.1.3.
111 Article 4(8) UNFCCC; Article 3(14) Kyoto Protocol.
112 Article 8, Paris Agreement.
113 See D Tilman, KG Cassman, PA Matson, R Naylor and S Polasky 2002 Agricultural sustainability and intensive production practices. *Nature,* 418(6898), 671–677; PM Glibert and JM Burkholder 2006 The complex relationships between increases in fertilization of the earth, coastal eutrophication and proliferation of harmful algal blooms. In *Ecology of harmful algae,* 189(1940), 341–354; E Granéli 2006 *Ecology of harmful algae* J. T. Turner (Ed.). Springer, 189, 341–354; FAO 2020 *Agriculture and climate change.* FAO. https://www.fao.org/3/cb1593en/CB1593EN.pdf, 167.
114 Basel Convention on the Control of Transboundary Movements of Hazardous Wastes and their Disposal 1989. http://www.basel.int/TheConvention/Overview/tabid/1271/Default.aspx.
115 Rotterdam Convention on the prior informed consent procedure for certain hazardous chemicals and pesticides in international trade1998 http://www.pic.int/TheConvention/Overview.
116 Stockholm Convention 2001 on Persistent Organic Pollutants (POPs) http://chm.pops.int/TheConvention/Overview/TextoftheConvention/tabid/2232/Default.aspx.
117 K Ripley 2017 The 2017 BRS COPs: Steps forward, but challenges persist. *Environmental Policy and Law,* 47(2), 66–77.
118 BRS 2021 Synergies. http://www.brsmeas.org/Decisionmaking/Overview/About Synergies.
119 The FAO international code of conduct for the effective use and management of fertilisers also complements other instruments that regulate the use of pesticides or chemicals which are dangerous to human health and the environment such as the ISO's Codex Alimentarius standards for maximum residue limits and the Globally Harmonized System of Classification and Labelling of Chemicals. See FAO 2007 The State of Food and Agriculture. http://www.fao.org/3/a1200e/a1200e.pdf.
120 FAO 2020 *Agriculture and climate change.* https://www.fao.org/3/cb1593en/CB1593EN.pdf, 100.
121 World Bank 2021 Climate-Smart Agriculture. https://www.worldbank.org/en/topic/climate-smart-agriculture.
122 See for example article 6 of the FAO international code of conduct for the effective use and management of fertilisers.
123 FAO 2014 The International Code of Conduct on Pesticide Management http://www.fao.org/3/I3604E/i3604e.pdf.
124 FAO 2020 *Agriculture an*
climate change. https://www.fao.org/3/cb1593en/CB1593EN.pdf, 172.
125 See Chapter 1, Section 1.2. In addition, see S Reid, B Smit, W Caldwell and S Belliveau 2007 Vulnerability and adaptation to climate risks in Ontario agriculture. In Mitigation *and adaptation strategies for global change*, 12(4), 609–637.
126 G Steier and KK Patel (Eds.) 2017 *International farm animal, wildlife and food safety law*. Springer International Publishing, 602.
127 W Brussaard and MR Grossman 1992 Agrarian land law in the Western World: Some reflections. In *Agrarian land law in the western world*. CAB International, 254–270.; NJ Chalifour, P Kameri-Mbote, LH Lye and JR Nolon (Eds.) 2006 *Land use law for sustainable development*. Cambridge University Press, 53.
128 G Steier and KK Patel (Eds.) 2017 *International Farm Animal, Wildlife and Food Safety Law.* Springer International Publishing, 601.
129 The CBD was negotiated in 1992, but entered into force in 1993.

130 Article 1. CBD. See also G Steier and KK Patel (Eds.) 2017 *International farm animal, wildlife and food safety law*. Springer International Publishing, 603, 606.
131 United Nations Department of Economic and Social Affairs 1992 https://sdgs.un.org/partnerships.
132 Section 14.72.
133 Cattle Grazing Act Cap 42, Laws of Uganda 2000.
134 Section 2 (2) (b).
135 See Chapter 1, Section 1.3.2.
136 DL Kershen 2010 Trade and commerce in improved crops and food: An essay on food security. *New Biotechnology*, 27(5), 623–627. The laws that govern supply chain production, marketing, and sales also vary in their form, scope, and impact, depending on geographical, socio-economic, and political context. See SA Schneider 2009 A reconsideration of agricultural law: A call for the law of food, farming, and sustainability. *William and Mary Environmental Law and Policy Review*, 34(3), 935.
137 See Alabrese et al 2017 Agricultural law from a global perspective: An introduction. In *Agricultural law current issues from a global perspective*, 1, 1–12; FAO 2020 Agriculture and climate change. https://www.fao.org/3/cb1593en/CB1593EN.pdf 17; SAAB 2016 *Leiden Journal of International Law*, 29(2), 503–528.
138 See Chapter 1 Section 1.2; K Gannon, DA Conway and P Curran 2018 *Policy brief: Climate risks to Hydro-Power Supply in Eastern and Southern Africa*. Grantham Research Institute on Climate Change and the Environment (LSE) 1–8; JM Huho and RC Kosonei 2014 Understanding extreme climatic events for economic development in Kenya. *IOSR Journal of Environmental Science, Toxicology and Food Technology*, 8(2), 14–24.
139 JC Sanchez and J Roberts 2014 *Transboundary water governance: Adaptation to climate change*, 23, 1–305.
140 In this regard, see the provisions of articles 2(a)(iii), 3 and 5 Kyoto Protocol 1997. See also P Martinez and P Hofwegen 2006 *Synthesis of the 4th world water forum*, 90.
141 JC Sanchez and J Roberts 2014 *Transboundary water governance: Adaptation to climate change*, 1–305.
142 R Adler 2012 Balancing compassion and risk in climate adaptation: US water, drought, and agricultural law. *Florida Law Review*, 64(1), 201.
143 R Iza and R Stein 2009 *RULE Reforming water governance*, International Union for Conservation of Nature. IUCN, Water and Nature Initiative (WANI), 1–126.
144 However, it is not possible to give a holistic overview of laws that provided the guiding principles of adaptive water management in this chapter, except few examples which are discussed below.
145 AE Boyle and C Redgwell 2021 *Birnie, Boyle, and Redgwell's International Law and the Environment*. Oxford University Press, 541.
146 The UNECE Water Convention is also known as the Helsinki Convention of 1992. The main objective of the Helsinki convention of 1992. See Convention on the Protection and Use of Transboundary Watercourses and International Lakes. https://unece.org/fileadmin/DAM/env/water/pdf/watercon.pdf.
147 *African Union* 2003 African Convention on the Conservation of Nature and Natural Resources. https://au.int/sites/default/files/treaties/41550-treaty-Charter_ConservationNature_NaturalResources.pdf
148 SADC Protocol on Shared Water System, 1995- https://www.samsa.org.za/Documents/SADC_Protocol_on_Shared_Watercourses.pdf
149 European Union Water Framework Directive of 2000 https://environment.ec.europa.eu/topics/water/water-framework-directive_en?prefLang=fi

In addition, the principle of equitable and reasonable utilisation and that of prevention of significant harm are supported by the Berlin Rules on Water Resources of 2004 and the 1992 Convention on the Protection and Use of Transboundary Watercourses and International Lakes and the 1997 UN Convention on the Law of the Non-Navigational Uses of International Watercourses, etc. However, these international legal instruments will not be discussed, as they are not within the scope of this book.

150 See for instance the Berlin Rules on Water Resources 2004, https://www.internationalwaterlaw.org/documents/intldocs/ILA/ILA_Berlin_Rules-2004.pdf, 71; AE Boyle and C Redgwell 2021 *Birnie, Boyle, and Redgwell's International Law and the Environment,* Oxford University Press, 541.

151 JC Sanchez and J Roberts 2014 *Transboundary water governance: Adaptation to climate change,* 1–305.; P Sands, J Peel, A Fabra and R MacKenzie 2012 *Principles of international environmental law* (3rd ed.). Cambridge University Press, 305.

152 See for example, Article 5 of the *Helsinki Rules of the International Law Association*. https://www.internationalwaterlaw.org/documents/intldocs/ILA/Helsinki_Rules-original_with_comments.pdf 3.

153 Some of these adaptation strategies were previously explained as rainwater harvesting, digging of pits, and the construction of diversions and water-storage systems. B Smit (Ed.) 1993 *Adaptation to climatic variability and change:Report of the task force on climate adaptation*. Prepared for the Canadian Climate Program, Department of Geography Occasional Paper No. 19, University of Guelph, Guelph, 1–28: WE Easterlin 1996 Adapting North American agriculture to climate change in review. *Agricultural and Forest Meteorology,* 80(1), 1–53.

154 JA Swart, HJ Van Der Windt and J Keulartz 2001 Valuation of nature in conservation and restoration. *Restoration Ecology,* 9(2), 230–238.

155 An ecosystem-based approach in water management is an approach that seeks to achieve sustainability and conservation of water systems through a cooperative and ecology-based management system that takes into consideration the complexity; interconnection of hydrological, ecological, and social systems; associated with water use. See *The Global Environment Facility* 2011 *GEF Action on Transboundary Water Resources,* 31. M Falkenmark and J Rockström 2006 The new blue and green water paradigm: Breaking new ground for water resources planning and management. *Journal of Water Resources Planning and Management,* 132(3), 129–132.

156 Article 3(1) (i) of the UNECE Convention on the Protection and Use of Transboundary Watercourses and International Lakes.

157 Article 2(2) (A) UNECE Water Convention.

158 Article 7 (3) RACCNNR 2003; BT Erinosho 2015 *Environmental Law in Africa.* Kraft Books Limited, 180.

159 Article 7, RACCNNR 2003. The functions of the Inter-State Commission also include the settlement of disputes among States arising from the use of shared natural resources.

160 Article 7 RACCNNR 2003; BT Erinosho 2015 *Environmental Law in Africa*. Kraft Books Limited, 179.

161 Article 7, RACCNNR 2003; BT Erinosho 2015 *Environmental Law in Africa*. Kraft Books Limited, 180.

162 See the Preamble to the 1995 SADC Protocol on Shared Watercourse Systems; SM Salman 2007 The Helsinki Rules, the UN Watercourses Convention and the Berlin Rules: Perspectives on international water law. *Water Resources Development,* 23(4), 625–640; BT Erinosho 2015 *Environmental Law in Africa*. Kraft Books Limited, 171.

163 Directive 2000/60/EC establishing a framework for the Community action in the field of water policy. https://eur-lex.europa.eu/resource.html?uri=cellar:5c835afb-2ec6-4577-bdf8-756d3d694eeb.0004.02/DOC_1&format=PDF.
164 J Verschuuren and J McDonald 2012 Towards a legal framework for coastal adaptation: Assessing the first steps in Europe and Australia. *Transnational Environmental Law,* 1(2), 355–379.
165 G Kallis and D Butler 2001 The EU water framework directive: Measures and implications. *Water Policy,* 3(2), 125–142.
166 See"The nation that destroys its soil destroys itself" G7: Examining the evidence. Plymouth Perspectives https://www.plymouth.ac.uk/news/pr-opinion/the-nation-that-destroys-its-soil-destroys-itself
167 RLS Castellano and J Moroney 2018 Farming adaptations in the face of climate change. *Renewable Agriculture and Food Systems,* 33 (3), 206–211; FAO 2020 *Agriculture and climate change.* https://www.fao.org/3/cb1593en/CB1593EN.pdf, 157.
168 Although these approaches were previously explained as on-farm adaptation strategies, without proper regulation they could also trigger climate change. See Chapter 1. Section 1.5.1.1.
169 It is important to note that land use and soil conservation laws at the national level are presented as land tenure rights. Well-defined land tenure rights are relevant in promoting sustainable land use such as agro-ecological zoning, which ensures that land is used in a manner that is suitable for its soil. As a corollary, poorly defined or unclear land tenure rights could heighten land degradation. On these remarks, see for example E Nkonya (Ed.) 2004 *Strategies for sustainable land management and poverty reduction in Uganda* (Vol. 133). Intl Food Policy Res Institute, 133.
170 FAO 2012. *Voluntary guidelines on the responsible governance of tenure of land, fisheries, and forests in the context of national food security*. First revision. Rome. https://doi.org/10.4060/i2801e.
171 FAO 2017. *Voluntary guidelines for sustainable soil management food and agriculture organization of the United Nations*. https://www.fao.org/3/bl813e/bl813e.pdf.
172 FAO 2015 World Soil Charter. https://www.fao.org/3/I4965E/i4965e.pdf. The World Soil Charter is a non-binding instrument prepared by the FAO. It was first adopted in 1981 and revised in 2015.
173 *African Union 2003* African Convention on the Conservation of Nature and Natural Resources. https://au.int/sites/default/files/treaties/41550-treaty-Charter_ConservationNature_NaturalResources.pdf
174 Section 1.1, VGGT, 2012.
175 Section 1.2, VGGT, 2012.
It has been previously explained in Chapter 1 of this book that some practices such as manuring, composting, and mulching could improve soil fertility while biological nitrogen fixation derived from leguminous crops could restore damaged agricultural lands. E Nkonya (Ed.) 2004 *Strategies for sustainable land management and poverty reduction in Uganda* (Vol. *133).* Intl Food Policy Res Institute, 133.
176 H Ginzky, E Dooley, IL Heuser, E Kasimbazi, T Markus, T Qin and OC Ruppel (Eds.) 2017 *International yearbook of soil law and policy*. Springer, 72.
177 Section 23 of the VGGT.
178 FAO 2020 *Agriculture and climate change*. https://www.fao.org/3/cb1593en/CB1593EN.pdf, 199.
179 FAO 2020 *Agriculture and climate change*, 99.

180 BW Boer, H Ginzky and IL Heuser 2017 International soil protection law: History, concepts and latest developments. *International Yearbook of Soil Law and Policy*, 2016, 52.

181 See Sections 3–11 of the World Soil Charter; FAO 2020 *Agriculture and climate change*. https://www.fao.org/3/cb1593en/CB1593EN.pdf, 99.

182 Article IV RACCNNR 2003.

183 Hannam and Boer 2019 *Land degradation and international environmental law*. In Response to land degradation. CRC Press, 429–440; E Kasimbazi 2019 Soil protection law in Uganda. In *International yearbook of soil law and policy 2018*. Springer, 322.

184 No. 165 of 2012.

185 No. 6 of 2012.

186 For example, Sections 11 and 19 of the Kenyan Land Act provide for the conservation of ecologically sensitive public land and the conservation of land-based natural resources. M Anarbaev 2018 Soil legislation and policy in the Kyrgyz Republic on the development of the law "on soil fertility protection of agricultural lands". In H Ginzky, E Dooley, I Heuser, E Kasimbazi, T Markus, T Qin (Eds.), *International yearbook of soil law and policy 2017*. Springer, 65.

187 Article 4(8) UNFCCC.

188 Article 3(14) Kyoto Protocol; article 8(4)(f) of the Paris Agreement.

189 J Linnerooth-Bayer, R Mechler and S Hochrainer-Stigler 2011 Insurance against losses from natural disasters in developing countries: Evidence, gaps and the way forward. *Journal of Integrated Disaster Risk Management*, 1.

190 Hence, as could be seen below, national governments such as Kenya, Nigeria, and South Africa have enacted laws or have formulated policies regulating agriculture insurance.

191 Act No. 12729, Jun. 3, 2014.

192 Article 1 of the AFDIA provides for the purpose of the law, which among others is "to contribute to the stability in management of agriculture and fisheries and the improvement of productivity in agriculture and fisheries…"

193 Law No 1617-VQ of the Republic of Azerbaijan.

194 See Sections 2, 3, and 5 of the Law.

References

Adger et al. 2007 Assessment of adaptation practices, options, constraints and capacity. In ML Parry, OF Canziani, JP Palutikof, PJ van der Linden, CE Hanson (Eds.) *Climate change 2007: Impacts adaptation and vulnerability contribution of Working Group II to the fourth assessment report of the Intergovernmental Panel on Climate Change*. Cambridge University Press, 717–743.

Adler R 2012 Balancing compassion and risk in climate adaptation: US Water, Drought, and Agricultural Law. *Florida Law Review*, 64(1), 201.

Afinowi O 2018 *An outline and critical assessment of the role of planning laws in the regulatory framework of climate change adaptation in South Africa and Nigeria*. PhD-thesis, University of Cape Town.

Alabrese et al 2017 Agricultural law from a global perspective: An introduction. In *Agricultural law*. Springer, 1–12.

Albrecht E, Schmidt M, Mißler-Behr M and Spyra SP (Eds.) 2014 *Implementing adaptation strategies by legal, economic and planning instruments on climate change*. Springer, 4.

Anarbaev M 2018 Soil legislation and policy in the Kyrgyz Republic on the development of the law "on soil fertility protection of agricultural lands". In H Ginzky, E

Dooley, I Heuser, E Kasimbazi, T Markus, T Qin (Eds.) *International yearbook of soil law and policy, 2017*. Springer, 65.
Barnett JR and Webber M 2010 Accommodating migration to promote adaptation to climate change. World Bank Policy Research Working Paper, 5270.
Berry RJ (Ed.) 2006 *Environmental stewardship (Vol. 56)*. Aand C Black.
Bodansky D 2016 The legal character of the Paris Agreement. *Review of European, Comparative and International Environmental Law*, 25(2), 142–150.
Bodansky D, Brunnée J and Rajamani L 2017 *International climate change law*. Oxford University Press, 118.
Boer BW, Ginzky H and Heuser IL 2017 *International soil protection law: History, concepts and latest developments* Springer International Publishing, 49–72.
Boyle AE and Redgwell C 2021 *Birnie, Boyle, and Redgwell's International Law and the environment*. Oxford University Press, 541.
Brussaard W and Grossman MR 1992 Agrarian land law in the Western World: Some reflections. In *Agrarian Land Law in the Western World*, CAB International, 254–270.
Bryant CR, Smit B, Brklacich M, Johnston TR, Smithers J, Chiotti Q and Singh B 2000 Adaptation in Canadian agriculture to climatic variability and change. In *Societal adaptation to climate variability and change*. Springer, 45, 181–201.
Bulkeley H, Andonova L, Bäckstrand K, Betsill M, Compagnon D, Duffy R and VanDeveer S 2012 Governing climate change transnationally: Assessing the evidence from a database of sixty initiatives. *Environment and Planning C: Government and Policy*, 30(4), 591–612.
Castellano RLS and Moroney J 2018 Farming adaptations in the face of climate change. *Renewable Agriculture and Food Systems*, 33(3), 206–211.
Chalifour NJ, Kameri-Mbote P, Lye LH and Nolon JR (Eds.) 2006 *Land use law for sustainable development*. Cambridge University Press, 53.
Chiotti QP and Johnston T 1995 Extending the boundaries of climate change research: A discussion on agriculture. *Journal of Rural Studies*, 11(3), 335–350.
Clare A, Fankhauser S and Gennaioli C 2017 The national and international drivers of climate change legislation. In A. Averchenkova, S Fankhauser and M Nachmany (Eds.), *Trends in Climate Change Legislation*. Cheltenham, UK: Edward Elgar Publishing, 1–27.
Contiades X and Fotiadou A 2012 Social rights in the age of proportionality: Global economic crisis and constitutional litigation. *International Journal of Constitutional Law*, 10(3), 660–686.
Craig RK 2010 Stationarity is dead-long live transformation: Five principles for climate change adaptation law. *Harvard Environmental Law Review*, 34, 9.
Dellapenna JW 2014 Patterns of water law. In *The global water system in the anthropocene: Challenges for Science and Governance*. Springer, 401–413.
Dobbs S and Van Staden C 2016 Motivations for corporate social and environmental reporting: New Zealand evidence. *Sustainability Accounting, Management and Policy Journal*, 7(3), 449–472.
Dong S, Wen L, Liu S, Zhang X, Lassoie JP, Yi S and Li Y 2011 Vulnerability of worldwide pastoralism to global changes and interdisciplinary strategies for sustainable pastoralism. *Ecology and Society*, 16(2), 1–23.
Driessen PP and van Rijswick HF 2011 Normative aspects of climate adaptation policies. *Climate Law*, 24, 559–581.
Druzin BH 2017 Why does soft law have any power anyway? *Asian Journal of International Law*, 7(2), 361–378.

Dzebo A and Stripple J 2015 Transnational adaptation governance: An emerging fourth era of adaptation. *Global Environmental Change*, 100(35), 423–435.

Easterlin WE 1996 Adapting North American agriculture to climate change in review. *Agricultural and Forest Meteorology*, 80(1), 1–53.

Ensor J and Berger R 2009 Community-based adaptation and culture in theory and practice. In *Adapting to climate change: Thresholds, values, governance*, Cambridge University Press, 227–239.

Erinosho BT 2015 *Environmental Law in Africa*, Kraft Books Limited, 171.

Falkenmark M and Rockström J 2006 The new blue and green water paradigm: Breaking new ground for water resources planning and management. *Journal of Water Resources Planning and Management*, 132(3), 129–132.

Fankhauser S, Gennaioli C and Collins M 2014 Domestic dynamics and international influence: What explain the passage of climate change legislation? Centre for Climate Change Economics and Policy Working Paper, 156.

Few R, Brown K and Tompkins EL 2007 Public participation and climate change adaptation: Avoiding the illusion of inclusion. *Climate Policy*, 7(1), 46–59.

Flatt VB 2012 Adapting laws for a changing world: A systemic approach to climate change adaptation. *Florida Law Review*, 64(1), 269.

Gannon K, Conway Da, and nd Curran P 2018 *Policy brief: Climate risks to hydropower supply in Eastern and Southern Africa*. Grantham Research Institute on Climate Change and Environment.

Ginzky H, Dooley E, Heuser IL, Kasimbazi E, Markus T, Qin T and Ruppel OC (Eds.) 2017 *International yearbook of soil law and policy 2016*. Springer, 52–72.

Glibert PM and Burkholder JM 2006 The complex relationships between increases in fertilization of the earth, coastal eutrophication and proliferation of harmful algal blooms. In *Ecology of harmful algae*. Springer, 189(1940), 341–354.

Granéli E 2006 *Ecology of harmful algae*. JT Turner (Ed.). Springer, 341–354.

Grasso M 2010 An ethical approach to climate adaptation finance. *Global Environmental Change*, 20(1), 74–81.

Hall N and Persson Å 2017 Global climate adaptation governance: Why is it not legally binding? *European Journal of International Relations*, 24(3), 540–566.

Hamilton ND 1990 The Study of agriculture law in the United States: Education, Organization and Practice. *Arkansas Law Review*, 43, 503–522.

Härtel I 2018 *Handbook of agri-food law in China*. European Union, 496.

Hornby AS, Cowie AP and Lewis JW 1994 *Oxford advanced learner's encyclopedic dictionary*. Oxford University Press, 841.

Huho JM and Kosonei RC 2014 Understanding extreme climatic events for economic development in Kenya. *IOSR Journal of Environmental Science, Toxicology and Food Technology*, 8(2), 14–24.

Iza R and Stein R 2009 *RULE–reforming water governance*, International Union for Conservation of Nature. IUCN, Water and Nature Initiative (WANI), 1–126.

Kallis G and Butler D 2001 The EU water framework directive: Measures and implications. *Water Policy*, 125–142.

Kasimbazi E 2019 Soil Protection Law in Uganda. In: H Ginzky, E Dooley, I Heuser, E Kasimbazi, T Markus and T Qin (Eds.) *International yearbook of soil law and policy*, vol 2018. Springer, Cham, 315–330.

Kaswan A 2012 Seven principles for equitable adaptation. *Sustainable Development Law and Policy*, 13(1), 41.

Kehoe DP 2006 Landlords and tenants. In D S Potter *A companion to the Roman Empire*, Blackwell Publishing Ltd, 298–311.

Kershen DL 2010 Trade and commerce in improved crops and food: An essay on food security. *New Biotechnology*, 27(5), 623–627.

Kibugi R 2017 Adaptation to climate change in smallholder agriculture in Kenya: The role of law. In *Research handbook on climate change and agricultural law*. Edward Elgar Publishing, 367–384.

Kiss A and Shelton D 2007 *Guide to international environmental law*. Brill, 69.

Knutti R, Rogelj J, Sedláček J and Fischer EM 2016 A scientific critique of the two-degree climate change target. *Nature Geoscience*, 9(1), 13–18.

Kulovesi K, Mehling M and Hollo EJ 2013 Introduction: Climate change and the law. In E Hollo, K Kulovesi, M Mehling (Eds.), *Climate Change and the Law. Ius Gentium: Comparative Perspectives on Law and Justice*, vol. 21. Dordrecht: Springer, 1–8.

Kumamoto M and Mills A 2012 What African countries perceive to be adaptation priorities: Results from 20 countries in the Africa adaptation programme. *Climate and Development*, 4(4), 265–274.

La Croix SJ 2002 *Land tenure: An introduction*. East-West Center, 1–44.

Lesnikowski A, Ford J, Biesbroek R, Berrang-Ford L and Heymann SJ 2016 National-level progress on adaptation. *Nature Climate Change*, 6(3), 261–264.

Linnerooth-Bayer J, Mechler R and Hochrainer-Stigler S 2011 Insurance against losses from natural disasters in developing countries: Evidence, gaps and the way forward. *Journal of Integrated Disaster Risk Management*, 1(1), 59–81.

Maguire R 2013 Foundations of International Climate Law: Objectives, Principles and Methods. In E Hollo, K Kulovesi, M Mehling (Eds.), *Climate Change and the Law. Ius Gentium: Comparative Perspectives on Law and Justice*, vol 21. Springer, Dordrecht, 83–110.

Martinez P and van Hofwegen P 2006 *Synthesis of the 4th world water forum synthesis*, 90.

McDonald J, Dew K and Cumming J 2010 Change and adaptation in families with young carers. *Australian Journal of Social Issues*, 45(4), 459–475.

McDonald Jan and Megan C 2014 Legal strategies for adaptive management under climate change. *Journal of Environmental Law*, 26(1), 25–53.

Mehling M 2015 The comparative law of climate change: A research agenda. *Review of European, Comparative and International Environmental Law*, 24(3), 341–352.

Mwebaza R and Kotzé LJ 2009 Environmental governance and climate change in Africa: Legal perspectives. *Institute for Security Studies Monographs*, 167, 283.

Nachmany M and Setzer J 2018 Policy briefs global trends in climate change legislation and litigation: 2018 snapshot. *Grantham Research Institute on Climate Change and the Environment*, 1–8. https://www.lse.ac.uk/GranthamInstitute/wp-content/uploads/2018/04/Global-trends-in-climate-change-legislation-and-*litigation*-2018-snapshot-1.pdf.

Nachmany M, Fankhauser S, Davidová J, Kingsmill N, Landesman T, Roppongi H and Townshend T 2015 *The 2015 global climate legislation study: A review of climate change legislation in 99 countries: Summary for policy-makers*. Grantham Research Institute on Climate Change and the Environment, GLOBE International, London, UK.

Ndoko S 2015 *Introducing the NAP process Commission on Genetic Resources for Food and Agriculture*. FAO UNFCCC Secretariat https://www.fao.org/fileadmin/*templates*/nr/documents/CGRFA/NAP_process.pdf, 1–6, accessed 31 October 2019.

Ngwadla X and El-Bakri S 2016 *The global goal for adaptation under the Paris agreement: Putting ideas into action*. Climate and Development Knowledge Network (CDKN). South Africa, https://policycommons.net/artifacts/1445643/the-global-goal-for-adaptation-under-the-paris-agreement/2077406/, 41

Nkonya E (Ed.) 2004. *Strategies for sustainable land management and poverty reduction in Uganda* (Vol. 133). Intl Food Policy Res Institute.

Paavola J and Adger WN 2006 Fair adaptation to climate change. *Ecological Economics*, 56(4), 594–609.

Pallangyo DM and Scholtz W 2015 Africa and climate change: Legal perspectives from the AU. In W Scholtz and J Verschuuren (Eds.), *Regional environmental law*. Edward Elgar Publishing, 51–71.

Puthucherril TG 2012 International law on climate change adaptation has the time come for a new protocol. *Macquarie Journal of International and Comparative Environmental Law*, 8(2), 42–74.

Rajamani L 2011 The Cancun climate Agreements: Reading the text, subtext and tea leaves. *International and Comparative Law Quarterly*, 60(2), 499–519.

Rajamani L 2016 The 2015 Paris Agreement: Interplay between hard, soft and non-obligations. *Journal of Environmental Law*, 28(2), 337–358

Ripley K 2017 The 2017 BRS COPs: Steps Forward, but challenges persist. *Environmental Policy and Law*, 47(2), 66–77.

Rowland P 2009 *National inventory of environmental management systems in Australian agriculture*. RIRDC.

Ruhl JB 2010 Climate change adaptation and the structural transformation of environmental law. *Environmental Law*, 40(2), 363.

Ruhl JB and Salzman J 2012 Climate change meets the law of the horse. *Duke Law Journal*, 62(5), 975–1027.

Rumble O 2019 Facilitating African climate change adaptation through framework laws. *Carbon and Climate Law Review*, 237–245.

Saab A 2016 Climate-resilient crops and international climate change adaptation law. *Leiden Journal of International Law*, 29(2), 503–528.

Salman SM 2007 The Helsinki Rules, the UN Watercourses Convention and the Berlin Rules: Perspectives on international water law. *Water Resources Development*, 23(4), 625–640.

Sanchez JC and Roberts J 2014 Transboundary water governance: Adaptation to climate change, 23, 1–305.

Sands P, Peel J, Fabra A and MacKenzie R 2012 *Principles of international environmental law* (3rd ed.). Cambridge University Press, 45.

Savaresi A 2016 The Paris Agreement: A new beginning? *Journal of Energy and Natural Resources Law*, 34(1), 16–26.

Schneider SA 2009 A reconsideration of agricultural law: A call for the law of food, farming, and sustainability. *William and Mary Environmental Law and Policy Review*, 34(3), 935.

Scotford E, Minas S and Macintosh A 2017 *Climate change and national laws across Commonwealth countries*. *Commonwealth Law Bulletin*, 43(3–4), 318–361.

Smit B 1994 Adaptation to climatic variability and change. Report of the task force on climate adaptation (No. GU/G--CE04610). Guelph University, 1–28.

Smit B, Burton I, Klein RJ and Street R 1999 The science of adaptation: A framework for assessment. In *Mitigation and adaptation strategies for global change*, 4, 199–213.

Smit B, Reid S, Caldwell W and Belliveau S 2007 Vulnerability and adaptation to climate risks in Ontario agriculture. *Mitigation and adaptation strategies for global change*, 12(4), 609–637.

Sovacool BK 2011 Hard and soft paths for climate change adaptation. *Climate Policy*, 11(4), 1177–1183.

Steier G, and Patel KK (Eds.) 2017 *International farm animal, wildlife and food safety law*. Springer International Publishing, 606.

Sussman E, Major DC, Deming R and Esterman PR 2010 Climate change adaptation: Fostering progress through law and regulation. *New York University Law School Environmental Law Journal*, 18, 55–60.

Swart JA, Van Der Windt HJ and Keulartz J 2001 Valuation of nature in conservation and restoration. *Restoration Ecology*, 9(2), 230–238.

Tilman D, Cassman KG, Matson PA, Naylor R and Polasky S 2002 Agricultural sustainability and intensive production practices. *Nature*, 418(6898), 671–677.

Tompkins EL and Adger WN 2004 Does adaptive management of natural resources enhance resilience to climate change? *Ecology and Society*, 9(2), 10.

Townshend T, Fankhauser S, Matthews A, Feger C, Liu J and Narciso T 2011 Legislating climate change on a national level. *Environment: Science and Policy for Sustainable Development*, 53(5), 5–17.

Truby J and Deehring M 2017 The Oxford handbook of International Climate Change Law. *Journal of Environmental Law*, 29(1), 197–201.

Usman H and Berti Suman A 2018 The value of a human rights-based approach to the climate change mitigation policy. In Daniels, JA (Ed), *Advances in Environmental Research*, Volume 63. New York: Nova Science Publishers.

Verheyen R 2002 Adaptation to the impacts of anthropogenic climate change-the international legal framework. *Review of European Community and International Environmental Law*, 11(2), 129–143.

Verschuuren J 2017 Climate change and agriculture under the United Nations framework Convention on Climate Change and related documents. In *Research handbook on climate change and agricultural law*. Edward Elgar Publishing, 21–46.

Verschuuren J and McDonald J 2012 Towards a legal framework for coastal adaptation: Assessing the first steps in Europe and Australia. *Transnational Environmental Law*, 1(2), 355–379.

Yamin F and Depledge J 2004 *The international climate change regime: A guide to rules, institutions and procedures*. Cambridge University Press, 9.

3 Kenyan policy and legal framework for adaptation and agriculture

3.1 Introduction

Chapter 1 discussed the impacts of climate change on agriculture, and explained how agricultural practices such as crop and livestock production partly contribute to climate change. Adaptation was explored, both as a coping strategy for agriculture against the impacts of climate change and as a strategy for protecting the environment from agricultural practices that are capable of contributing to climate change. Chapter 2 explored the role of the law in facilitating adaptation, both in a broad context and in the specific context of agriculture. This chapter will build on the conceptual foundations developed by Chapter 1 and Chapter 2 of the book to explore the role of the Kenyan climate change regulatory framework and agricultural regulatory framework in facilitating adaptation in the context of agriculture.

In order to broadly facilitate adaptation, the Government of Kenya formulated some policies and enacted several laws pursuant to the guidelines that were provided by the provisions of the United Nations Framework Convention on Climate Change (UNFCCC), the Kyoto Protocol, and the Paris Agreement. In addition, the Government of Kenya formulated several agriculture-based policies, and enacted agricultural laws pursuant to the country's *Constitution* to ameliorate the problems posed by agriculture to the environment.[1] However, the question remains: how do Kenyan climate change and agriculture-related policies and laws provide for adaptation measures in the context of agriculture, and are there observable gaps in this regulatory framework?

Answering the research sub-question above requires a structured analysis of the policy and legal framework for adaptation, as well as key agricultural policies and laws in Kenya. The main aim of the analysis is to determine whether Kenyan climate change policies and laws, as well as agricultural policies and laws, offer provisions that support adaptation in the context of agriculture.

3.2 Regulatory framework for climate change adaptation

As noted earlier, Kenya's commitment in adapting to the impacts of climate change is demonstrated by several policies and laws that were formulated

DOI: 10.4324/9781003536437-3

and enacted pursuant to the guiding principles provided by article 4 of the UNFCCC, article 10 of the Kyoto Protocol, and article 7 of the Paris Agreement,[2] as well as in accordance with the provisions of the country's *Constitution*.[3] In this regard, examples of climate change adaptation policies that were formulated by the Government of Kenya include the National Environment Action Plan (NEAP),[4] National Climate Change Response Strategy (NCCRS),[5] National Climate Change Action Plan (NCCAP),[6] and the National Adaptation Plan (NAP).[7] It has been explained in Chapter 2 that policies generally serve as the basis for the enactment of specific laws, while laws are meant to support the implementation and operationalisation of policies.[8] Hence, in the context of Kenya's climate change adaptation governance, the aforementioned climate change policies led to the enactment of the *Environment Management and Coordination Act* (*EMCA*) of 1999,[9] the inclusion of provisions for climate change adaptation in the *Constitution*, and the enactment of a specific climate change law in 2016.[10] Considering that policies precede laws, discussion of the regulatory framework for adaptation will commence with a discussion of policies and then an analysis of the laws. However, before discussing climate change policies in Kenya, it is necessary to discuss the *Constitution of Kenya* as the basis for all policies in Kenya including climate change adaptation.

3.2.1 The Constitution of Kenya

The *Constitution of Kenya*, 2010 is the latest Constitution which operates in the country.[11] As with most national constitutions, the *Constitution of Kenya*, 2010 (hereinafter referred to as the *Constitution*) serves as the basis for all the policies and laws of Kenya including environmental governance.[12] This means that all other laws in Kenya derive their validity from the country's *Constitution*, and accordingly, any law that is inconsistent with the *Constitution* is null and void to the extent of the inconsistency.[13] The *Constitution* through its preamble makes environmental protection one of its main objectives. This is something which most constitutions do not have. Part of the preamble to the *Constitution* provides:

> [w]e the people of Kenya are respectful of the environment as a national heritage and are determined to sustain it for the benefit of future generations.[14]

Sustaining the environment could mean taking all necessary measures, including relevant adaptation measures, such as the construction of ocean walls or coastal defences as strategies against sea-level rise, or maintaining tree cover plantation, as measures for protecting the environment against degradation and climate change.[15] Protection of the environment through adaptation is a constitutional right which has been provided by article 42 of the *Constitution*, in the following words:

> Every person has the right to a clean and healthy environment, which includes the right—(a) to have the environment protected for the benefit of present and future generations through legislative and other measures, particularly those contemplated in article 69...[16]

In a similar vein, article 69 of the *Constitution* makes it obligatory for the Government of Kenya and stakeholders to implement measures that could support the sustainable exploitation and conservation of the environment.[17] By inference, these measures could also include adaptation measures. Article 69 also mandates the State to utilise the environment and its components, such as land and water, for the benefit of the people of Kenya.[18]

Land is an important example of a natural resource that is vital in supporting agriculture, but at the same time, it is impacted by climate change. By implication and as previously stated, adaptation strategies such as the planting of cover crops and conservation tillage, construction of water purification plants, sprinkler irrigation methods, and so on could also serve to protect and conserve components of the environment, especially land and water.[19] Against this background, adaptive land management practices are provided for under article 60 in the *Constitution*, which provides:

> Land in Kenya shall be held, used and managed in a manner that is equitable, efficient, productive and sustainable, and in accordance with the following principles—(a) equitable access to land; (b) security of land rights; (c) sustainable and productive management of land resources; (d) transparent and cost-effective administration of land; (e) sound conservation and protection of ecologically sensitive areas...[20]

These requirements of article 60 are closely aligned with the principles of international environmental law, which also form part of the Kenyan law.[21] In practical terms, the sustainable exploitation, utilisation, and conservation of land, for instance, could only be possible when land is equitably allocated. Uneven allocation of land could lead to the over-utilisation of land and over-concentration of agricultural activities, for instance over-grazing of cattle and by extension more emission of greenhouse gas (GHG). The concept of adaptive land management practices, which was previously explained in Chapter 1, also includes spacing of grazing reserves and bush fallowing.[22] These strategies could be effectively implemented when land is equitably allocated in accordance with the provisions of article 69 of the Constitution.

Given that the *Constitution* is the basis for all policies in Kenya,[23] article 69 of the Constitution provides the required stimulus for formulating the NCCRS, the NCCAP, and a Climate-Smart Agriculture (CSA) Strategy Framework that seeks to facilitate adaptation in the country.[24] However, it should be noted that these climate change policies, response strategies, and action plans are not equivalent to hard binding laws, but (as discussed previously) they are soft laws or tools for coordinating climate change institutional and governance actions, including adaptation.[25]

3.2.2 Policies for adaptation in Kenya

As stated above, Kenya formulated climate change-related policies based on the guidelines provided by the UNFCCC, the Kyoto Protocol, and the Paris Agreement, and pursuant to the provisions of the *Constitution*. The provisions of the UNFCCC, Kyoto Protocol, and Paris Agreement dealing with adaptation were discussed in Chapter 2.[26] The provisions of the *Constitution* that focus on adaptation, in a broad context and in the context of agriculture, were explained above. This section will discuss environmental-related and climate change policies in Kenya in order to ascertain whether the policies have sufficient content for adaptation.

3.2.2.1 National Environment Action Plan

The NEAP of 1994 was the first environmental action plan developed in Kenya. The development of the NEAP was influenced by the outcome of the proceedings of the United Nations Conference on Environment and Development (UNCED) of 1992.[27] Additionally, the NEAP was developed out of the need to address the negative impacts of environmental problems such as droughts, desertification, and climate change on Kenya's economic development.[28] However, the NEAP did not expressly provide for the means and methods of addressing the negative impacts of environmental problems, which include adaptation strategies, such as adaptive land management or water-resource management techniques. Nonetheless, the NEAP provides the platform for the mainstreaming of climate change concerns into the national economic planning and development of Kenya.[29] On this basis, the NEAP was described as:

> [t]he most comprehensive and ambitious environmental blueprint, serving as a platform for the formulation of policies and strategies, enactment of statutes, and establishment of institutions that touch on environment and climate change in Kenya.[30]

Examples of additional policies and laws that were formulated and enacted to address the impacts of climate change in Kenya based on the recommendations of the NEAP include the NCCRS of 2010, NCCAP of 2012 and 2018, the *EMCA* of 1999, and the *Climate Change Act* of 2016 (hereinafter, referred to as the *Climate Change Act*).[31] These policies are briefly analysed below.

3.2.2.2 National Climate Change Response Strategy

Pursuant to article 4 of the UNFCCC,[32] many African countries have developed (or are in the process of developing) their NCCRS.[33] Most of these NCCRS include either only adaptation measures, or a combination of both adaptation and mitigation measures.[34] The NCCRS of Kenya, 2010 is an example of a national climate change policy response strategy which caters for both adaptation and mitigation measures in addressing the impacts of climate change.[35] As a policy strategy, the NCCRS does not have the binding force of law,[36] but is nevertheless guided by some basic principles derived from the *Constitution*, which have been explained above (especially the principles of land policy).[37]

The NCCRS identified key sectors of the Kenyan economy (such as agriculture, forestry, rangelands, health, and tourism) that are susceptible to the impacts of climate change, and worthy of receiving adaptation and mitigation responses.[38] In terms of adaptation, the NCCRS is aimed at:

> [p]romoting focused adaptation strategies against changing climate by ensuring commitment and engagement of all stakeholders while taking into account the vulnerable nature of our natural resources and society as a whole.[39]

Notable adaptation strategies specified by the NCCRS include, but are not limited to, the production and promotion of crop varieties that could withstand the impacts of climate change (such as droughts and pests, especially invasive and alien species) as well as the production and promotion of early maturing crops.[40] In addition, as part of agricultural adaptation strategies, the NCCRS makes provision for the breeding of animal varieties that could adapt to changing climate.[41] As part of further support for adaptation in agricultural practices, the NCCRS recognises "special livestock insurance schemes to spread and transfer risks from climate change".[42] Insurance schemes have been firmly established as a coping strategy against climate change-induced risk on agriculture, such as droughts and lack of fodder.[43] Hence, a special livestock insurance scheme with low premium rate could provide an opportunity for the restocking of cattle herds lost due to climate change-induced starvation or diseases.[44]

The NCCRS also aims to facilitate adaptation against the impacts of climate change on the resources that support agriculture – for instance, in the case of water, through provisions which promote irrigation strategies. Some examples of irrigation strategies provided by the NCCRS include the construction of water basins and pans to allow for the utilisation of marginal-quality water in cases of declined water cycles due to climate change.[45] Other irrigation strategies include the construction of waterways to capture and retain rainwater and the construction of strategic boreholes to boost water availability in dry seasons.[46]

Apart from providing guidance on adaptation strategies, the NCCRS was also formulated to provide "a conducive and enabling policy, legal and institutional framework to combat climate change in Kenya".[47] In this regard, the NCCRS served as the basis for reviewing existing legislation such as the EMCA and the drafting of new legislation, for instance, the *Climate Change Act* (discussed below).[48] Additionally, the NCCRS facilitated the formulation of a NCCAP (explained below) that also provides for adaptation in some important sectors of the Kenyan economy, particularly agriculture and tourism.[49] Although the NCCRS sets the tone for sectoral adaptation in Kenya, an additional policy, especially the NCCAP is still required for providing a more focused implementation.[50]

3.2.2.3 National Climate Change Action Plan

As a follow-on to the NCCRS, the Government of Kenya developed its NCCAP in 2012.[51] The NCCAP covers a plethora of climate change governance themes, such as low-carbon development, adaptation and mitigation strategies, and climate finance.[52] As a climate change action plan, the NCCAP makes adaptation a topmost priority. In the context of agriculture, the aim of the NCCAP is to promote strategies that could boost crops and livestock production services through strategies such as integrated harvesting techniques, introduction of drought-tolerant crops, integrated soil-fertility management, and crop insurance.[53] The NCCAP was partly formulated to support the operationalisation of the NCCRS from 2013 to 2017 and to consolidate Kenya's adaptation governance.[54] Further, the NCCAP was designed to serve as a platform for enacting and developing comprehensive and coherent climate change laws in Kenya.[55] The NCCAP was amended in 2018 to cater for climate change adaptation and mitigation actions in Kenya from 2018 to 2022.[56]

One of the objectives of the NCCAP of 2018–2022 is to pursue low-carbon and climate-resilient development pathways,[57] while aiming to address the risks that climate-related disasters such as droughts and floods pose to communities and infrastructure.[58] Additionally, the NCCAP of 2018–2022 is aimed at increasing food and nutrition security by enhancing productivity and resilience of agricultural systems.[59] At the same time, the goal of the NCCAP of 2018–2022 is to enhance the resilience of water systems, especially for the purposes of agriculture and manufacturing.[60]

Equally, the NCCAP of 2018–2022 aims to promote adaptation and mitigation actions by providing guidelines in the areas of capacity building and knowledge sharing, technology transfer, climate finance, measurement, reporting, and verification.[61] These guidelines are intended to coordinate adaptation and mitigation actions at all tiers of governments (both national and county governments), the private sector and civil society.[62]

As well, the NCCAP demonstrates Kenya's determination to mainstream climate change concerns into national development plans, where "both national and county governments (under the supervision of the National Climate Change Council (NCCC)) are tasked with the responsibility of integrating climate change responses into development planning, decision-making, and implementation in all sectors of the economy".[63]

Based on the foregoing discussion, the Kenyan NCCAP of 2018–2022 acts as a framework for Kenya to deliver on its Nationally Determined Contributions (NDCs) under the Paris Agreement.[64] As noted by Schwarte,[65] a State's ability and readiness to addressing global climate change is usually articulated in its NDC. In other words, NDCs are recognised as implementation tools for the Paris Agreement, which is legally binding on signatory countries that ratified the Agreement, including Kenya.[66] As part of its commitment to addressing the impacts of climate change, and as required by the Paris Agreement, the Government of Kenya (even though a small emitter of GHGs) prepared and submitted its NDC to the UNFCCC Secretariat for the first time in 2015.[67]

The Kenyan NDC is prepared in line with the country's NAP (elaborated upon below) and is a combination of both the mitigation and adaptation objectives of the country.[68] The NDC of Kenya further affirms the country's vulnerability to climate change impacts (especially in critical sectors such as agriculture and tourism), while at the same time emphasising the "need for a stronger focus on climate adaptation and the reduction of vulnerability, particularly in the smallholder agriculture sector".[69] Therefore, Kenya's NDC (in line with the NCCAP) seeks to enhance resilience and adaptive capacity in the agricultural sector by mainstreaming adaptation into medium-term development plans. [70]

However, the NCCAP has been criticised for falling short in providing the required institutional support or direction on how to boost resilience.[71] For example in the agriculture sector, the NCCAP either failed to recognise or presented a poor understanding of the impacts and vulnerabilities of climate change on the sector.[72] As a result, coordinated programmes that could promote CSA are equally missing from the NCCAP.[73] Similarly, the NCCAP (as an adaptation action plan) has been faulted for not encouraging water conservation, which has been explained as an adaptation strategy against climate change-induced water shortages.[74] Based on the criticisms advanced against the NCCAP (which by extension also applies to the NCCRS), the Government of Kenya developed a NAP for the country.

3.2.2.4 National Adaptation Plan

Grounded in the foundations laid by the NCCRS and NCCAP, and supported by the provisions of the 2010 *Constitution*, the Government of Kenya developed its NAP in 2015.[75] The NAP is designed to operate for a period of 15 years, within which the programme is expected to consolidate Kenya's vision on adaptation through the promotion of low-carbon and climate-resilient development in line with the goals of the Paris Agreement.[76] Low-carbon and climate-resilient development is expected to be achieved in Kenya in the short term and medium term, which involves, among others, an assessment of existing and planned low-carbon climate infrastructural assets against the risk and vulnerability of climate change impacts.[77]

To achieve its objectives, the Kenyan NAP identifies both current and future dimensions and peculiarities of the country's climate change and associated vulnerabilities, which were previously identified in Chapter 1.[78] Also, in a bid to promote low-carbon and climate-resilient development, the NAP coordinates the required governance and institutional arrangements, and monitors and evaluates adaptation processes in Kenya.[79] For example, based on the NAP, the Government of Kenya was able to identify 20 socio-economic sectors (including agriculture) that require protection against the impacts of climate change, either through short-term, medium-term, or long-term adaptation measures.[80]

As a policy plan for adaptation, the objectives of Kenya's NAP are in alignment with other environmental protection-based laws that were enacted before the NAP, for instance, the *EMCA* of 1999, as well as adaptation laws which came after the NAP such as the *Climate Change Act* of 2016. These adaptation laws will now be discussed in the next section.

3.2.3 Adaptation-related Laws

As already explained, the *Constitution* is the centrepiece for adaptation legal framework in Kenya. As explained above, article 42 of the *Constitution* makes environmental protection a right available to every person in Kenya. It has also been stated above that adaptation is one of the means of protecting the environment against destruction. In accordance with article 72 of the *Constitution*, the parliament of Kenya has the power to make laws (including adaptation laws) for the protection of the environment.[81]

Article 2 of the *Constitution* also affirms the role and position of international law, treaties, and conventions ratified by Kenya as an important source of Kenyan law.[82] This is particularly relevant and important for adaptation governance in Kenya, which is guided by the provisions of the UNFCCC, the Kyoto Protocol, and the Paris Agreement as well as the *Constitution*. Examples of laws derived from the *Constitution* that support adaptation are the EMCA and the *Climate Change Act*.

3.2.3.1 The Environment Management and Coordination Act of 1999

The *EMCA* of 1999 is a law that deals with environmental management, including climate change adaptation in Kenya.[83] The *EMCA* became operational in 2000 as a framework law that governs matters such as policy planning and designing of strategies and tools, as well as the establishment of environmental institutions both at the national and local levels of government in Kenya.[84] The National Environment Management Authority is an example of an institution that was established based on the recommendation of the *EMCA*.[85] Similarly, climate change adaptation is among the strategies recommended by the *EMCA* for the protection and conservation of the environment. To this effect, a provision of the *EMCA* reads as follows:

> The Authority [National Environment Management Authority] shall, in consultation with the relevant lead agencies, develop, issue and implement regulations, procedures, guidelines and measures for the sustainable use of hill sides, hill tops, mountain areas and forests and such regulations, guidelines, procedures and measures shall control the harvesting of forests and any natural resources located in or on a hill side, hill top or mountain area so as to protect water catchment areas, prevent soil erosion and regulate human settlement.[86]

The provision of the *EMCA* discussed above seeks to regulate GHG-intensive farming activities that take place near hillsides and on hill tops, in mountain areas and forests, which have the potential to destroy sensitive resources that support agriculture, such as water-catchment areas and soil components. Measures for the sustainable use of hillsides, hill tops, mountain areas, and the control of harvesting of forests, as indicated by the provisions of the *EMCA* above, also include adaptation strategies already discussed in Chapter 1, such as conservation tillage, land contouring, and terrace farming, which have acknowledged benefits in adjusting the impacts of climate change manifesting as water imbalance, water runoff, and soil moisture reduction.[87]

Other processes to support adaptation measures provided by the *EMCA* for controlling GHG-intensive farming activities near hillsides, hill tops, mountain areas, and forests, include the Strategic Environment Assessment (SEA) and the Environmental Impact Assessment (EIA).[88] According to the *EMCA*, the SEA is "the process of subjecting public policy, programmes and plans to test for compliance with sound environmental management"[89], while an EIA is "a systematic examination conducted to determine whether or not a programme, activity or project will have any adverse impacts on the environment".[90] As a precautionary measure, the EIA is done purposely in order to determine impacts of development activities (including farming) on the environment.[91] The EMCA outlines the requirements and methods of conducting an EIA.[92]

As a process for assessing proposed developmental activities, the SEA could be embarked on for the purposes of the protection and conservation of natural resources and biodiversity.[93] For example, the SEA is conducted to prevent farming near hillsides or on hill tops that could result in the emission of more carbon, which by extension would lead to climate change-induced flooding and erosion. Another benefit of the SEA and EIA to adaptation as supported by the *EMCA* according to Onyango and Schmidt[94] is that they provide important practical tools for mainstreaming climate change concerns into Kenya's development agenda.[95] In the context of agriculture and climate change, for instance, a well-conducted SEA and EIA could aid in integrating less GHG-intensive agricultural activities into medium and long-term national development plans.

In spite of the well-drafted provisions of the *EMCA* on EIA and SEA, the implementation of EIA and SEA remains a challenge. Mainly due to lack of trained workforce, it is difficult to use EIA and SEA to assess projects or monitor the development of some projects.[96] For example, EIA and SEA permits disallowing marine farming within marine protected areas, made pursuant to *Regulation 21 of the Riverbanks, Lake Shores and Sea Shore Management Regulations,*[97] appear to be mere requirements in theory, and it is possible to find farmers cultivating crops on marine protected areas such as river banks, lake shores, and sea shores in contravention to the *Regulations*.[98] This is evident in the case of Takaungu and Uyombo villages in Kilifi District on the Kenyan coast, where the cultivation of crops such as sorghum, pearl millet, and cowpeas takes place.[99] The implication of this is that adaptation strategies intended for such areas, (particularly those provided by the provisions of the *Wetlands, Riverbanks, Lake Shores and Sea Shore Management Regulations*) including conservation of water catchments and the control of floods, are often impossible.[100]

Notwithstanding these concerns, certain provisions of the *EMCA* were reaffirmed by provisions of the *Constitution* which deal with environmental management. [101] For example, the provisions of section 9 of the *EMCA*[102] are supported by the principles of land policy in the *Constitution*.[103] As well, the provisions of article 62 of the *Constitution* designating public lands as lands worthy of being protected validated the provisions of the *EMCA* Regulations on *Wetlands, Riverbanks, Lake Shores and Sea Shore Management*.[104] In the same vein, article 69 of the *Constitution* which provides for obligations in respect of the environment restated the provisions of *EMCA*.

As stated earlier, the *EMCA* and the *Climate Change Act* derived their validity from the *Constitution* of Kenya. Having discussed the *EMCA*, the *Climate Change Act* is now discussed.

3.2.3.2 Climate Change Act of 2016

The Kenyan *Climate Change Act* is the first national law dedicated to climate change in Africa.[105] Coordinated by a high-level advisory body known as the NCCC,[106] the *Climate Change Act* was enacted to provide the requisite legal and institutional framework for both mitigation and adaptation actions in Kenya.[107] In order to achieve its objectives, the *Climate Change Act* is guided by certain values and principles, which are largely drawn from the *Constitution* such as the principles of economic and social rights (article 43), principles of land policy (article 60) as well as the principles of transparency, accountability, and sustainable development (article 10).[108] These values and principles, as already explained above, are intended to provide the legal and policy incentives and obligations for achieving low-carbon and climate-resilient development in the country, which is also possible through adaptation actions.

With regard to adaptation, section 13 of the *Climate Change Act* provides that:

> The Cabinet Secretary shall, in accordance with Article 10 of the Constitution and section 3 of this Act, and through public consultation, formulate a National Climate Change Action Plan ... The National Climate Change Action Plan shall prescribe measures and mechanisms —for adaptation to climate change...[109]

The reviewed NCCAP of 2018 serves as a notable example of a NCCAP that has been formulated pursuant to the provisions of section 13 of the *Climate Change Act*.[110] Based on the requirements of section 13 of the *Climate Change Act*, the NCCAP of 2018 is expected to further:

> [p]rescribe measures and mechanisms—(a) to guide the county toward the achievement of low -carbon climate resilient sustainable development; (b) to set out actions for mainstreaming climate change responses into sector functions; (c) for adaptation to climate change...[111]

The *Climate Change Act* is also considered as a cornerstone for principles of adaptation because it makes provision for policy coordination and oversight functions through:

> [t]he establishment of a high-level advisory body; a dedicated access to information provision, the operationalization of a decentralized governance framework; a clear role of a financial mechanism; the utilization of a climate change mainstreaming approach, while at the same time emphasizing on a low- carbon and climate resilient development pathway.[112]

Serving as a legal instrument that guides national and sub-national responses to climate change in Kenya, the *Climate Change Act* is also expected to promote the minimisation of climate change risk in all sectors of the Kenyan economy through the integration of climate change concerns into national development plans and processes.[113] The process of integration (mainstreaming) in this sense, or simply just as previously described,[114] implies the:

> [i]ncorporation and prioritization of climate change (adaptation) considerations into all areas of government regulation and planning for development.[115]

In other words, some provisions of the *Climate Change Act* offer guidance and directions on how adaptation actions against the risk of climate change in all sectors of the Kenyan economy could be mainstreamed into national and county development plans and processes.[116] The concept of mainstreaming as provided by the *Climate Change Act* also includes the building of resilience and adaptive capacity of human and natural systems against the impacts of climate change.[117]

In particular, the *Climate Change Act* 2016 is much more specific in terms of institutional arrangements for climate change coordination because it provides for mechanisms and measures intended to achieve low-carbon climate development and related goals.[118] These mechanisms and measures are contained in the objects and purpose of the Act and they include, among others, mainstreaming climate change responses into development planning and decision-making and implementation, and formulating programmes and plans to enhance the resilience and adaptive capacity of human and ecological systems to the impacts of climate change.[119]

In spite of the progressive provisions of the *Climate Change Act* which alludes to its main objectives of achieving "a herculean task of creating a legal framework that places climate change concerns at the front and centre of the development process in Kenya",[120] the law appears to contain some gaps. In particular, agriculture was not specifically considered by the *Climate Change Act*. Rather, agriculture-focused adaptation is inferred from the general provisions of the Act. The lack of specific agriculture-focused adaptation provisions is seen as a barrier to the implementation of an ideal adaptation strategy for agriculture in Kenya.[121] In the same context, some institutional and sectoral compartmentalisation could operate to undermine the purposes of adaptation either in a general context or in the context of agriculture as envisaged by Kenya's regulatory framework for adaptation.

In sum, the *Climate Change Act,* as well as the generic regulatory framework for climate change in Kenya, could be described as comprehensive as far as adaptation in a broad context is concerned. However, in terms of agriculture-focused adaptation, the generic regulatory framework for climate change only provides a few meaningful provisions. Therefore, it is necessary to discuss agriculture and agriculture-related policies and laws in order to determine whether they facilitate adaptation in the context of agriculture.

3.3 Regulatory framework for agriculture

Considering the linkages between agriculture and climate change, agricultural policies and laws must not only promote supply chain production, marketing and sales, but must also aim to protect the environment from agriculture-induced climate change.[122] As well, agricultural policies and laws must be applied to promote coping strategies for agriculture against the impacts of climate change. Consequently, agricultural policies and laws have been applied in "equipping the Kenyan government and its various agencies with effective tools to respond to the complex challenges of climate change to agriculture".[123] These policies and laws are enacted and formulated in accordance with international standards and guidelines issued by international organisations such as the Food and Agricultural Organization (FAO), and with reference to the Kenyan *Constitution*.[124] Even though most of these agricultural policies and laws are not directly focused on climate change, they nonetheless address climate change concerns in the context of agriculture.[125] These agricultural policies and laws are discussed below, not chronologically according to their year of formulation and enactment, but based on their relevance.

Given that CSA strategies are required in coping with the impacts of climate change on agriculture, and in adjusting agricultural practices that contribute to climate change; also, relying on the assertion that effective CSA practices at the national level are generally dependent on well-established institutional mechanisms, which are supported by robust policy and legal frameworks[126]; a starting point for discussing the regulatory framework for agricultural adaptation in Kenya is the Kenya CSA Strategy.

3.3.1 Kenyan Climate-Smart Agriculture Strategy of 2017

The Government of Kenya devised a Climate-Smart Agriculture Strategy Framework in 2017.[127] The CSA Strategy Framework seeks to promote a sustainable agricultural system, which could increase productivity and income, build resilience, and address GHG emissions in the country.[128] Dinesh et al.[129] believe that the CSA Strategy Framework has been modelled on the guidelines issued by article 4 of the UNFCCC, which has been previously emphasised.[130] Hence, the status of the CSA Strategy Framework is that of a policy which has no binding legal force, but is merely a guide for stakeholders towards the implementation of related policies. In other words, stakeholders in the agriculture sector both at the local and national level of government look to the CSA Strategy Framework for directions in identifying the most suitable agricultural production methods in given circumstances.[131] For example, agricultural training institutions such as Kenya Water Institute, Kenya School of Agriculture, Kenya Institute of Organic Farming, and the Kenya Meat Training Institute are guided by the Kenyan CSA Strategy Framework in operationalising integrated farming and pastoral production systems which for instance consist of agroforestry practices and conservation agriculture.[132]

The CSA Strategy Framework has broad and specific objectives. Broadly, the CSA Strategy Framework seeks to promote agriculture-focused adaptation by enhancing the resilience of the agricultural sector "through measures that could reduce the vulnerabilities of farmers, pastoralists, and fisher-folk to changing temperature regimes and precipitation patterns".[133] Another broad objective of the CSA Strategy Framework is to minimise the emission of GHGs emanating from agricultural practices, without affecting food security and income from agriculture.[134]

Further and more specifically, the objectives of the CSA Strategy Framework are to:

> create an enabling regulatory and institutional framework; and address cross-cutting issues that adversely impact climate-smart agriculture in Kenya.[135]

To achieve its broad and specific objectives, the CSA Strategy Framework identifies focal areas such as water, temperature, and land that require urgent attention against the impacts of climate change.[136] These focal areas have been previously explained as supporting resources for agriculture.[137] Subsequent sections of this chapter will explain the regulatory framework (starting with the policies, then laws) for the resources that support agriculture in Kenya in order to determine how they facilitate adaptation in the context of agriculture.

3.3.2 *Agricultural policies*

As noted above, the *Constitution* is the basis for the formulation of all policies and law concerning all matters in Kenya, including that of agriculture.[138] For example, the *Constitution* makes provision for the development of a national land policy which is expected to be reviewed regularly.[139] In addition, a draft national livestock policy and a draft national agricultural insurance policy were formulated in Kenya pursuant to the *Constitution* in order to partly reduce the vulnerability of agriculture to the impacts of climate change. These policies are explained in the following order.

3.3.2.1 *National Land Policy*

Sustainable and equitable use of natural resources including land is part of good adaptive agricultural practices, previously explained as CSA.[140] In Kenya, sustainable and equitable use of land is also provided by the National Land Policy of 2007,[141] which has been revised in 2009[142] pursuant to the provisions of the *Constitution*.[143] The National Land Policy of 2007 promotes sustainable and equitable use of land by focusing on actions that seek "to address environmental problems such as land degradation, soil erosion and pollution".[144] Some of the strategies formulated for addressing land degradation include adaptive land management techniques such as equitable land allocation that aims to avoid over-concentration of agricultural activities on land – one of the drivers of climate change.[145]

The reviewed National Land Policy of 2009 focused more on how to avoid land deterioration, especially loss of soil nutrients emanating from unconventional farming practices (especially GHG-intensive cultivation methods).[146] Compared to the National Land Policy of 2007, the National Land Policy of 2009 accommodates sustainable land practices, which also employ soil- conservation strategies such as conservation tillage and crop-cover techniques.[147]

3.3.2.2 *Draft National Livestock Policy of Kenya*

The role of livestock in Kenya's economic development, how it contributes to climate change, and the manner by which it is impacted by climate change have been previously highlighted.[148] The draft National Livestock Policy of

Kenya, 2020 (hereinafter referred to as the draft Policy) has the mandate of defining roles and monitoring responsibilities of major stakeholders in the livestock sub-sector.[149]

The draft Policy, once fully implemented, is expected to safeguard livestock production and supportive resources (water, soil, and fodder) from the impacts of climate change, especially droughts and desertification.[150] Furthermore, the objectives of the draft Policy are to promote food security and contribute to better livelihoods, without significantly affecting the environment.[151] Since livestock production has also been identified as a contributing factor to climate change,[152] the draft Policy also seeks to manage livestock interaction with the environment (especially natural resources such as land and water) through strategies that, among others, include monitoring and controlling the allocation and use of range land and other sustainable land-management techniques.[153]

To achieve the aforementioned objectives, the draft Policy formulates and seeks to implement effective and climate-friendly breeding strategies (feeding and nutrition support), disease (genetic disorders and pests) control, and management services at the national and county levels of government.[154] The draft Policy provides the necessary guidance and mandate for both national and county governments in steering the development of the livestock industry as an important component of Kenya's economy, which has been affirmed by Vision 2030 (Kenya's blueprint for national development).[155]

In a nutshell, the draft Policy identifies and facilitates appropriate measures that could enable the livestock sector to:

> [e]nhance its contribution to food and nutritional security, provide raw materials for agro-based industries and contribute to improved livelihoods in the country.[156]

An example of such measures is the proposal for the establishment of country-wide livestock insurance schemes based on a public-private-partnership model, which will safeguard the livestock sub-sector from the risks of climate change such as droughts and shortage of fodder.[157] This point prompts a further discussion on agricultural insurance governance in Kenya.

3.3.2.3 Draft National Agricultural Insurance Policy

Insurance has been identified as a means of coping with the impacts of climate change, whereby farmers usually resort to either the indemnity-based insurance or the index-based insurance in recouping their lost or damaged crops and livestock against climate change-based risk and vulnerability.[158] Consequently, the Government of Kenya formulated a draft National Agricultural Insurance Policy in 2021 (hereinafter referred to as the draft Insurance Policy) as a starting point for agriculture-focused insurance services.[159] The draft Insurance Policy recognises the indemnity-based insurance and index-based insurance as two major categories of insurance for the agriculture sector.[160] Further, the draft Insurance Policy is expected to:

> [f]oster the development and growth of agriculture, promote sustainable food and nutritional security through enhanced agricultural productivity and profitability.[161]

When fully implemented, it is anticipated that the draft Insurance Policy will serve as a means of unlocking more potential in the agricultural sector, particularly a risk-free and equitable utilisation of natural resources such as land.[162] Equitable utilisation of land has been explained as a means of promoting adaptation in land use,[163] meaning that the availability of adequate land for the planting of crops and the grazing of cattle could ease the emission of GHGs, and could also allow degraded lands affected by erosion to recover.

Furthermore, the draft Insurance Policy is relevant for adaptation in the context of agriculture as:

> [i]t has been identified as one of the ways to de-risk the agriculture sector and reduce vulnerability of farmers and other value chain actors.[164]

Hence, reading through the main objective of the draft National Agricultural Insurance Policy, which is to promote sustainable food security by enhancing agricultural productivity, it can be concluded the adaptation is probably facilitated by the Policy.[165]

However, it has been observed that the draft Insurance Policy has neglected some basic requirements of a comprehensive insurance for agriculture. For example, those basic requirements that are relevant in "expanding agriculture risk management options, enhancing productivity capacity, stabilising agricultural income, and reducing food insecurity"[166] are missing.

Given that policies are often translated into binding laws, as explained in Chapter 2,[167] the gaps that were identified in the draft National Livestock Policy and the draft National Agricultural Policy could be remedied, at least in principle, by some agriculture-based legislation.

3.3.3 Agriculture-related Laws

In accordance with the provisions of the *Constitution*, the Government of Kenya had enacted a set of laws in order to boost agricultural productivity in the country.[168] These agricultural laws also seek to enhance the resilience and adaptive capacity, as well as reduce the vulnerability of agricultural productivity against risk and disasters such as climate change-induced droughts, floods, and pests. Examples of these agriculture-based laws (that will be further discussed below) include the *Water Act*,[169] *National Drought Management Authority Act* (*NDMA Act*),[170] *Kenyan Agriculture and Livestock Research Act* (*KALRA*),[171] *Crops Act*,[172] *Land Act*,[173] and the *Agriculture and Food Authority Act* (*AFA Act*).[174]

3.3.3.1 Water Act

As explained earlier, water is one of the most important resources for agricultural practices, particularly crops and livestock production. However, water is severely affected by the impacts of climate change and, without adaptive water strategies (especially irrigation), crop yield and livestock productivity would drastically decline.[175] Adaptive water strategies in Kenya are regulated by water resources laws. Therefore, the *Water Act* of 2016 is the most recent law that regulates water resources in Kenya, and for the purposes of clarification, the law describes water resources as:

> [a]ny lake, pond, swamp, marsh, stream, watercourse, estuary, aquifer, artesian basin or other body of flowing or standing water, whether above or below the ground, and includes sea water and transboundary waters within the territorial jurisdiction of Kenya.[176]

The *Water Act* was enacted as a successor to the *Water Institute Act* of 2001,[177] which is also a water regulatory framework that clearly delineates roles and responsibilities in water governance, at the same time promoting the decentralisation of water and sewerage services.[178] The *Water Act* seeks to regulate the use, management, and development of water resources in Kenya in accordance with certain constitutional principles.[179] These constitutional principles related to water management are the same principles previously described as relevant in supporting broad adaptation strategies, and they include the principles of national values and governance; principles of economic and social right; and the principles of land policy.[180]

By extension, the *Water Act* seeks to protect the ecosystem in Kenya (against climate change-induced water damages such as erosion and flooding) through the conservation, control, and management of water resources. Section 2 of the *Water Act* defines water resource management as "the conservation, including soil and water conservation, protection, development and utilization of water resources". [181] Section 22 of the Act emphasises the need for the conservation and protection of water from disasters such as climate change through specific measures such as prohibition of activities that are capable of damaging catchment areas.[182] Consequently, and for the purposes of conservation and protection of water resources in Kenya, the *Water Act* makes provision for an integrated water services strategy plan known as the National Water Resource Strategy (NWRS).[183] The plans of the government for the protection, conservation, control, and management of water resources as required by the *Water Act*[184] especially delineating the water sector into catchment areas are provided through this NWRS.[185]

Another significant aspect of the *Water Act* is that it provides for the establishment of an institution known as the Water Resources Authority to oversee the use and management of water resources.[186] The Water Resources Authority (preceded by the Water Resources Management Authority) is an institution that was established to carry out specific functions, which among others include the formulation and enforcement of standards and procedures as well as regulations relating to the use of water resources.[187] As well, the functions of the Water Resources Authority are to inform and advise the Cabinet Secretary on how to formulate policies and strategies for the control and management of floods in Kenya.[188] The provisions of the *Water Act* and the functions of the Water Resources Authority (as identified and discussed above) indicate that adaptation is also anticipated as a strategy against water shortage and droughts occurring because of climate change.

3.3.3.2 National Drought Management Authority Act

Crop yield and livestock productivity are impacted by droughts, and in turn, droughts are linked to climate change, which means that adaptation is required as a strategy for coping with droughts.[189] Hence, due to the devastating impacts of droughts on the environment across the Horn of Africa in general, the Government of Kenya enacted the *NDMA Act* in 2016 with provisions to oversee the management of drought-related disasters in the country.[190] The provisions of the *NDMA Act* are implemented by the NDMA, through strategic projects and programmes.[191] These strategic projects and programmes include mapping of zones vulnerable to droughts, the incorporation of risk reduction and climate adaptation into planning, and the implementation of social protection programmes in food-insecure populations.[192]

Given that drought management is an adaptation option as discussed previously,[193] the *NDMA Act* also makes provision for strategies that could improve drought preparedness in the country, such as early warning information systems and Ending Drought Emergency (EDE) strategies.[194] Accordingly, these strategies, especially the EDE strategies, are expected to reduce the overall impacts of drought in Kenya by the year 2022, and to also aid in "... strengthening the institutional and financing framework for drought risk management in Kenya".[195]

Other than making provisions for strategies aimed at adjusting the impacts of droughts in Kenya, the *NDMA Act* also serve as a future policy formulation instrument for drought management and adaptive water management in Kenya. In other words, the *NDMA Act* mandates the NDMA to:

> [p]romote knowledge on drought management and climate change adaptation activities and contribute to policy formulation on matters that align with its mandate.[196]

Apart from crop production services, other agricultural sub-sectors such as livestock services also benefit from the general provisions of the *NDMA Act* in the management of droughts. In other words, and as discussed below, the *NDMA Act* supports adaptation efforts in other sectors particularly the livestock services. It should be noted that Kenya's livestock-production services are also guided by the *KALRA*. The next section will discuss this law.

3.3.3.3 Kenyan Agriculture and Livestock Research Act

The livestock sub-sector has been noted as an important contributor to Kenya's GDP, and it is not only heavily affected by the impacts of climate change (especially droughts) but also contributes to climate change.[197] Therefore, adaptation strategies such as migration for livestock and the rearing of animal breeds that can withstand the impacts of climate change were identified.[198] These adaptation strategies for the livestock sub-sector are partially supported by the *KALRA*.[199] In other words, section 5(2) of the *KALRA* is to the effect that:

> [f]or the purpose of carrying out its functions under subsection (1), the Organisation [Kenya Agricultural and Livestock Research Organization] shall-(a) formulate policy and make policy recommendations to the Cabinet Secretary on agricultural research.

By extension, the draft National Livestock Policy which outlines Kenya's livestock adaptation strategies is an example of a policy that was prepared in 2020 pursuant to the *KALRA*. The draft National Livestock Policy has already been discussed above[200] and, therefore, does not require further analysis here. However, apart from the *KALRA*, another significant legislation regulating agricultural practice in Kenya is the *Crops Act*.

3.3.3.4 Crops Act of 2013

The *Crops Act* is a law enacted in order to consolidate all existing laws dealing with the growth and development of crops in Kenya.[201] In particular, the *Crops Act* is concerned with the "promotion of the production, processing, marketing, and distribution of crops in suitable areas of the country ..."[202] By virtue of section 6 of the *Crops Act*, national and county governments are obliged to provide an enabling environment for the development of the crop sub-sector. Accordingly, the *Crops Act* mandates both the national and county governments in Kenya "to determine and promote the implementation of agricultural policies and measures in a manner designed to promote, support and enhance productivity in the crop subsector".[203] These measures for the promotion, support, and enhancement of crop productivity could mean all forms of adaptation measures aimed at addressing the impacts of climate change on crops. Examples of adaptation measures that are relevant for crop productivity, which are provided by the *Crops Act*, include soil and water conservation strategies.[204] The importance of soil and water to crop production was discussed in Chapter 1.[205]

The *Crops Act* also has a pivotal role in facilitating agricultural insurance in Kenya – a form of adaptation in the context of agriculture that has been previously discussed.[206] The *Crops Act* provides for the formulation of a national policy for agricultural insurance. According to a provision of the *Crops Act*:

> The Authority [Agriculture, Fisheries and Food Authority] may recommend the establishment of subsidiary bodies to discharge specific

> aspects, in particular—(a) food security; (b) value addition, marketing and export; (c) offer extension services for irrigation farming; (d) pest and disease control; (e) crop insurance; (f) marketing; and (g) any other aspect relating to crop development.[207]

Based on this provision, the Government of Kenya formulated a draft National Agricultural Insurance Policy in 2021 (already discussed above).[208] It is worth noting that the establishment of a specific insurance for agriculture by the *Crops Act* did not altogether exclude the administration of agricultural insurance under the *Insurance Act* of 1984. [209] The *Insurance Act* generally served as a framework for the regulation, supervision, and development of the insurance industry in Kenya, and agricultural services are included under miscellaneous matters.[210]

3.3.3.5 Land Act of 2012

The 2009 National Land Policy served as a basis for the Government of Kenya to enact a more focused national land law in a bid to promote adaptive land management, known as the *Land Act*.[211] As part of efforts to facilitate adaptive land management, the *Land Act* recognises the need to protect public lands, and it makes provision in that regard. A provision of the *Land Act* reads as:

> The [National Land] Commission shall identify ecologically sensitive areas that are within public lands and demarcate or take any other justified action on those areas and act to prevent environmental degradation and climate change.[212]

Justified action in the context of climate change governance could mean both mitigation and adaptation strategies. With regard to adaptation, the rationale for the protection of public land is to moderate activities that are capable of inducing more GHG emissions, for instance. This would entail using previously explained land-management strategies such as controlled grazing and other restrictive crop-cultivation methods which include conservation tillage and cover cropping.[213]

Essentially, the *Land Act* is a law that revises and harmonises major land policies and laws in order to have a unified and coherent framework for the sustainable management of land and land-based resources.[214] A useful example of a law harmonised by the *Land Act* is the *Agriculture Act* which hitherto provided for adaptive land management strategies.[215] Section 48 of the *Agriculture Act* provides for land preservation rules that encapsulate soil conservation methods and the prevention of the effects of soil erosion on lands (especially agricultural lands). The rules on land preservation seek to avoid:

> (i) the breaking or clearing of land for the purposes of cultivation; (ii) the grazing or watering of livestock; the firing, clearing or destruction of vegetation...[216]

Regarding the *Land Act*, it must be reemphasised that the law was enacted pursuant to the *Constitution of Kenya*, 2010.[217] As a result, the *Land Act* accommodates relevant values and principles that guide state organs, state officers, and all other persons in implementing actions that could promote "the sustainable and productive management of land resources; conservation and protection of ecologically sensitive areas in Kenya".[218] As explained previously, sustainable land uses are part of adaptation strategies in the context of agriculture.[219]

While the *Land Act* contains some provisions for the protection of ecologically sensitive areas against climate change from which adaptation strategies could be inferred, there are concerns that these provisions of the *Land Act*, especially section 11, did not provide sufficient direction on the type of justified adaptation actions, or how they could be implemented.[220] On the basis of this, the contents of the *Land Act* are not sufficient as far as adaptation in the context of agriculture is concerned.

3.3.3.6 Agriculture and Food Authority Act

As part of efforts to promote adaptation in the context of agriculture, the Government of Kenya enacted the *AFA Act*.[221] The *AFA Act* was mainly enacted in order to consolidate all laws and policies that promote crops and food production (that is, cultivation of land and animal husbandry) in Kenya, especially the *Lands Act* and the *Crops Act*.[222] In order to effectively administer the *Lands Act* and *Crops Act*, the *AFA Act* makes provision for the establishment of a body known as the Agriculture and Food Authority, which defines the roles of government at the national and county levels in promoting crops and food production in Kenya. [223]

Considering that land is one of the important resources of agricultural production, which if not properly utilised could lead to more GHG emissions, adaptive land management is required.[224] Therefore, the *AFA Act* seeks to facilitate adaptive land management through the *Lands Act*. In this context, section 21 of the *AFA Act* provides for the formulation of policy guidelines on the development, preservation, and utilisation of agricultural land, which could be interpreted as guidelines for adaptation. In other words, section 21 provides for:

> the adoption of such system of management or farming practice or other system in relation to land in question (including the execution of such work and the placing of such things in, on or over the land, from time to time) as may be necessary for the proper development of land for agricultural purposes.[225]

Most, or all of adaptive land management – such as conservation tillage, changing the topography of land, cover cropping, etc., previously explained in Chapter 1 – could serve as illustrations for the aforementioned measures of the *AFA Act*.

Also, given that livestock production is one of the main sources of GHG emission and climate change in Kenya,[226] section 22 of the *AFA Act* provides for certain rules that specify the manner and method by which livestock can be kept on agricultural land.[227] These rules for livestock management are required in order to reduce the vulnerability of the environment to livestock-derived GHG emissions.

Regarding the administration of the *Crops Act*, the *AFA Act* prohibits some agricultural activities such as clearing or destruction of vegetation for the purposes of farming, which could lead to more GHG emission.[228] By extension, the *AFA Act* seeks to protect water catchment areas and soil fertility as vital components of the environment and resources for agriculture. In this regard, the *AFA Act* provides for adaptation measures which, among others, include the construction, maintenance or repair of drains, gullies, contour banks, terraces, and diversion ditches as well as the salination of the soil.[229] These adaptation strategies were previously discussed in detail.[230]

In addition to promoting adaptation in the context of agricultural production through the administration of the *Crops Act*, the *AFA Act* makes provision for guidelines on the control and management of invasive weeds.[231] As explained previously, climate change serves as a transmission mechanism for invasive weeds, and invasive weeds lead to poor crop yield.[232]

As far as adaptation in the context of agriculture is concerned, the *AFA Act* seems to have sufficient provisions. However, a major lacuna that has been identified in the *AFA Act* is that the required cattle herds to be kept on agricultural land is not indicated by section 22 of the *AFA Act*. In other words, a farmer may circumvent the requirements of section 22 of the *AFA Act*, resulting in a barrier to adaptation in the context of agriculture. This lacuna prompts further discussion on the legal and policy challenges for agriculture-focused adaptation in Kenya.

3.4 Gaps in the Kenyan policy and legal frameworks for climate change adaptation and agriculture

As discussed, Kenya is committed to adapting to the impacts of climate change through various policies and laws targeting critical sectors of interest, such as agriculture. The *Constitution* and the NCCRS, NCCAP, NAP, *EMCA*, and the *Climate Change Act* allude to the interest and commitment of Kenya to adaptation governance. However, it appears that most of these climate change policies and laws of Kenya have some shortcomings in facilitating agriculture-focused adaptation. This section will summarise the gaps in the policy and legal frameworks for climate change adaptation and agriculture in Kenya.

For a start, the NEAP was the first action plan developed to address environmental problems in Kenya. Equally important, the NEAP operates as a platform for the formulation of additional climate change policies and strategies, enactment of statutes, and establishment of climate change institutions in Kenya. However, the NEAP scarcely promotes adaptation in the context of agriculture. In other words, relevant adaptation strategies that could enhance the resilience of crops or livestock against the impacts of climate change such as droughts or desertification are not included in the NEAP. Likewise, adaptation strategies for resources (especially water) that support crops and livestock productions such as irrigation are not catered for by the NEAP.

Turning to a generic and framework environmental law such as the *EMCA*, it has been observed that the implementation of EIAs and SIAs as practical tools that support adaptation strategies is difficult due to the lack of a trained workforce, thus rendering EIAs and SIAs as mere declaratory requirements. Additionally, since the *EMCA* is a generic and framework environmental law, it cannot be expected to have specific provisions for agriculture-focused adaptation.[233] The implementation of a law with insufficient provisions for agriculture-focused adaptation could pose challenges for the integration of adaptation and agricultural development in Kenya.[234] In addition, legislation that lacks (or has insufficient contents) on agriculture-focused adaptation could result:

> [i]n the sprawling of mandates among institutions involved in regulating adaptation and by extension could lead to poor coordination and collaboration among institutions and stakeholders in agriculture focused adaptation.[235]

As for the *Climate Change Act*, a gap that has been identified could be the lack of specific provisions dealing with adaptation in the context of agriculture. Most of the provisions on adaptation in the *Climate Change Act*, especially section 13, are broad provisions without a specific focus on agriculture. This omission could defeat the very purpose of the *Climate Change Act* in driving low-carbon and climate-resilient development for an agriculture-dependent-GDP country such as Kenya.

Another important observation is the lack of synchronisation of scientific solutions with legal solutions to climate change in the *Climate Change Act*. Without proper synchronisation, it is believed that there could be a tendency of not including:

> [w]ell trained personnel who are required in translating scientific data into information that can be used to track implementation and effective compliance of the *Climate Change Act*.[236]

An implication of this is the possibility that essential adaptation strategies targeting critical sectors such as agriculture cannot be effectively implemented.

In the same way, some of Kenya's agricultural laws and policies, especially the *AFA Act*, appear to have provisions that are insufficient in linking actors and agencies responsible for adaptation.[237] This, according to experts, "could lead to detachment and inefficiency in implementation of programs".[238] By implication, laws such as the *AFA Act* are considered to be part of the reason for the weak adaptation management of climate change impacts on agriculture in Kenya.[239]

Regarding the institutional framework for adaptation in Kenya, the current organisational arrangement appears to be diffused and uncoordinated. For example, the activities of the NCCC (as an advisory body for climate change in Kenya) appear to be detached from other sectoral bodies involved in adaptation – such as the Agriculture and Food Authority, Water Resources Authority, and the NDMA. This regulatory concern could have also fed into other challenges such as "jurisdictional ambiguities and conflicts" in Kenya's legal and policy framework for adaptation.[240]

In a nutshell, some Kenyan policies for climate change (the NEAP) and laws (the *EMCA* and the *Climate Change Act*):

> [d]o not provide for a well-coordinated, coherent and effective response to the challenges of climate change on agriculture, and are not responsive to global opportunities.[241]

Accordingly, these gaps could lead to policy incoherence, as well as pose a risk to the success of the goals of adaptation and agriculture in Kenya.[242] In short, the current legal and policy framework for adaptation and agriculture in Kenya requires review, and possibly an overhaul.

3.5 Conclusion

Agriculture has been identified as an important sector of the Kenyan economy to boosts the country's GDP, as well as a means of providing employment opportunities for a majority of the population in Kenya. However, being a climate-sensitive sector, agriculture is also impacted by climate change and often suffers from periods of extended drought and flooding, which impairs agricultural yield and productivity.[243]

Cognisant of the fact that climate change is a global challenge that requires local solutions, Kenya acceded to key international legal instruments (the UNFCCC, Kyoto Protocol, and Paris Agreement) in a bid to adapt to the impacts of climate change locally. These international and regional laws and policies influenced Kenyan adaptation policies and laws, which are generally tailored towards the realisation of low -carbon and climate-resilient development pathways in the country.

This chapter reviewed relevant climate change adaptation as well as agricultural policies and laws of Kenya and found that most of the policies and laws are sophisticated and relevant to adaptation. However, at the same time, some of the policies are weak and the laws did not contain the type of provisions that could enable and promote adaptation in the context of agriculture. The problems of the regulatory framework are further compounded by poor coordination and lack of collaboration among institutions and stakeholders.

These challenges make it difficult to implement adaptation policies or enforce adaptation laws in Kenya without creating discrepancies in mandates among institutions and stakeholders involved in agriculture-focused adaptation. Simply put, existing climate change laws and policies have neither addressed the impacts of climate change on agriculture through adaptation nor do they mainstream agriculture-focused adaptation into development planning. Having discussed how the Kenyan climate change and agriculture policy and legal frameworks support adaptation in agriculture, it is imperative to discuss the policy and legal frameworks for climate change and agriculture in Nigeria in order to determine the extent to which they cater to agriculture-focused adaptation in the country.

Notes

1 M Radeny, C Mungai, D Amwata, J Osumba and D Solomon 2020 Climate change, agriculture, food and nutrition security policies and frameworks in Kenya. CCAFS Working Paper. http://repository.seku.ac.ke/handle/123456789/6184 .
2 It must be noted that, as previously explained, the UNFCCC, Kyoto Protocol, and Paris Agreement do not entirely regulate climate change adaptation, but they only provide guidelines for formulating policies and enacting legal solutions to climate change at the national and regional level. See Chapter 2, Section 2.3.1.
3 What this means is that Kenya does not have a single or consolidated climate change adaptation law. Instead, adaptation is found in the provisions of different laws.
4 Kenya National Environment Action Plan 1994.
5 Government of Kenya 2010 https://cdkn.org/wp-content/uploads/2012/04/National-Climate-Change-Response-Strategy_April-2010.pdf; PO Ongugo, D Langat, VO Oeba, JM Kimondo, B Owuor, J Njuguna and AJ Russell 2014 A review of Kenya's national policies relevant to climate change adaptation and mitigation: Insights from Mount Elgon Centre for International Forestry Research Working Paper, 10–21.
6 The NAP of Kenya has been acclaimed internationally as one of the best action plans in Africa. See for instance, J-E Parry 2016 Review of current and planned adaptation action in Kenya. CARIAA Working Paper No. 16. International Development Research Centre.
7 Government of Kenya 2016 Kenya National Adaptation Plan: 2015-2030, Government of Kenya, July 2016. https://www4.unfccc.int/sites/NAPC/Documents%20NAP/Kenya_NAP_Final.pdf.
8 See generally, Chapter 2, Section 2.2.
9 No. 8 of 1999.
10 O Rumble 2019 (January) Climate change legislative development on the African continent. In P Kameri-Mbote, A Paterson, O C Ruppel, B B Orubebe, E D Kam Yogo (Eds.), *Law environment Africa*, Publication of the 5th Symposium | 4th Scientific Conference | 2018 of the Association of Environmental Law Lecturers from African Universities in cooperation with the Climate Policy and Energy Security Programme for Sub-Saharan Africa of the KonradAdenauer-Stiftung and UN Environment, 31–60.
11 *Constitution of Kenya, 2010* https://kenyalaw.org/kl/fileadmin/pdfdownloads/TheConstitutionOfKenya.pdf
12 JK Bosek 2014 Implementing environmental rights in Kenya's new constitutional order: Prospects and potential challenges. *African Human Rights Law Journal*, 14(2), 489–508.

13 See article 2 of the *Constitution of Kenya, 2010*. In addition, article 2 of the *Constitution* provides for the supremacy of the *Constitution* over all laws, persons, and organs and levels of government in Kenya, and the legality of the *Constitution* is excluded from challenge by all before any established court or organ of state with competent authority.
14 See Preamble to the *Constitution of Kenya*, 2010 https://kenyalaw.org/kl/fileadmin/pdfdownloads/TheConstitutionOfKenya.pdf
15 These strategies were previously discussed. See for example, Chapter 1, Section 1.4.
16 Article 42, *Constitution of Kenya*. It should be noted that the scope of article 42 generally extends to the protection of biodiversity, natural resources, ecosystems, and plant and animal species. See T Chore 2019 Reconceptualising the right to a clean and Healthy environment in Kenya: The need to Move from an Anthropocentric view to a Bicentric view. *Strathmore Law Review*, 4(1), 71–88.
17 Article 69, *Constitution of Kenya*.
18 See article 69, *Constitution of Kenya*.
19 See Chapter 1, Section 1.5.1.
20 Article 60(1), *Constitution of Kenya*, 2010.
21 Article 2(5), *Constitution of Kenya*, 2010.
22 See Chapter 1, Section 1.5.1.1.
23 Article 2 of the *Constitution of Kenya*, 2010.
24 Kenya Climate Smart Agriculture Strategy-2017–2026, https://adaptation-undp.org/sites/default/files/resources/kenya_climate_smart_agriculture_strategy.pdf 20; Government of Kenya 2010 NCCRS https://faolex.fao.org/docs/pdf/ken163118.pdf 88; I Maina, A Newsham and M Okoti 2013 Agriculture and climate change in Kenya: Climate chaos, policy dilemmas. *Future Agricultures*, 70.
25 See Chapter 2, Section 2.2.
26 See Chapter 2, Sections 2.3.1, 2.3.2 and 2.3.3.
27 The UNCED of 1992 (dubbed the "Earth Summit") emphasised that, since the environment is interconnected with social and economic factors, the success of the environment is dependent on the sustenance of the other two sectors. As such, "a broad agenda and a new blueprint for international action on environmental and development issues that would help guide international cooperation and development policy in the twenty-first century is required". See UNO 1992 https://www.un.org/en/conferences/environment/rio1992
28 Climate change is addressed by the NEAP under "Environmental Pollution and Waste Management", where both mitigatory and adaptive strategies have been recommended as response actions to climate change. See for example, MO Oulu 2011 Mainstreaming climate adaptation in Kenya. *Climate Law*, 2(3), 375–394.
29 Government of Kenya 1994The Kenya National Environment Action Plan (NEAP). https://documents1.worldbank.org/curated/en/961121468753321847/pdf/multi0page.pdf 1-207.; SK Mutiso 1994 Kenya's National environmental action plan. *Waterlines*, 13(2),11–14.
30 MO Oulu 2011 Mainstreaming climate adaptation in Kenya. *Climate Law*, 2(3), 375–394.
31 No. 11 of 2016.
See National Environmental Action Plan 2009–2013. https://faolex.fao.org/docs/pdf/ken147884.pdf. 28.
32 See Chapter 2, Section 2.3.1.
33 R Kibugi 2010 *New Constitutional Environmental Law in Kenya: Changes in 2010*, 1–7.
34 P Mwendwa and RA Giliba 2012 Climate change impacts and adaptation strategies in Kenya. *Chinese Journal of Population Resources and Environment*, 10(4), 22–29.

35 Government of Kenya 2010 National Climate Change Response Strategy. https://cdkn.org/sites/default/files/files/National-Climate-Change-Response-Strategy_April-2010.pdf, 50.
It should also be noted that the NCCRS was the first response strategy exclusively developed to deal with climate change in Kenya. What existed prior to the formulation of the NCCRS in 2010 were documents as well as fragments of statutes and environmental policies such as *the Environmental Management and Coordination Act of 1999* and the National Environment Action Plan of 1994 (explained below), which only addressed aspects of climate change remotely, as a subset of environmental problems. See MO Oulu 2011 Mainstreaming climate adaptation in Kenya. *Climate Law,* 2(3), 375–394.
36 The role of policies (described as soft laws) in guiding law and governance has been explained previously. See Chapter 2, Section 2.2.
37 Article 60.
38 Government of Kenya 2010 https://cdkn.org/wp-content/uploads/2012/04/National-Climate-Change-Response-Strategy_April-2010.pdf, 7 and 44. It should be noted that adaptation, especially for critical sectors such as agriculture, has been placed on a higher priority than mitigation by the Kenyan NCCRS.
39 Government of Kenya 2010 https://cdkn.org/wp-content/uploads/2012/04/National-Climate-Change-Response-Strategy_April-2010.pdf, 45.
40 Government of Kenya 2010 https://cdkn.org/wp-content/uploads/2012/04/National-Climate-Change-Response-Strategy_April-2010.pdf 7.
41 Government of Kenya 2010 https://cdkn.org/wp-content/uploads/2012/04/National-Climate-Change-Response-Strategy_April-2010.pdf 7. In this regard, the Government of Kenya, through the assistance of the Kenya Livestock Breeding Board and other relevant institutions, developed a special livestock insurance scheme that will provide an opportunity to spread and transfer climate change risks, e.g., to restock herds after decimation by starvation caused by droughts. See Government of Kenya 2010 https://cdkn.org/wp-content/uploads/2012/04/National-Climate-Change-Response-Strategy_April-2010.pdf, 51.
42 Similarly, Equity Bank, UAP and ILRI have established such a scheme, called Drought Insurance. See Government of Kenya 2010 https://cdkn.org/wp-content/uploads/2012/04/National-Climate-Change-Response-Strategy_April-2010.pdf, 51.
43 See Chapter 1, Section 1.5.1.3.
44 P Rowlinson 2008 Adapting livestock Production Systems to climate change temperate zones. In *Livestock and global change conference proceeding,* 61–63.
45 Government of Kenya 2010 https://cdkn.org/wp-content/uploads/2012/04/National-Climate-Change-Response-Strategy_April-2010.pdf, 50.
46 Government of Kenya 2010 https://cdkn.org/wp-content/uploads/2012/04/National-Climate-Change-Response-Strategy_April-2010.pdf, 52.
47 Government of Kenya 2010 https://cdkn.org/wp-content/uploads/2012/04/National-Climate-Change-Response-Strategy_April-2010.pdf, 45.
48 Government of Kenya 2010 https://cdkn.org/wp-content/uploads/2012/04/National-Climate-Change-Response-Strategy_April-2010.pdf, 7.
49 Government of Kenya 2010c National Climate Change Response Strategy: Executive Brief. https://cdkn.org/sites/default/files/files/National_Climate_Change_Response_Strategy_Executive_Brief.pdf, 1–30, 50. O Rumble 2019 (January) Climate change legislative development on the African continent. In *Law environment Africa*, 31–60.
50 JM Njoroge, BM Ratter and L Atieno 2017 Climate change policy-making process in Kenya: Deliberative inclusionary processes in play. *International Journal of Climate Change Strategies and Management,* 9(4), 535–554.

51 Republic of Kenya National Climate Change Action Plan 2013– 2017 https://cdkn.org/sites/default/files/files/Kenya-National-Climate-Change-Action-Plan.pdf.

52 T Mikunda and MI Halstead 2014 Prioritising mitigation and adaptation options as part of the development of a National Climate Change Action Plan in Kenya. *Policy Studies*, 2017.

53 Republic of Kenya National Climate Change Action Plan 2013–2017 https://cdkn.org/sites/default/files/files/Kenya-National-Climate-Change-Action-Plan.pdf, 60.

54 PO Ongugo, D Langat, VO Oeba, JM Kimondo, B Owuor, J Njuguna and AJ Russell 2014 A review of Kenya's national policies relevant to climate change adaptation and mitigation: Insights from Mount Elgon. Centre for International Forestry Research Working Paper, 155.

55 The *Climate Change Act* of 2016 is a good example of a comprehensive and coherent stand-alone law that was enacted in line with the recommendation of the NCCAP. As explained below, the 2016 *Climate Change Act of Kenya* is a law that was enacted in order to coordinate climate change policies and to promote actions that are necessary in mainstreaming climate change actions across all sectors of national development in Kenya.

56 The Kenyan Ministry of Environment and Forests facilitated the formulation of the NCCAP in 2018 through extensive consultation among stakeholders, consisting of members of parliament, representatives of interest and civil groups, heads of ministries, parastatals, and departments, etc. See Government of Kenya 2018 National Climate Change Action Plan. https://www.lse.ac.uk/GranthamInstitute/wp-content/uploads/2018/10/8737.pdf, 9.

57 A low-carbon and climate-resilient development is any national development pathway which considers sustainable development and gives high priority to adaptation, while at the same time taking cognisance of surrounding inequalities and vulnerability of weak populations such as persons with disabilities, women and marginalised groups. See for instance, M Bagheri, S H Delbari, M Pakzadmanesh and C A Kennedy. 2019 City-integrated renewable energy design for low-carbon and climate-resilient communities *Applied Energy,* 239(C), 1212–1225.

58 Government of Kenya 2018 National Climate Change Action Plan. https://www.lse.ac.uk/granthaminstitute/wp-content/uploads/2018/10/8737.pdf, 17.

59 S Mutimba, SW Simiyu, TL Lelekoiten, and AV Ospina 2019 *sNAPshot,* 1–9.

60 Government of Kenya 2018 https://napglobalnetwork.org/wp-content/uploads/2022/01/napgn-en-2022-kenya-NCCAP-2018-2022-Implemantation-Status-Report.pdf, 19.

61 Government of Kenya 2018 https://napglobalnetwork.org/wp-content/uploads/2022/01/napgn-en-2022-kenya-NCCAP-2018-2022-Implemantation-Status-Report.pdf, 15.

62 Government of Kenya 2018 National Climate Change Action Plan https://www.lse.ac.uk/granthaminstitute/wp-content/uploads/2018/10/8737.pdf, 17.
By involving all tiers of governments (national and county governments), the private sector, and civil societies in adaptation, the NCCAP simply affirms the provisions of article 69 of the *Constitution of Kenya* 2010, which portrays climate change governance as a shared responsibility.

63 Government of Kenya 2018 National Climate Change Action Plan https://www.lse.ac.uk/granthaminstitute/wp-content/uploads/2018/10/8737.pdf, 67.
In this regard, it should be noted that the NCCC places more obligation on the national government than county governments, where the national government is expected to design most of the strategies that could be operationalised in a

country-wide context. Government of Kenya 2018 https://napglobalnetwork.org/wp-content/uploads/2022/01/napgn-en-2022-kenya-NCCAP-2018-2022-Implemantation-Status-Report.pdf; W Lafferty and E Hovden 2003 Environmental policy integration: Towards an analytical framework. *Environmental Politics,* 12(3), 1–22.

64 G Cunliffe, C Holz, K Mbeva, P Pauw and H Winkler 2019 *Comparative Analysis of the NDCs of Canada, the European Union, Kenya and South Africa from an Equity Perspective,* Cape Town: Energy Research Centre, University of Cape Town. https://doi.org/10.5281/zenodo.8387347, 21–30.

65 C Schwarte 2017 Legislating the Paris Agreement in developing countries. Climate Law and Governance Working Paper Series, No1/2017. Centre for International Sustainable Development Law, 8; FAO 2020 *Agriculture and Climate Change*. http://www.fao.org/3/cb1593en/CB1593EN.pdf, 1-376.

66 Article 4 (2) Paris Agreement. This has been previously explained. See Chapter 2, Section 2.3.3.

67 M Dal Maso, KH Olsen, Y Dong, MB Pedersen, and MZ Hauschild 2020 Sustainable development impacts of nationally determined contributions: Assessing the case of mini grids in Kenya. *Climate Policy*, 20(17), 815–831; UNFCCC 2015 Kenya's Intended Nationally Determined Contribution (INDC). https://www4.unfccc.int/sites/submissions/INDC/Published%20Documents/Kenya/1/Kenya_INDC_20150723.pdf.

68 Though much priority is given to adaptation as could be found below. See Government of Kenya 2015 https://www4.unfccc.int/sites/ndcstaging/PublishedDocuments/Kenya%20First/Kenya_NDC_20150723.pdf.

69 Government of Kenya 2015 https://www4.unfccc.int/sites/ndcstaging/PublishedDocuments/Kenya%20First/Kenya_NDC_20150723.pdf, 6.

70 Government of Kenya 2015 https://www4.unfccc.int/sites/ndcstaging/PublishedDocuments/Kenya%20First/Kenya_NDC_20150723.pdf, 7.

71 JM Njoroge, BM Ratter and L Atieno 2017 Climate change policy-making process in Kenya: Deliberative inclusionary processes in play. *International Journal of Climate Change Strategies and Management,* 9(4), 535–554.

72 D Murphy and M Harris 2014 Policy and regulatory barriers in Kenya's National Climate Change Action Plan: Areas for Private Sector Action International Institute for Sustainable Development. https://cdkn.org/sites/default/files/files/Policy-Regulation-Review.pdf, 13.

73 D Murphy and M Harris 2014 Policy and regulatory barriers in Kenya's National Climate Change Action Plan: Areas for Private Sector Action International Institute for Sustainable Development. https://cdkn.org/sites/default/files/files/Policy-Regulation-Review.pdf, 14.

74 See Chapter 1, Section 1.5.1.2. See also D Murphy and M Harris 2014 Policy and regulatory barriers in Kenya's National Climate Change Action Plan: Areas for Private Sector Action International Institute for Sustainable Development https://cdkn.org/sites/default/files/files/Policy-Regulation-Review.pdf, 10.

75 The development of the Kenyan NAP was done through a process of consultations and cooperation that involved stakeholders from the public sector, private sector, and international organisations. See Government of Kenya 2016 Kenya National Adaptation Plan: 2015-2030,https://www4.unfccc.int/sites/NAPC/Documents%20NAP/Kenya_NAP_Final.pdf, 10.

76 Government of Kenya 2016 Kenya National Adaptation Plan: 2015-2030, https://www4.unfccc.int/sites/NAPC/Documents%20NAP/Kenya_NAP_Final.pdf, 3.

77 I Karani 2018 Development of national and sub-national adaptation metrics: Lessons from Kenya. *Adaptation metrics: Perspectives on measuring, aggregating and comparing adaptation results,* 113.

78 See Chapter 1, Section 1.3.1. See also Government of Kenya 2016 https://www4.unfccc.int/sites/NAPC/Documents%20NAP/Kenya_NAP_Final.pdf, 15.
79 Government of Kenya 2016 https://www4.unfccc.int/sites/NAPC/Documents%20NAP/Kenya_NAP_Final.pdf, 8.
80 Government of Kenya 2016. https://www4.unfccc.int/sites/NAPC/Documents%20NAP/Kenya_NAP_Final.pdf, 21.
81 The provisions of article 72 should be read alongside with article 261(1) and the Fifth Schedule to the Constitution of Kenya, which mandates the Parliament to enact a new law or amend existing laws for the protection of the environment in every four years following the adoption of the 2010 *Constitution*.
82 Article 2 of the *Constitution of Kenya, 2010* provides that: (5) the general rules of international law shall form part of the law of Kenya. (6) Any treaty or convention ratified by Kenya shall form part of the law of Kenya under this *Constitution*.
83 L Juma 2000 Environmental Protection in Kenya: Will the *Environmental Management and Co-ordination Act* (1999) Make a difference. *South Carolina Environmental Law Journal,* 9, 181.
84 *The Environmental Management and Co-Ordination Act* No. 8 of 1999. https://kenyalaw.org/kl/fileadmin/pdfdownloads/Acts/EnvironmentalManagementandCo-ordinationAct_No8of1999.pdf.
85 The task of monitoring, coordinating, and implementing all policy matters affecting the environment (including climate change) is principally vested in the National Environmental Management Authority (NEMA). Similarly, it has a crucial role to play in mainstreaming "environmental considerations into development policies, plans, programmes and projects". Section 40 EMCA; MO Oulu 2011 Mainstreaming climate adaptation in Kenya. *Climate Law,* 2(3), 375–394.
86 Section 44.
87 See Chapter 1, Section 1.5.1.1.
88 Both the SEA and the EIA are recommended by the provisions of article 4 of the UNFCCC as national measures for addressing the adverse effects of climate change on the economy, public health, and the quality of the environment. See Chapter 2, Section 2.3.1.
89 Section 2 *EMCA*.
90 Section 2 *EMCA*.
91 Article 69 of the Constitution of Kenya, 2010 also provides for EIA as a means of facilitating adaptation.
92 Sections 58–67 of the *EMCA*.
93 Section 42 of the Environmental (Impact Assessment and Audit) Regulations, Legal Notice No 101, 2003 and section 42 EMCA.
94 V Onyango and M Schmidt 2007 Towards a strategic environment assessment framework in Kenya: Highlighting areas for further scrutiny. *Management of Environmental Quality: An International Journal*, 18(3), 309–328.
95 Kenya's mainstreaming approach is equally accommodated in other laws and policies, which are explained below.
96 F Alemayehu 2016 Challenges and gaps in the existing laws and policies in marine-related resource use and conservation in Watamu Mida Creek, Kenya. *Environmental Management and Sustainable Development,* 6(1), 10766.
97 *Wetlands, Riverbanks, Lake Shores, and Sea Shore Management Regulations*, Legal Notice No. 19, 2009. https://www.nema.go.ke/images/Docs/Regulations/Wetlands%20regulations-1.pdf.
98 Regulation 21 of the *Wetlands, Riverbanks, Lake Shores, and Sea Shore Management Regulations* provides that: "(1) A developer intending to a undertake a project which may have a significant impact on a wetland, river bank, lake shore, or the sea shore shall carry out an EIA in accordance with the provisions

of the Act. (2) The developer referred to in sub-regulation (1) shall carry out an environmental audit as provided for by the Act, and the Authority to monitor such activities in accordance with the Act."

99 See GN Morara, F Hassan and MK Osore 2015 Sustaining local livelihoods through coastal fisheries in Kenya. *Sustainability Indicators in Practice,* 99; N Versleijen and J Hoorweg 2008 From farming to fishing: Marine resource conservation and a new generation of fishermen. *Western Indian Ocean Journal of Marine Science,* 7(1), 1–14.

100 See Regulation 17 of the Wetlands, Riverbanks, Lake Shores, and Sea Shore Management Regulations for measures of conserving river banks, lake shores, and the seashore, which among others include bunding, terracing, mulching, tree planting or agro forestry, and grassing. These measures were discussed as on farm adaptation strategies. See Chapter 1, Section 1.5.1.1.

101 JM Njoroge, BM Ratter and L Atieno 2017 Climate change policy-making process in Kenya: Deliberative inclusionary processes in play. *International Journal of Climate Change Strategies and Management,* 9(4), 535–554.

102 Section 9(2) of the *EMCA* provides that "(2) Without prejudice to the generality of the foregoing, the Authority [EMCA] shall—(a) co-ordinate the various environmental management activities being undertaken by the lead agencies and promote the integration of environmental considerations into development policies, plans, programmes and projects with a view to ensuring the proper management and rational utilization of environmental resources on a sustainable yield basis for the improvement of the quality of human life in Kenya".

103 See article 60 of the *Constitution of Kenya,* 2010.

104 See for example Regulation 8 of The *Environmental Management and Co-Ordination (Wetlands, River Banks, Lake Shores, and Sea Shore Management) Regulations,* 2009 which provides that "The Minister may, by notice in the Gazette, declare an area to be a protected wetland where such area has national and international significance due to its- (a) biological diversity; (b) ecological importance; (c) landscape; (d) natural heritage, or (e) aesthetic value".

105 C Wambua 2019 *The Kenya Climate Change Act* 2016: Emerging Lessons from a Pioneer Law. *Carbon & Climate Law Review,* 13(4), 257–269.

106 Section 5.

107 Section 3.

108 According to article 42 of the *Constitution*, the state is obliged to apply these values and principles contained in article 10 which guides the parliament in promoting and safeguarding the right to a clean and healthy environment (as provided by Article 69 of the *Constitution*).

109 Section 13 *Climate Change Act.*

110 Government of Kenya 2018 National Climate Change Action Plan https://www.lse.ac.uk/granthaminstitute/wp-content/uploads/2018/10/8737.pdf.

111 Section 13(3) *Climate Change Act.*

112 O Rumble 2019 (January) Climate change legislative development on the African continent. *Law Environment Africa,* 31–60.

113 C Wambua 2019 *The Kenya Climate Change Act* 2016: Emerging Lessons from a Pioneer Law. *Carbon & Climate Law Review,* 13(4), 257–269.

114 See Chapter 1, Section 1.4.4. It should be noted that all forms of strategies aimed at addressing the impacts of climate change discussed under this sub-heading are suitable for mainstreaming with development.

115 RD Lasco, FB Pulhin, PA Jaranilla-Sanchez, RJP Delfino, R Gerpacio and K Garcia 2009 Mainstreaming adaptation in developing countries: The case of the Philippines. *Climate and Development,* 1(2), 130–146. The act of mainstreaming climate change adaptation consideration into Kenyan national development planning is discussed below.

116 For example, article 3(2) of the *Climate Change Act* provides that "... this Act shall be applied in all sectors of the economy by the national and county governments to— (a) mainstream climate change responses into development planning, decision making and implementation ..."
117 Article 3(2) (c).
118 Kenyan *Climate Change Act* 2016. https://kenyalaw.org/kl/fileadmin/pdfdownloads/Acts/ClimateChangeActNo11of2016.pdf.
119 Art.3 (2) Kenyan *Climate Change Act* 2016.
120 C Wambua 2019 *The Kenya Climate Change Act* 2016: Emerging Lessons from a Pioneer Law. *Carbon & Climate Law Review,* 13(4), 257–269.
121 OO Olashore 2019 Implementation of the international legal framework regarding climate change in developing countries; A review of Nigeria, Kenya, and Botswana's environmental provisions governing climate change. *Environmental Law Review,* 21(3), 189–209.
122 FAO 2020 *Agriculture and Climate Change.* http://www.fao.org/3/cb1593en/CB1593EN.pdf, 3.
123 R Kibugi 2017 Adaptation to climate change in smallholder agriculture in Kenya: The role of law. In *Research handbook on climate change and agricultural law.* Edward Elgar Publishing, 367; C Wambua 2019 *The Kenya Climate Change Act* 2016: Emerging Lessons from a Pioneer Law. *Carbon & Climate Law Review,* 13(4), 257–269. See also the provisions of Section 13(3)(b) of the *Climate Change Act* 2016.
124 See for example, PO Ongugo, D Langat, VO Oeba, JM Kimondo, B Owuor, J Njuguna and AJ Russell 2014 A review of Kenya's national policies relevant to climate change adaptation and mitigation: Insights from Mount Elgon Centre for International Forestry Research Working Paper, 155; J Dell'Angelo PF McCord, E Baldwin, M E Cox, D Gower, K Caylor and TP Evans . 2014 Multilevel governance of irrigation systems and adaptation to climate change in Kenya.*The global water system in the Anthropocene*: *Challenges for Science and Governance,* 323–341; JE Parry, D Echeverria, J Dekens and J Maitima 2012 Climate Risks, Vulnerability and Governance in Kenya: A review. Commissioned by climate risk management technical assistance support project (CRM TASP), joint initiative of bureau for crisis prevention and recovery and bureau for development policy of UNDP, 83, 78–83.
125 FAO 2020 Agriculture and Climate Change. http://www.fao.org/3/cb1593en/CB1593EN.pdf, 4.
126 Chapter 1, Section 1.5.1.4; FAO 2020 *Agriculture and Climate Change.* http://www.fao.org/3/cb1593en/CB1593EN.pdf, 60.
127 TO Williams, ML Mul, OO Cofie, J Kinyangi, RB Zougmoré, G Wamukoya and BM Campbell 2015 *Climate smart agriculture in the African context,* 1–26. The legal satatus of a strategy in the context of climate change governance has been previously discussed. See Chapter 2, Section 2.2.
128 *Kenya Climate Smart Agriculture Strategy-2017–2026.* https://www.adaptation-undp.org/sites/default/files/resources/kenya_climate_smart_agriculture_strategy.pdf, 1–53.
129 D Dinesh, P Aggarwal, A Khatri-Chhetri, AML Rodríguez, C Mungai, L Sebastian and RB Zougmore 2017 The rise in Climate-Smart Agriculture strategies, policies, partnerships and investments across the globe. *Agriculture for Development,* 30, 4–9.
130 See for example Chapter 2, Section 2.3.1.
131 Marcus Taylor 2018 Climate-smart agriculture: What is it good for? *The Journal of Peasant Studies,* 45(1), 89–107. See also *Kenya Climate Smart Agriculture Strategy-2017–2026.*https://www.adaptation-undp.org/sites/default/files/resources/kenya_climate_smart_agriculture_strategy.pdf, 23.

132 *Kenya Climate Smart Agriculture Strategy-2017–2026*. https://www.adaptation-undp.org/sites/default/files/resources/kenya_climate_smart_agriculture_strategy.pdf, 33.
133 *Kenya Climate Smart Agriculture Strategy-2017–2026*. https://www.adaptation-undp.org/sites/default/files/resources/kenya_climate_smart_agriculture_strategy.pdf, 22, 57.
134 *Kenya Climate Smart Agriculture Strategy-2017–2026*. https://www.adaptation-undp.org/sites/default/files/resources/kenya_climate_smart_agriculture_strategy.pdf, 23.
135 *Kenya Climate Smart Agriculture Strategy-2017–2026*. https://www.adaptation-undp.org/sites/default/files/resources/kenya_climate_smart_agriculture_strategy.pdf, 57.
136 *Kenya Climate Smart Agriculture Strategy-2017–2026*. https://www.adaptation-undp.org/sites/default/files/resources/kenya_climate_smart_agriculture_strategy.pdf, 57.
137 See Chapter 1, Section 1.2.
138 See Fourth Schedule, Item 29, *Constitution of Kenya*, 2010.
139 Article 60(2), *Constitution of Kenya*, 2010.
140 See Chapter 1, Section 1.5.1.4.
141 JM Klopp and O Lumumba 2016 The State of Kenya's Land Policy and Law Reform: A political institutional analysis. *Washington, DC: 2016 World Bank Conference on Land and Policy*, 5.
142 Kenya Government 2009 *Sessional Paper No.3 of 2009* on National Land Policy.
143 Article 60(2) of the *Constitution* provides: "these principles shall be implemented through a national land policy developed and reviewed regularly by the national government and through legislation".
144 *Kenya Climate Smart Agriculture Strategy* 2017–2026, https://www.adaptation-undp.org/sites/default/files/resources/kenya_climate_smart_agriculture_strategy.pdf, 18. These principles of sustainable land management were subsequently incorporated into the *Constitution of Kenya*, 2010. See article 60 *Constitution of Kenya*, 2010.
145 Equitable land allocation is reflected in the *Constitution of Kenya*, 2010. See the preamble to the constitution of Kenya, 2010 and article 60, *Constitution of Kenya*, 2010.
146 Kenya Government 2009 Sessional Paper No.3 of 2009 on National Land Policy.
147 These are on-farm adaptation strategies that were explained previously. See Chapter 1, Section 1.5.1.1.
148 See Chapter 1, Section 1.3.1 above. In addition, see E Odoyo, B Bett and AS Murenga Mwimali 2019 Challenges of and responses to climate variability: A case study of peri-urban female farmers in Kenya. *World Journal of Agricultural Research*, 7(2), 61–68.
149 Government of Kenya 2019 Draft National Livestock Policy. http://repository.kippra.or.ke/bitstream/handle/123456789/483/Draft-reviewed-National-Livestock-Policy-February-2019.pdf, 73.
150 Government of Kenya 2019. http://repository.kippra.or.ke/bitstream/handle/123456789/483/Draft-reviewed-National-Livestock-Policy-February-2019.pdf, 24.
151 Government of Kenya 2019. http://repository.kippra.or.ke/bitstream/handle/123456789/483/Draft-reviewed-National-Livestock-Policy-February-2019.pdf, 32.
152 See generally Chapter 1, Section 1.2.
153 Government of Kenya 2019. http://repository.kippra.or.ke/bitstream/handle/123456789/483/Draft-reviewed-National-Livestock-Policy-February-2019.pdf, 2 and 39.

154 Government of Kenya 2019. http://repository.kippra.or.ke/bitstream/handle/123456789/483/Draft-reviewed-National-Livestock-Policy-February-2019.pdf, 14.
155 Government of the Republic of Kenya 2007 Vision 2030. https://vision2030.go.ke/wp-content/uploads/2018/05/Vision-2030-Popular-Version.pdf, 17.
156 *Kenya Climate Smart Agriculture Strategy-2017–2026*. https://www.adaptation-undp.org/sites/default/files/resources/kenya_climate_smart_agriculture_strategy.pdf, 1–53.
157 *Kenya Climate Smart Agriculture Strategy* 2017–2026. https://www.adaptation-undp.org/sites/default/files/resources/kenya_climate_smart_agriculture_strategy.pdf, 2–53.
158 See Chapter 1, Section 1.5.1.3.
159 Government of the Republic of Kenya 2021 draft-National Agriculture Insurance Policy. https://www.treasury.go.ke/wp-content/uploads/2021/04/Draft-National-Insurance-Policy.pdf.
160 Government of the Republic of Kenya 2021. https://www.treasury.go.ke/wp-content/uploads/2021/04/Draft-National-Insurance-Policy.pdf, 8.
161 PO Alila and R Atieno 2006 Agricultural policy in Kenya: Issues and processes. *Nairobi: Institute of Development Studies*, 1–41.
162 J Morton 2017 Climate change and African agriculture: Unlocking the potential of research and advisory services. In *Making climate compatible development happen*. Routledge, 87–113. It is believed that a secured and equitable land tenure system will not only support agricultural practices that are beneficial to the environment but will also ensure that agricultural practices which contribute to climate change such as overgrazing are eliminated.
163 See Chapter 1, Section 1.5.1.1.
164 Government of the Republic of Kenya 2021. https://www.treasury.go.ke/wp-content/uploads/2021/04/Draft-National-Insurance-Policy.pdf, 4.
165 Government of the Republic of Kenya 2021. https://www.treasury.go.ke/wp-content/uploads/2021/04/Draft-National-Insurance-Policy.pdf, 5.
166 Government of Kenya 2021. https://www.treasury.go.ke/wp-content/uploads/2021/04/Draft-National-Insurance-Policy.pdf, 17.
167 See Chapter 2, Section 2.2.
168 See item 1, part 2, Fourth Schedule to the Constitution of Kenya 2010. However, it should be noted that the powers of the National Government to make law on agricultural matters are delegated to the County Government.
169 The *Water Act* No. 43 of 2016.
170 *National Drought Management Authority Act*, No. 4 of 2016.
171 The *Kenya Agriculture and Livestock Research Act*, No. 17 of 2013.
172 *Crops Act*, No. 16 of 2013.
173 The *Land Act*, No. 6 of 2012.
174 *Agriculture, Fisheries and Food Authority Act*, No 13 of 2013.
175 See Chapter 1, Section 1.2.
176 Section 2 of the *Water Act*.
177 *Kenya Water Institute Act*, of 2001 (Cap. 372A).
178 It should be noted that the launching of the Kenyan National Water Master Plan 2030 in 2014 was informed by the *Water Institute Act*. See https://wasreb.go.ke/wp-content/uploads/2023/08/National-Water-Master-Plan-2030-Exec.-Summary-Vol.-1-Main-1.pdf
179 See Section 3 of the *Water Act* which specifies the purpose of the Act as "... the regulation, management and development of water resources and water and sewerage services in line with the Constitution".
180 These were discussed previously. See articles 10, 43, and 60 of the *Constitution of Kenya*.

181 Section 2(1), *Water Act*.
182 See Section 22 of the *Water Act*.
183 Item 3, Third Schedule to the *Water Act*.
184 Section 10(2).
185 Catchment areas are formed taking into consideration a list of locations containing all surface water and ground water, an inventory of potential water users, purpose of use, the volume of water used, as well as the chemical and bacteriological elements of each class of water. See the National Water Resources Management Strategy of Kenya. https://waterfund.go.ke/watersource/Downloads/006.%20Water%20Resources%20Management%20Strategy.pdf, 8.
186 Section 11, *Water Act*.
Other similar institutions to the Water Resources Authority among others include: The National Water Harvesting and Storage Authority (replacing the National Water Conservation and Pipeline Corporation); the Water Services Regulatory Authority; Water Works Development Agencies; the Water Sector Trust Fund and the Water Tribunal (charged with dispute resolution), etc. E Gachenga 2019 (January) *Kenya's Water Act* (2016): Real devolution or simply the "same script, different cast". In *Law environment Africa*. Nomos Verlagsgesellschaft mbH and Co. KG, 429–452.
187 Section 12, *Water Act*.
188 Section 12, *Water Act*.
189 See Chapter 1, Section 1.2.
190 Government of Kenya 2017 National Drought Management Authority. https://ndma.go.ke/download/strategic-plan-2018-2022/#
191 See Government of Kenya 2017 National Drought Management Authority. https://ndma.go.ke/download/strategic-plan-2018-2022/#, 6.
192 Government of Kenya 2017 National Drought Management Authority. https://ndma.go.ke/download/strategic-plan-2018-2022/#, 25; C King-Okumu, VA Orindi and L Lekalkuli 2019 Drought management in the drylands of Kenya: What have we learned? In *Current directions in water scarcity research*. Elsevier, 2, 277–294.
193 See Chapter 1, Section 1.5.1.4.
194 Section 5 *NDMA Act*.
195 Government of Kenya 2013. https://vision2030.go.ke/wp-content/uploads/2018/05/Second-Medium-Term-Plan-2013-2017.pdf, 41. See also Government of Kenya 2017 National Drought Management Authority. https://ndma.go.ke/download/strategic-plan-2018-2022/#, 25–26; PA Mugabe, F Mwaniki, KA Mamary and HM Ngibuini 2019 An assessment of drought monitoring and early warning systems in Tanzania, Kenya, and Mali. *Current Directions in Water Scarcity Research*, 2, 211–219.
196 DW Ndegwa and JM Kinyua 2018 Strategic measures employed by the National Drought Management Authority for drought mitigation in Kenya. *International Journal of Management and Commerce Innovations*, 6(1), 802–812.
197 See E Odoyo, B Bett and AS Murenga Mwimali 2019 Challenges of and responses to climate variability: A case study of peri-urban female farmers in Kenya. *World Journal of Agricultural Research*, 7(2), 61–68.
198 See Chapter 1, Section 1.5.1.
199 The *Kenya Agriculture and Livestock Research Act*, No. 17 of 2013.
200 See Section 3.3.2.2.
201 The *Crops Act* (No. 16) of 2013. The *Crops Act* is mainly administered by the Agriculture, Fisheries and Food Authority which is established by the *Agriculture, Fisheries, and Food Authority Act*, 2013 (discussed below).
202 Section 3 *Crops Act*.
203 Section 6(3).

204 Section 6(1) b.
205 See Chapter 1, Section 1.2.
206 See Chapter 1, Section 1.5.1.3.
207 See Section 34 (1) (e).
208 See Section 3.3.2.3 above.
209 *Kenyan Insurance Act* Cap 487, 1984. http://www.kenyalaw.org/kl/fileadmin/pdfdownloads/Acts/Insurance_Act_Cap_487.PDF.
In addition to its indigenous insurance law (*Insurance Act* of 1984), Kenya is also a signatory to global and regional conventions and instruments on insurance, aimed at reducing vulnerability and strengthening the resilience of agricultural systems against endemic disasters including the impacts of climate change. Some of these global and regional conventions and instruments include the Sendai Framework for Disaster Risk Reduction, 2015–2030 and the Africa Regional Strategy for Disaster Risk Reduction. It is important to note that the Africa Regional Strategy for Disaster Risk Reduction is administered by the Africa Union as part of AU Agenda 2063. See Sendai Framework for Disaster Risk Reduction 2015-2030. https://www.undrr.org/media/16176/download?startDownload=20241116; African Union 2004 https://www.preventionweb.net/files/4038_africaregionalstrategy1.pdf, 1–18.
210 *Kenyan Insurance Act* Cap 487, 1984. http://www.kenyalaw.org/kl/fileadmin/pdfdownloads/Acts/Insurance_Act_Cap_487.PDF. Apart from the *Kenyan Insurance Act* of 1984, agricultural insurance has also been featured by other legislation such as the *Deposit Insurance Act* No.10, of 2012 and the *Re-insurance Corporation Act*, No. 9 of 1997.
211 *Land Act* No.6 of 2012. Other land-use laws in Kenya include but are not limited to, the *Government Lands Act* (CAP 280, 2010); *Registration of Titles Law* (No. 6, 2010); *Land Titles Act* (CAP 282, 2010); *Land Consolidation Act* (CAP 283, 2012); *Land Control* Act (CAP 302, 2012); and the *Land Registration Act* (No. 3 of 2012). However, these laws do not accommodate provisions regulating GHG emission and adaptation. As such they do not require further discussion.
212 Section 11 (2).
213 See Chapter 1, Section 1.5 generally.
214 Section 4 *Land Act* No. 6 of 2012.
215 *Agriculture Act*, Cap 318, 1986.
216 Section 48 (1) (a) *Agriculture Act*.
217 Articles 60–68 on land and environment.
218 Kenya Climate Smart Agriculture Strategy-2017–2026. https://www.adaptation-undp.org/sites/default/files/resources/kenya_climate_smart_agriculture_strategy.pdf, 1–53.
219 See Chapter 1, Section 1.5.1.4.
220 OO Olashore 2019 Implementation of the international legal framework regarding climate change in developing countries; A review of Nigeria, Kenya, and Botswana's environmental provisions governing climate change. *Environmental Law Review*, 21(3), 189–209.
221 Act 13 of 2013.
222 See Long Title to the *AFA Act*, 13 of 2013.
223 Long title to the *AFA Act* 13 of 2013.
224 See Chapter 1, Section 1.5.1.1.
225 Section 21 (3). See also section 23(1) of the *AFA Act* for an expansive guideline on land preservation.
226 See Chapter 1, Section 1.3.1.
227 Section 22 of the *AFA Act* mandate the Cabinet Secretary to make rules for preservation, utilisation, and development of agricultural land by specifying good animal husbandry rules for farmers.

228 Section 23 *AFA Act*.
229 Section 23(1) (b) *AFA Act*.
230 See generally, Chapter 1, Section 1.5.1.
231 See section 24 of the *AFA Act*. The provisions of this section also prescribe punitive measures against any person who fails to clear land containing invasive weeds within a time specified by a notice under the *AFA Act*.
232 See Chapter 1, Section 1.5.1.3.
233 JM Njoroge, BM Ratter and L Atieno 2017 Climate change policy-making process in Kenya: Deliberative inclusionary processes in play. *International Journal of Climate Change Strategies and Management,* 9(4), 535–554.
234 Kenya Climate Smart Agriculture Strategy-2017–2026, 28. In like manner, economic growth and sustainable livelihood in Kenya will be affected by the implementation of policies and the enforcement of legislation with insufficient content for agriculture-focused adaptation. See for example, JM Njoroge, BM Ratter and L Atieno 2017. Climate change policy-making process in Kenya: Deliberative inclusionary processes in play. *International Journal of Climate Change Strategies and Management,* 9(4), 535–554.
235 Kenya Climate Smart Agriculture Strategy 2017–2026. https://www.adaptation-undp.org/sites/default/files/resources/kenya_climate_smart_agriculture_strategy.pdf, 46, 53, 60.
236 OO Olashore 2019 Implementation of the international legal framework regarding climate change in developing countries; A review of Nigeria, Kenya, and Botswana's environmental provisions governing climate change. *Environmental Law Review*, 21(3), 189–209.
237 I Maina, A Newsham and M Okoti 2013 Agriculture and climate change in Kenya: Climate chaos, policy dilemmas. *Future Agricultures*, 70.
238 *Kenya Climate Smart Agriculture Strategy-2017–2026*. https://www.adaptation-undp.org/sites/default/files/resources/kenya_climate_smart_agriculture_strategy.pdf, 1–53.
239 Kenya Climate Smart Agriculture Strategy-2017–2026. https://www.adaptation-undp.org/sites/default/files/resources/kenya_climate_smart_agriculture_strategy.pdf, 20; Government of Kenya 2010 https://cdkn.org/sites/default/files/files/National-Climate-Change-Response-Strategy_April-2010.pdf ,88; I Maina, A Newsham and M Okoti 2013 Agriculture and climate change in Kenya: Climate chaos, policy dilemmas. *Future Agricultures, 70.*
240 C Schwarte 2017 Legislating the Paris Agreement in developing countries. Climate Law and Governance Working Paper Series, No1/2017. Centre for International Sustainable Development Law, 8.
241 MJ Naeku 2020 Climate change governance: An analysis of the climate change legal regime in Kenya. *Environmental Law Review*, 22(3), 170–183.
242 M Faling 2020 Framing agriculture and climate in Kenyan policies: A longitudinal perspective. *Environmental Science and Policy*, 106, 228–239.
243 See *Kenya Climate Smart Agriculture Strategy 2017–2026* https://www.adaptation-undp.org/sites/default/files/resources/kenya_climate_smart_agriculture_strategy.pdf, 17.

References

Alemayehu F 2016 Challenges and gaps in the existing laws and policies in marine related resource use and conservation in Watamu Mida Creek, Kenya. *Environmental Management and Sustainable Development*, 6(1) 10766.

Alila PO and Atieno R 2006 *Agricultural policy in Kenya: Issues and processes*. Institute of Development Studies.

Bosek JK 2014 Implementing environmental rights in Kenya's new constitutional order: Prospects and potential challenges. *African Human Rights Law Journal*, 14(2), 489–508.

Chore T 2019 Reconceptualising the right to a clean and healthy environment in Kenya: The need to move from an anthropocentric view to a bicentric view. *Strathmore Law Review*, 4(1), 71–88.

Cunliffe G, Holz C, Mbeva K, Pauw P and Winkler H 2019 Comparative analysis of the NDCs of Canada, the European Union, Kenya and South Africa from an equity perspective, 21–30.

Dal Maso M, Olsen KH, Dong Y, Pedersen MB, and Hauschild MZ 2020 Sustainable development impacts of nationally determined contributions: Assessing the case of mini grids in Kenya. *Climate Policy*, 20(17), 815–831.

Dinesh D, Aggarwal P, Khatri-Chhetri A, Rodríguez AML, Mungai C, Sebastian L and Zougmore RB 2017 The rise in climate-smart agriculture strategies, policies, partnerships and investments across the globe. *Agriculture for Development*, 30, 4–9.

Faling M 2020 Framing agriculture and climate in Kenyan policies: A longitudinal perspective. *Environmental Science and Policy*, 106, 228–239.

Gachenga E 2019 (January) Kenya's Water Act (2016): Real devolution or simply the 'same script, different cast'. In P Kameri-Mbote, A Paterson, O C Ruppel, B B Orubebe, E D Kam Yogo (Eds.), *Law environment Africa*. Nomos Verlagsgesellschaft mbH and Co. KG, 429–452.

Juma L 2000 Environmental Protection in Kenya: Will the environmental management and co-ordination act (1999) Make a difference. *South Carolina Environmental Law Journal*, 9, 181.

Karani I 2018 Development of national and sub-national adaptation metrics: Lessons from Kenya. In *Adaptation metrics: Perspectives on measuring, aggregating and comparing adaptation results*, 113.

Kibugi R 2010 *New constitutional environmental law in Kenya: Changes in 2010*, 1–7.

Kibugi R 2017 Adaptation to climate change in smallholder agriculture in Kenya: The role of law. In M J Angelo and A Du Plesis (Eds.), *Research handbook on climate change and agricultural law*. Edward Elgar Publishing, 367.

King-Okumu C, Orindi VA and Lekalkuli L 2019 Drought management in the drylands of Kenya: What have we learned? In E Mapedza, D Tsegai, M Bruntrup, R Mcleman (Eds.) *Current directions in water scarcity research*. Elsevier, 277–294.

Klopp JM and Lumumba O 2016 The State of Kenya's Land Policy and Law Reform: A political institutional analysis. *Washington, DC: 2016 world bank conference on land and policy*. <https://www.conftool.com/landandpoverty2016/index.Php>, 5 (accessed June 14, 2022).

Lafferty W and Hovden E 2003 Environmental policy integration: Towards an analytical framework. *Environmental Politics*, 12(3), 1–22.

Lasco RD, Pulhin FB, Jaranilla-Sanchez PA, Delfino RJP, Gerpacio R and Garcia K 2009 Mainstreaming adaptation in developing countries: The case of the Philippines. *Climate and Development*, 1(2), 130–146.

Maina I, Newsham A and Okoti M 2013 *Agriculture and climate change in Kenya: Climate chaos, policy dilemmas*. Future Agricultures.

Mikunda T and Halstead MI 2014 Prioritising mitigation and adaptation options as part of the development of a National Climate Change Action Plan in Kenya. *Policy Studies*, 2017.

Morara GN, Hassan F and Osore MK 2015 Sustaining local livelihoods through coastal fisheries in Kenya. In A E Latawiec and D Agol (Eds.) *Sustainability Indicators in Practice*. De Gruyter Open Ltd, Warsaw/Berlin, 99-122.

Morton J 2017 Climate change and African agriculture: Unlocking the potential of research and advisory services. In Fiona Nunan (Eds.) *Making climate compatible development happen*. Routledge, 87–113.

Mugabe PA, Mwaniki F, Mamary KA and Ngibuini HM 2019 An assessment of drought monitoring and early warning systems in Tanzania, Kenya, and Mali. *Current Directions in Water Scarcity Research*, 211–219.

Murphy D and Harris M 2014 Policy and regulatory barriers in Kenya's National Climate change action plan: Areas for private sector action international institute for sustainable development https://cdkn.org/sites/default/files/files/Policy-Regulation-Review.pdf13 accessed 24 July 2021.

Mutimba S, Simiyu SW, Lelekoiten TL, and Ospina AV 2019 *sNAPshot*, 1–9.

Mutiso SK 1994 Kenya's National environmental action plan. *Waterlines*, 11–14.

Mwendwa P and Giliba RA 2012 Climate change impacts and adaptation strategies in Kenya. *Chinese Journal of Population Resources and Environment*, 10(4), 22–29.

Naeku MJ 2020 Climate change governance: An analysis of the climate change legal regime in Kenya. *Environmental Law Review*, 22(3), 170–183.

Ndegwa DW and Kinyua JM 2018 Strategic measures employed by the National Drought Management Authority for drought mitigation in Kenya. *International Journal of Management and Commerce Innovations*, 6(1), 802–812.

Njoroge JM, Ratter BM and Atieno L 2017 Climate change policy-making process in Kenya: Deliberative inclusionary processes in play. *International Journal of Climate Change Strategies and Management*, 9(4), 535–554.

Odoyo E, Bett B and Murenga Mwimali AS 2019 Challenges of and responses to climate variability: A case study of peri-urban female farmers in Kenya. *World Journal of Agricultural Research*, 7(2), 61–68.

Olashore OO 2019 Implementation of the international legal framework regarding climate change in developing countries: A review of Nigeria, Kenya, and Botswana's environmental provisions governing climate change. *Environmental Law Review*, 21(3), 189–209.

Ongugo PO, Langat D, Oeba VO, Kimondo JM, Owuor B, Njuguna J and Russell AJ 2014 A review of Kenya's national policies relevant to climate change adaptation and mitigation: Insights from Mount Elgon. Centre for International Forestry Research Working Paper, 155.

Onyango V and Schmidt M 2007 Towards a strategic environment assessment framework in Kenya: Highlighting areas for further scrutiny. *Management of Environmental Quality: An International Journal*, 18(3), 309–328.

Oulu MO 2011 Mainstreaming climate adaptation in Kenya. *Climate Law,* 2(3), 375–394.

Parry J-E 2016 Review of current and planned adaptation action in Kenya. CARIAA Working Paper No. 16. International Development Research Centre.

Parry JE, Echeverria D, Dekens J and Maitima J 2012 Climate risks, vulnerability and governance in Kenya: A review. Commissioned by climate risk management technical assistance support project (CRM TASP), joint initiative of bureau for crisis prevention and recovery and bureau for development policy of UNDP, 1–83.

Radeny M, Mungai C, Amwata D, Osumba J and Solomon D 2020 Climate change, agriculture, food and nutrition security policies and frameworks in Kenya. CCAFS

Working Paper, https://repository.seku.ac.ke/handle/123456789/6184 (accessed 9 November 2021).

Rowlinson P 2008 Adapting livestock production systems to climate change–temperate zones. In *Livestock and global change conference proceeding*, 61–63.

Rumble O 2019 (January) Climate change legislative development on the African continent. In *Law environment Africa*, 31–60.

Schwarte C 2017 Legislating the Paris Agreement in developing countries. Climate Law and Governance Working Paper Series, No. 1/2017. Centre for International Sustainable Development Law, 8.

Taylor Marcus 2018 Climate-smart agriculture: What is it good for? *The Journal of Peasant Studies*, 45(1), 89–107.

Versleijen N and Hoorweg J 2008 From farming to fishing: Marine resource conservation and a new generation of fishermen. *Western Indian Ocean Journal of Marine Science*, 7(1), 1–14.

Wambua C 2019 The Kenya Climate Change Act 2016: Emerging lessons from a Pioneer Law. *Carbon & Climate Law Review*, 257–269.

Williams TO, Mul ML, Cofie OO, Kinyangi J, Zougmoré RB, Wamukoya G Nyasimi M, Mapfumo P, Speranza CI, Amwata D, Frid-Nielsen S, Partey S, Girvetz E, Rosenstock T, Campbell BM. 2015. *Climate Smart Agriculture in the African Context*. Background Paper. Feeding Africa Conference 21–23 October 2015.

4 Nigeria

The policy and legal framework for adaptation and agriculture

4.1 Introduction

In line with the approach that was followed in Chapter 3, this chapter will investigate some climate change-related and agriculture-based policies and laws in Nigeria, with a view to determining whether the policies and laws are able to promote adaptation in the context of agriculture.

In a bid to adapt to the impacts of climate change, Nigeria is guided by the provisions of the United Nations Framework Convention on Climate Change (UNFCCC), the Kyoto Protocol, and the Paris Agreement, and other national climate change policies, action plans, and strategies that were formulated in accordance with the *Constitution of the Federal Republic of Nigeria* 1999 (as amended) (hereinafter referred to as the CFRN).[1] Examples of these national climate change policies include the National Adaptation Strategy and Plan of Action for Climate Change in Nigeria (NASPA-CCN),[2] the Nigeria Climate Change Policy Response and Strategy (NCCPRS),[3] National Policy on Environmental Conservation (NPEC),[4] and the National Adaptation Plan (NAP).[5] In addition to these policies, other laws such as the *National Environmental Standards and Regulations Enforcement Agency* (NESREA) *Act*[6] and the *Climate Change Act*[7] were enacted to promote adaptation pursuant the CFRN.

Considering that climate change is partly induced by agricultural practices (especially greenhouse gas (GHG)-intensive crops and livestock production methods), the Federal Government of Nigeria formulated and enacted certain agricultural-based policies and laws to enable the environment to cope with agriculture-induced climate change, as well as to enable agriculture to adapt to the impacts of climate change. These agricultural-based policies include the Nigerian Agricultural Policy,[8] the Nigerian Agricultural Resilience Framework (NARF),[9] the National Water Policy,[10] and National Drought Plan,[11] as well as the National Policy on Erosion and Flood Control (NPEFC).[12] In support of these agricultural policies, there are other laws which seek to facilitate adaptation in the context of agriculture. These laws include the *Water Resources Act,*[13] *River Basin Development Authority Act,*[14] *Land Use Act,*[15] *National Environmental (Soil Erosion and Flood Control) Regulations* (*NESEFC Regulations*),[16] and the *Nigerian Agricultural Insurance Corporation (NAIC) Act.*[17]

DOI: 10.4324/9781003536437-4

Therefore as done in the last chapter, the policies and laws are analysed below, based on their relevance (not chronologically according to their year of formulation and enactment), to determine whether they contain the type of provisions that could enable and promote adaptation in the context of Nigeria's agricultural sector. The analysis is guided by the research sub-question: how do Nigerian climate change and agriculture-related policies and laws provide for adaptation measures in the context of agriculture, and are there observable gaps in this regulatory framework?

4.2 Regulatory framework for climate change adaptation in Nigeria

As a party to the UNFCCC, the Kyoto Protocol, and the Paris Agreement, and as provided by the CFRN,[18] Nigeria is required to take actions that will address the impacts of climate change on the environment through various measures, including adaptation. For example, in order to realise the terms and objectives of article 4 of the UNFCCC, article 10 of the Kyoto Protocol, and article 7 of the Paris Agreement, as well as the provisions of the CFRN on environmental protection (discussed below), the Federal Government of Nigeria established a special climate change unit (SCCU) under the Federal Ministry of Environment. The SCCU is complemented by the Department of Climate Change in the Federal Ministry of Environment and the Climate Change section of the office of the Vice President of Nigeria.[19] These climate change institutions are also guided by Key climate change policies (which are also relevant for adaptation in the context of agriculture) and they include the NASPA-CCN, NCCPRS, NPEC, NAP (as mentioned above),[20] as well as Nigeria's Nationally Determined Contributions (NDCs) to climate change governance.[21] Having already discussed the provisions of the UNFCCC, Kyoto Protocol, and the Paris agreement in facilitating adaptation in a national context,[22] the constitutional provision that laid the foundation for adaptation policies in Nigeria will now be explained.

4.2.1 The constitution

The CFRN, 1999 (as amended) is the prevailing constitution in the country.[23] The provisions of the CFRN are supreme over any other law, person, or authority in Nigeria.[24] In effect, all laws and policies in Nigeria, including environmental laws and policies, derive their validity from the CFRN. Consequently, and in specific terms, section 20 of the CFRN mandates the State (Federal Government and State Governments of Nigeria):

> [t]o protect and improve the environment and safeguard the water, air, land, forest and wildlife in Nigeria.[25]

In acknowledgment of the opinion that "the constitution is a living tree that could be interpreted in the light of emerging realities",[26] an interpretation of section 20 of the CFRN could mean the deployment of all measures and strategies, including adaptation, to protect and improve the environment.[27] In a strict sense, section 20 of the CFRN could also be applied in providing a constitutional basis for the regulation of adaptation in the context of agriculture, and (as previously discussed) related resources such as water and land.[28]

Further, and as part of efforts to protect the environment through adaptation, section 12(1) of the CFRN serve as the basis for the implementation of important Multilateral Environmental Agreements (MEAs) such as the UNFCCC, Kyoto Protocol, and the Paris Agreement that were signed and ratified by Nigeria.[29] It should be noted, however, that although Nigeria signed and ratified the UNFCCC earlier, at the time of drafting the CFRN as well as when it came into force in 1999, adaptation was not considered as part of the solutions to climate change.[30] It is based on inference to the provisions of section 20 of the CFRN that the Federal Government of Nigeria formulated policies and enacted several environmental protection laws that could drive climate change adaptation processes.[31]

4.2.2 Adaptation Policies in Nigeria

Based on the provisions of section 20 of the CFRN, the Federal Government of Nigeria made efforts to boost the adaptive capacity of the environment against the impacts of climate change. The first effort was to consolidate adaptation planning into the then-existing Millennium Development Goals (MDGs),[32] and to formulate a national environmental policy which is implemented through a set of policies and action plans for adapting to climate change.[33] These policies and action plans, as highlighted above, include the NASPA-CCN, NCCPRS, NPEC, and the NAP as well as Nigeria's NDCs.[34] These policies and action plans, which are further discussed below, target components of the environment (such as water and soil) that are critical for agricultural activities, but yet vulnerable to the impacts of climate change. In other words, the policies provide a set of foundational principles that are important in guiding climate change responses (especially adaptation) in Nigeria.[35]

4.2.2.1 National Adaptation Strategy and Plan of Action for Climate Change in Nigeria

To define Nigeria's adaptation priorities and to integrate existing adaptation initiatives and priorities into all components of economic sectors, the Federal Government of Nigeria formulated a NASPA-CCN in 2011.[36] The goal of NASPA-CCN is to facilitate adaptation actions that will reduce vulnerability and increase the resilience and sustainable well-being of all Nigerians against the impacts of climate change.[37] The NASPA-CCN did not only defines Nigeria's adaptation priorities but also harmonises existing initiatives and priorities for future action.[38] The NASPA-CCN considers adaptation as an all-inclusive and integrated component of all economic sectors and sustainable development. The goal of NASPA-CCN is to take action to adapt to climate change by reducing vulnerability to climate change impacts and increasing the resilience and sustainable well-being of all Nigerians.[39]

In other words, adaptation is described by the NASPA-CCN as a means of "leveraging new opportunities and facilitating collaboration inside Nigeria and with the global community".[40] Parts of new opportunities envisaged by the NASPA-CCN include an array of options aimed at facilitating adaptation strategies for agriculture, water resources, fisheries, forestry, etc.[41] These adaptation strategies were previously discussed in Chapter 1.[42]

Equally, the NASPA-CCN makes provisions for certain community-based projects (which also stimulate wide stakeholders' participation) that could foster Nigeria's adaptive capacity in line with the country's commitments as mandated by the international adaptation regulatory framework (UNFCCC, Kyoto Protocol, and the Paris Agreement).[43] Examples of adaptation projects facilitated by the NASPA-CCN, which also stimulate wide stakeholders' participation, include those carried out in towns such as Kwaikong, Falgore, Dashe, Billiri and others involving the Federal, State and Local Governments in Nigeria as well as civil society organisations and international development partners.[44] The projects in those towns saw the implementation of some adaptation strategies that could enhance the adaptive capacity of agricultural production in Nigeria, such as the adoption of improved seeds and early maturing crops, application of rainwater-harvesting techniques, dry-season irrigation, the use of weather forecasting tools, and so on.[45] Therefore, for all intents and purposes, it could be submitted that the NASPA-CCN promotes adaptation in the context of agriculture. However, consistent with modern trends and developments in climate change governance, the NASPA-CCN was upgraded by another climate change policy known as NCCPRS.

4.2.2.2 Nigeria Climate Change Policy Response and Strategy

Guided by the provisions of article 4 of the UNFCCC, the Federal Government of Nigeria in 2012 approved NCCPRS as a climate change policy instrument as well as "a framework for implementing climate activities in the country".[46] In order to effectively reduce Nigeria's vulnerability to climate change impacts and to enhance its resilience and adaptive capacity, the NCCPRS enjoins and guides stakeholders to participate and collaborate on suitable response actions to climate change, be it for a short-term, medium-term, or longer-term.[47] The NCCPRS targets critical sectors such as energy, water resources, and food production that appear vulnerable to climate change impacts[48].

The NCCPRS aims to promote a low-carbon and climate-resilient society, by strengthening adaptive capacity in (for instance) the energy, water resources, and food production sectors of Nigeria.[49] By using a multi-sectoral approach,[50] the NCCPRS serves as a platform for mainstreaming climate change concerns into key economic sectors particularly energy, agriculture, water resources, forestry and wildlife, education, health, security, and transportation.[51] It has been previously explained that a multi-sectoral approach to mainstreaming could take either a vertical or a horizontal form.[52] In this regard, the Federal Government of Nigeria has adopted both vertical and horizontal forms of mainstreaming in adapting to climate change.

On one hand, the vertical approach to mainstreaming in Nigeria is represented in a top-down approach taken by the Government, whereby efforts were made by the Department for Climate Change to integrate the Paris Agreement into plans and policies at the national level, then further bringing down to sectoral programmes, and finally, closer to the grassroots at the sub-national level. Similarly, a vertical approach is exemplified by the Department for Climate Change as part of efforts to engage states to collaborate with local governments on climate change issues. This vertical arrangement to mainstreaming is usually supported by Nigeria's NAP, which is implemented to "ensure that emerging issues from international platforms, and especially the UNFCCC, are integrated into development activities and programmes at the sectoral levels". In the same regard, the development and implementation of Nigeria's Intended Nationally Determined Contribution (INDC) takes vertical arrangement to mainstream into consideration, where the activities of federal ministries, state ministries, and local government authorities are cascaded and linked by the Department for Climate Change into a Joint Planning Board of the Federal Ministry of Budget and National Planning.[53]

On the other hand, a horizontal form of mainstreaming in Nigeria could be seen in the coordinated integration of climate change concerns between the Ministry of Environment and other ministries such as the Ministries of Agriculture, Water Resources and Rural Development, Budget and National Planning, Science and Technology, and Health. In facilitating a horizontal approach to sectoral mainstreaming, the Government of Nigeria established an Inter-Ministerial Committee on Climate Change (ICCC). The ICCC, in collaboration with the Department for Climate Change, oversees the programming and integration of adaptation in Ministries, Departments, and Agencies at the federal (national) level.[54]

As emphasised in Chapter 2, parties to the Paris Agreement are required to submit and periodically update their adaptation plans and response strategies.[55] In line with this requirement, the Federal Government of Nigeria reviewed the NCCPRS in 2021, so as to conform to recent developments in global adaptation governance.[56] In addition, the review of the NCCPRS was moved by the need to "have a new holistic framework to guide the country's response to the challenges of climate change".[57] Another rationale for reviewing the NCCPRS (in addition to the promotion of a low-carbon and climate-resilient development pathway) was to enhance Nigeria's adaptive capacity especially with regard to agriculture and related resources (land and water) through international scientific and technological cooperation, research and development as well as public-private partnerships.[58]

The enhancement of Nigeria's adaptive capacity in protecting the environment and its components (agriculture, land, and water) is equally facilitated by another related policy – the NPEC.

4.2.2.3 *National Policy on Environmental Conservation*

In line with emerging trends in global environmental regulations, the Federal Government of Nigeria revised an already existing environmental policy on environmental conservation in 2016.[59] The revised policy has been re-named as the NPEC.[60] The NPEC is relevant for adaptation especially in the context of agriculture, as expressed in its goals. In other words, the goals of the NPEC are to cater for sustainable development while facilitating measures that could:

> [s]ecure an environment adequate for good health and well-being; promote sustainable use of natural resources; restoration and maintenance of the ecosystem and ecological processes and preservation of biodiversity; public awareness and understanding of linkages between environment and development; and cooperation with government bodies and other countries and international organizations on environmental matters.[61]

Agriculture or (in a strict sense) components of the environment that support agriculture (land and water) could be considered as part of natural resources which need to be secured and preserved against factors such as climate change that could lead to their depletion. Land and water are examples of natural resources that are required to be used in a sustainable manner. Adaptation is one of the means of securing and preserving natural resources and ensuring that they are used in a sustainable manner.[62]

Therefore, to facilitate adaptation, the NEPC is also expected to strategically aid the implementation of other related policies on environmental conservation that also support adaptation in the context of agriculture.[63] Those policies related to the NEPC include the National Water Policy, National Drought Plan, and the NPEFC.[64] Some approaches provided by the NEPC 2016 in aiding the implementation of policies related to environmental conservation (for instance, the National Water Policy) include:

> [e]ncourage community participation in the design, establishment, operation and maintenance of freshwater management programmes and infrastructure while promoting the integration of land and water management; provide water in adequate quantity and quality to meet domestic, industrial, agricultural, recreational and other needs.[65]

Most importantly and in connection with enhancing Nigeria's agricultural adaptive capacity, the NEPC 2016 also aims to develop and incorporate other policy instruments related to adaptation, such as the NAP and Nigeria's NDC, which are discussed in the following sections.[66]

4.2.2.4 National Adaptation Plan

As already stated in Chapter 2, countries that are signatories to the UNFCCC have established their NAPs as recommended by the Cancun Adaptation Framework (CAF).[67] Through NAPs, countries affected by climate change are able to identify and develop their medium- and long-term adaptation needs.[68] In other words, signatories to the UNFCCC are guided by their NAPs, in effectively monitoring and implementing their adaptation strategies.[69] Nigeria also developed its NAP as a long-term adaptation plan, in line with the guidelines set by the UNFCCC and the CAF.[70]

The broad objectives of the Nigerian NAP (similar to most NAPs) are to enhance adaptive capacity, boost resilience, and to reduce the vulnerability of human and natural systems to the impacts of climate change.[71] In addition, Nigeria's NAP seeks to facilitate low-carbon and climate-resilient development pathways in the country by integrating key adaptation policies into relevant but existing policies or new development policies and planning processes such as agricultural development.[72] That is to say, Nigeria's NAP is an amalgamation of major climate change policy in the country, which consist of the NASPA-CCN, the NCCPRS, and the NPEC (already discussed), and the NDC (which is discussed below).

To achieve its stated objectives, Nigeria's NAP is further supported by a framework which is described as the NAP Framework of 2020. The NAP Framework is equally guided by the provisions of the UNFCCC.[73] Accordingly, the UNFCCC requires that any NAP which is implemented under a framework be:

> [u]ndertaken in accordance with the Convention; follow a country-driven, gender-sensitive, participatory and fully transparent approach, taking into consideration vulnerable groups, communities and ecosystems; be based on and guided by the best available science and, as appropriate, traditional and Indigenous knowledge, and by gender-sensitive approaches, with a view to integrating adaptation into relevant social, economic and environmental policies and actions, where appropriate; and not prescriptive, nor result in the duplication of efforts undertaken in-country, but facilitate country-owned, country-driven actions.[74]

However, the above UNFCCC guidelines for a NAP Framework appear too broad and possibly overlooked agriculture-focused adaptation, which could only be identified in guidelines by way of inference. By way of explanation, the NAP Framework of 2020 adopts a sectoral approach to the planning and implementation of the adaptation process.[75] A sectoral approach to adaptation planning and implementation also takes into consideration the specific roles and responsibilities of stakeholders in different sectors of the economy, including agriculture. [76]

Additionally, a sectoral approach to adaptation planning and implementation will not only avoid duplication of adaptation programmes and projects but will also promote collaboration and experience sharing among ministries and parastatals. In other words, a sectoral approach "will facilitate the alignment of adaptation programmes and projects with the country's overall development agenda".[77] For example, a sectoral approach to adaptation planning and implementation has resulted in an alignment of the goals of NAP with other sectoral policies such as the Water Policy of 2004 (discussed below). Both NAP and the Water Policy promote similar goals for climate change adaptation and agriculture, such as the provisions for irrigation measures during water scarcity (aggravated by climate change). This is an important strategy that could allow agricultural practices in many parts of Nigeria to adapt to the impacts of climate change.[78] Lastly, the NAP Framework supports the actualisation of Nigeria's NDC, which is also an important policy component of adaptation as required by the Paris Agreement.[79]

4.2.2.5 Nigeria's Nationally Determined Contributions

All country parties to the Paris Agreement, through their NDCs, are required to contribute to the success of the global action on climate change.[80] In fulfilment of this obligation, Nigeria formulated its first NDCs in 2015, when the country signed the Paris Agreement.[81] Based on the requirements of the Paris Agreement, Nigeria is expected, after every five years, to submit a report of its NDCs to the UNFCCC secretariat alongside in-depth review summaries and a progress report on the country's NAP, in addition to a Global Climate Observing System Report.[82] Consequently, Nigeria submitted an updated NDCs to the UNFCCC on 2 July 2021.[83] The 2021 updated NDCs of Nigeria indicates an enhanced climate action ambition for the country compared to the 2015 NDCs. For example, emissions reductions from the waste sector have been included in the updated 2021 NDCs, with a target reduction of 20% below business-as-usual by 2030.[84]

Specifically, Nigeria's updated NDCs identify climate change priorities, targets, as well as "strategies for addressing the causes of climate change and responding to its effects in the country from 2015 and even beyond".[85] These priorities and targets also include sectors such as agriculture and its supporting resources (land and water). In these aspects, the NDC seeks to promote climate-smart agriculture (CSA) by encouraging less carbon-intensive agricultural practices, including better soil management practices, non-energy intensive irrigation methods (such as conservation tillage and the use of organic fertilisers), as well as the production and promotion of drought-tolerant crops and livestock feeds.[86]

To an extent, Nigeria's NDCs recognise and support sector-specific adaptation strategies, programmes, and policies.[87] For example, agriculture development strategies, programmes, and policies such as the National Agricultural Policy, the NARF, National Water Policy, National Drought Plan, and the

NPEFC were integrated into Nigeria's NDCs in furtherance to the goals of the Paris Agreement.[88] The integration of the NDCs with other sectoral programmes and policies is also done in order to:

> [i]mprove awareness and preparedness for climate change impacts, mobilize communities for climate change adaptation actions, reduce the impacts of climate change on key sectors and vulnerable communities, integrate climate change adaptation into national, sectoral, State and Local Government planning and into the plans of universities, research and educational organizations, civil society organizations, the private sector and the media.[89]

Based on the above objectives and strategies, the NDCs has provided the required policy stimulus in Nigeria for "appropriate responses with respect to erosion, pollution, sustainable livestock and crop production, among others".[90] In other words, just like other adaptation policies in Nigeria such as the NCCPRS, NASPA-CCN, and the NAP, the NDCs also underscore the need for the promotion of a CSA as a means of "improving agricultural productivity as well as the physical environment of the country".[91] By way of a review, all the previously discussed policies (the NASPA-CCN, NCCPRS, and the NAP) and the NDCs seek to:

> [b]uild resilience of the people in the agro-ecosystems to ensure their capacity to provide ecosystem services; ensure the integration of the agriculture sectors as well as food security and nutrition concerns in climate change strategies and policies; integrate strategies related to water management, land and natural resource management, rural development and social protection.[92]

Apart from the improvement of agricultural productivity and the physical environment, it is envisioned that Nigeria's NDCs will be able to stimulate improvement in standards of living and promote clean energy access among others, while also being instrumental in keeping global temperature increase well below 2°C, which are all vital for agriculture and adaptation.[93] In essence, the overall strategic goal of the NDCs and other adaptation policies in Nigeria (discussed above) are to "foster low-carbon, high-growth economic development and build a climate-resilient society".[94]

In terms of alignment, considerable overlap seems to exist between the components of Nigeria's NDCs, and other adaptation policies such as the NCCPRS, NASPA-CCN, and the NAP.[95] The alignment between the NDCs and other adaptation policies in Nigeria is a clear indication of attempts in mainstreaming climate change concerns with national development. Mainstreaming can only be effective when implemented in the light of relevant policies and laws. Having discussed those relevant policies above, it is pertinent to also discuss the legal framework that broadly promotes adaptation in Nigeria.

4.2.3 Adaptation-related Laws

An effective climate change law is a prerequisite for achieving a low-carbon and climate-resilient development (through mainstreaming adaptation options) in an oil-dependent and agriculture-driven economy like Nigeria, which is also vulnerable to climate change.[96] The *NESREA Act* and the *Climate Change Act* are laws that directly provide for the protection and improvement of the environment in Nigeria from all kinds of environmental problems (including climate change). It must be reiterated that these laws (as analysed below) are enacted pursuant to the provisions of the CFRN.[97]

4.2.3.1 National Environmental Standards and Regulations Enforcement Agency Act

The *NESREA Act* was enacted in 2007 in accordance with the constitutional mandate of the Federal Government of Nigeria to provide standards for environmental protection.[98] In other words, the *NESREA Act* serves as a framework:

> [f]or the protection and development of the environment, biodiversity conservation and sustainable development of Nigeria's natural resources in general and environmental technology, including coordination and liaison with relevant stakeholders within and outside Nigeria on matters of enforcement of environmental standards, regulations, rules, laws, policies and guidelines.[99]

The functions of NESREA (the Agency) also include the enforcement of compliance with international agreements, protocols, conventions, and treaties on the environment (including climate change), which Nigeria is a party. In specific terms, a provision of the *NESREA Act* provides that:

> The Agency [NESREA] shall enforce compliance with the provisions of international agreements, protocols, conventions and treaties on the environment, including climate change, biodiversity, conservation, desertification, forestry, oil and gas, chemicals, hazardous wastes, ozone depletion, marine and wildlife, pollution, sanitation and such other environmental agreements as may from time to time come into force.[100]

Such international agreements, protocols, conventions, and treaties include the UNFCCC, Kyoto Protocol, and the Paris Agreement, and they were previously discussed.[101]

Furthermore, the *NESREA Act* provides standards on the preservation and the enhancement of the quality of air,[102] water,[103] and the ozone layer[104] among others, which could help the environment (including agricultural practices) to adapt to the impacts of climate change. For example, given that water is required in reasonable quantity and quality for domestic and agricultural needs, the *NESREA Act* makes provision for standards that could promote the preservation and enhancement of water bodies in Nigeria.[105] Measures adopted by the Agency in conserving and improving water quality in Nigeria include, but are not limited to, supporting the efforts of State governments in combating water pollution at the State level; adopting standards in water management as provided by conventions, treaties, protocols, and agreements on water to which Nigeria is a signatory;[106] as well as other control mechanisms that could promote water quality, especially the integrated water-resources management techniques that were previously explained.[107]

These measures of the *NESREA Act* on the conservation and enhancement of water standards also form the basis of the goals of national water resources management policies, especially the National Water Policy of 2004, which is "the provision of water in adequate quality and acceptable quality to meet domestic, industrial, agricultural and recreational needs".[108] An in-depth analysis of the National Water Policy is provided below.

Reading through the general provisions of section 7 of the *NESREA Act*, it could be concluded that adaptation has been provided for. However, as a generic law, the NESREA is not best placed to go into specificities on climate adaptation. In other words, the provisions of section 7 do not provide sufficient direction on the types of adaptation strategies that should be adopted for most sectors of the economy including agriculture. Hence, adaptation for agriculture (especially crop and livestock productions) has not been sufficiently provided for by the *NESREA Act*, although resources that support agriculture (such as land and water) are catered for by the Act. Considering the gap in the *NESREA Act* in providing sufficient direction for adaptation in the context of agriculture, recourse should be made to a more specific climate change law for guidance. In this context, the provisions of the *Climate Change Act* of 2021 would be referred.

4.2.3.2 Climate Change Act 2021

The *Climate Change Act* came into force in 2021, just when the COP 26 meeting was taking place in Glasgow, Scotland.[109] The *Climate Change Act* was enacted in accordance with the provisions of section 20 of the CFRN, 1999 (as amended) after more than a decade-long, rigorous law-making process.[110] The *Climate Change Act* is a law which provides the required legal basis for the coordination of Nigeria's long-term goals for climate change actions, as well as the legal stimulus needed to promote climate-resilient socio-economic development in Nigeria.[111] The Act regulates actions, measures, and interventions by public entities (ministries, departments, and agencies) and private entities (businesses and civil society organisations) operating in the country.[112] The Ministry for Agriculture and departments under it (for instance crops and livestock) are included in the application of the provisions of the *Climate Change Act*.[113]

In sum, the objectives of the *Climate Change Act* as provided in section 1 of the Act are to support low GHG-emission growth and sustainable development in Nigeria by identifying risks and vulnerabilities; building resilience and strengthening existing adaptive capacities to the impacts of climate change; mainstreaming climate change policies and actions with the national development agenda; as well as ensuring that private and public entities comply with climate change strategies and targets set out in the NAP.[114]

Therefore, in order to achieve its objectives, the *Climate Change Act* is coordinated by a National Council on Climate Change (NCC),[115] supported by a Carbon Budget[116] and operationalised through a National Climate Change Action Plan.[117] The functions of the NCC, as provided by section 4 of the *Climate Change Act*, include complementing the role of the Department for Climate Change (under the Ministry for Environment) in coordinating the implementation of a country-wide National Climate Change Action Plan,[118] the administration of the Climate Change Fund, and steering the process of mainstreaming climate change concerns in national development. In addition, the NCC is entrusted with the task of formulating climate change policies and programmes. It is on this basis that CSA policies are formulated as medium-term solutions to the impacts of climate change.

Given that agriculture is a sector of the economy that is particularly vulnerable to climate change risk, the sector is expected to benefit from the NCC, Carbon Budget, and National Climate Change Action Plan. In a bid to facilitate adaptation, especially for agriculture, section 27 of *the Climate Change Act* makes provision for nature-based solutions to climate change problems.[119] These nature-based solutions include conservation tillage, restorative agriculture, and regrowing clear-cut forests.[120] Therefore, with regard to adaptation in the context of agriculture, it could be stated that the provisions of the *Climate Change Act* have provided the required legal stimulus, but not in any explicit terms.

So far, the roles of the *Climate Change Act*, the *NESREA Act*, and other climate change policies in facilitating adaptation in a broad context and in the context of agriculture have been discussed. While the regulatory framework for climate change adaptation could be described as comprehensive in facilitating adaptation in a broad context, the same could not be stated in the context of agriculture. As explained in the preceding sections of this chapter, the NAP, NPEC, the *NESREA Act*, and *Climate Change Act* do not offer sufficient guidance for adaptation in the context of agriculture. Therefore, it is essential to review relevant agriculture-based policies and laws in Nigeria to determine whether they contain the type of provisions that could enable and promote adaptation in the context of agriculture. Before proceeding to examine agricultural policies and laws in Nigeria, it will be necessary to highlight that, as far as adaptation is concerned, agricultural policies and laws are expected to drive a process that is generally understood and previously explained as CSA.[121]

4.2.4 CSA Strategy

Unlike in Kenya where CSA is implemented under a framework,[122] in Nigeria, CSA is implemented as a mere strategy, and this has some implications which are explained below. Originally, even without a framework, poor and small-scale rural farmers in Nigeria (to some extent) practised CSA methods based on Indigenous and experiential knowledge in coping with the impacts of climate change on crops and livestock production.[123] Examples of CSA practices implemented by poor small-scale rural farmers in boosting the adaptive capacity of crops include minimum tillage and integrated crop management systems (crop rotation, cover cropping, intercropping, and planting of seed varieties that could withstand droughts, floods, and pests) while rearing disease-resistant animal breeds with high yield in enhancing the adaptive capacity of livestock against the impacts of climate change.[124]

These CSA practices are associated with numerous benefits. For example, using crop rotation as a CSA strategy could reduce the application of costly agrochemicals and non-organic fertilisers, which have been previously attributed to GHG emission.[125] In the same manner, adoption of minimum tillage and integrated crop-management systems as forms of CSA strategies tend to promote sustainable land-use practices. These strategies are noted for their roles in minimising the emission of GHGs.[126]

Indigenous knowledge-based CSA practices were replaced with formalised CSA practices in Nigeria, and these are regulated through climate change policies such as the NASPA-CCN and the NCCPRS. An example of formal CSA practices is the Pro-Resilience Action (PROACT) project that was implemented in Adamawa and Borno States (located in North East) of Nigeria.[127] In addition to the impacts of climate change, which persistently occur as an increase in temperature, floods, droughts, and desertification, Adamawa and Borno States were affected by prevailing insurgency and insecurity.[128] The PROACT CSA projects have strengthened the resilience of farmers in Adamawa and Borno States, who are mostly engaged in the farming of crops (such as cassava, yam, sweet potatoes, groundnut, cotton, rice, and maize) and the rearing of cattle, poultry, and goats.[129]

The PROACT CSA project in Adamawa State targeted four local government areas – Mubi South, Fufore, Song, and Guyuk. The project mainly focused on the provision of farm inputs and training of about 700 farmers' groups. In addition, farmers in the four local government areas were supplied with farm inputs (water pumps, improved seeds) and fertilisers, and were supported with access to flexible financing services. Following the successful launch of the PROACT CSA project, almost 100% of rice farmers recorded a bumper harvest, especially during the dry-season rice production.[130] While in Borno State, the PROACT CSA projects promoted terrace farming as a strategy for boosting farming in the rocky areas of southern and northern Borno State. Farmers in the area confirmed that the benefits of terrace farming facilitated by the PROACT CSA project include the control of soil erosion, preservation of soil moisture contents, as well as improved nutrients in crops such as sorghum and maize. In addition, the PROACT CSA project aided a better management of weeding and spraying of farms in the southern part of Borno State.[131]

However, there are several factors hindering the full adoption of CSA among farmers in Nigeria. According to the Food and Agricultural Organization (FAO), the acceptance of modern CSA practices is still low in Nigeria compared to other African countries such as Kenya because of lack of a supporting framework and the poor level of awareness.[132] It has been observed that most farmers in Nigeria are rural-based with less formal education, and they are barely conversant with modern CSA practices; even the few farmers that are aware of the innovations offered by CSA lack the skills needed for using CSA technologies.[133] Some CSA technologies are complex and cannot be easily understood within a short period of time.[134]

Similarly, as a result of culture and traditional beliefs, farmers are skeptical of moves to replace traditional practices with CSA technology.[135] Also, the cost involved in practising CSA makes it not viable for the majority of farmers in Nigeria, who are poor.[136] These challenges make it necessary for the Federal Government of Nigeria to formulate policies and enact laws in support of CSA.

4.3 Regulatory framework for agriculture

In addition to the broad climate change-adaptation policies and laws that have been analysed above, there are other agriculture-based policies and laws that also provide for adaptation in the agriculture sector. The Nigerian Agricultural Policy and the NARF are examples of agricultural policies that underscore the need for enhancing the adaptive capacity and resilience of agriculture against climate change, as well as enabling the environment to cope with the impacts of agriculture-induced climate change.[137] The Nigerian Agricultural Policy served as the basis for the formulation and implementation of other policies related to agriculture, such as the National Water Policy and the NPEFC. Apart from these policies, there are laws such as the *Water Act* and *Land Use Act*, *NESEFC Regulations,* and the *NAIC Act* that facilitate adaptation in the context of agriculture. The following sections will first analyse agriculture-related policies, and then explain the provisions of agriculture-related laws in order to determine how they facilitate adaptation in the context of agriculture.

4.3.1 Agricultural-based Policies

In line with the provisions of section 20 of the CFRN, which provide the basis for the formulation of policies dealing with environmental protection (including adaptation), the Federal Government of Nigeria formulated two agricultural policies in Nigeria to cater for adaptation. These policies, as already mentioned in the introduction to this chapter include the Nigerian Agricultural Policy of 2001 and the NARF of 2014. These agricultural policies are aimed at promoting a vibrant agricultural system in Nigeria that could also adapt to the impacts of climate change.[138] In other words, both the Agricultural Policy and the NARF seek to boost the resilience of crops and livestock against the impacts of climate change, by promoting sustainable soil-management practices (such as conservation tillage, integrated soil-nutrient management, etc.) and other on-farm adaptation strategies previously discussed.[139]

4.3.1.1 Nigerian Agricultural Policy of 2001

As was pointed out in Chapter 1, agriculture is important to the Nigerian economy, but it is susceptible to climate change, and at the same time, it has a tendency to drive climate change.[140] On this basis, the Federal Government of Nigeria formulated an Agricultural Policy in 2001 to promote the:

> [r]ational utilization of agricultural resources, improved protection of agricultural land resources from drought, desert encroachment, soil erosion and flood, and the general preservation of the environment for the sustainability of agricultural production. [141]

Taken together, the above objectives would seem to suggest that the Agricultural Policy promotes adaptation in the context of agriculture to some extent.

First, the Agricultural Policy serves as the basis for increasing water supply for crops and livestock production as well as domestic needs in Nigeria through the construction of boreholes, dams, dykes, and polder wells for irrigation purposes in strategic locations in Nigeria affected by shortfall in water supply.[142] Commensurately, the Agricultural Policy also recommends the construction of dams and dykes as adaptation strategies against erosion and floods as well as a means to facilitate afforestation schemes.[143] These adaptation strategies provided by the Agricultural Policy are operationalised through the retention and upgrading of existing water resources development agencies, especially River Basin Development Authorities, to make their operations more efficient and effective.[144] In addition, the adaptation strategies for water supply as provided by the Agricultural Policy are also realised by enacting or amending national water-resource laws for the regulation of the development and exploitation of water resources.[145] It is against this background that the Water Resources Act of 1993 was amended in 2004.[146]

Second, the Agricultural Policy advocates for appropriate land-use strategies, which perfectly fits with what has been previously explained as adaptive land-management strategies.[147] For example, in order to achieve its objective of "rational utilisation of agricultural resources", the Policy recommends the expansion of livestock-grazing reserves in Nigeria.[148] The expansion of livestock-grazing reserves is an appropriate adaptation strategy for land use that could minimise the over-concentration of cattle on a particular land, which had been described as one of the factors that increase the emission of methane and climate change in Nigeria.[149] For related purposes, the Agricultural Policy also recommended the rearing of livestock breeds that can cope with the impacts of climate change, in the following words:

> In recognition of the natural ecological zoning of the country which supports the raising of different kinds of livestock in different parts of the country, government will continue to encourage the production of various classes of livestock in those regions of the country with proven comparative geographical and production advantage. [150]

Also, with regard to livestock production, the Agricultural Policy promotes adaptation measures that could support animals' fodder against the impacts of climate change such as droughts and the decline of soil moisture. The Agricultural Policy specifically provides that:

> Government will ensure that livestock farmers are encouraged to practice various forms of fodder conservation techniques. It will in addition, provide adequate storage facilities for grains and strategic supplementary feed reserves as part of efforts to alleviate the effects of incessant drought. [151]

Third, given that the application of industrial fertilisers (such as NH3 and NH4) and non-industrial fertilisers (such as manure derived from fermentative digestion by cows, horses, and sheep) are contributing factors to GHG emission and climate change,[152] the Agricultural Policy recommends the use of alternative and local organic-based raw materials for the production of fertilisers for farming.[153] Precisely, the Agricultural Policy seeks to promote:

> [t]he development of appropriate technologies for the optimum utilisation of farm input such as fertilizers, herbicides, pesticides and so on, to increase production potentials without adverse effects on man and his environment.[154]

This policy aspiration is intended to be implemented through a law that would generally control the quality and quantity of agro-chemicals, including fertilisers, applied to land in order to protect it against degradation and pollution.[155] These measures could serve as an adaptation strategy that helps the environment to cope with a changing climate caused by the application of fertilisers.

Fourth, the Agricultural Policy recognises and supports agricultural insurance, which is also a form of an adaptation strategy that has been discussed previously.[156] In light of this, the objective of the Agricultural Policy is to make provision for a mechanism such as soft loans that would indemnify farmers engaged in crop and livestock production in all the ecological zones of

Nigeria against risk and loss resulting from natural hazards including climate change.[157] To this effect, a strategy employed by the Agricultural Policy to ensure that farmers are protected against climate change-induced risk and loss is to make insurance cover mandatory for all categories of farmers in Nigeria (self-financed farmers, small-to-medium scale farmers, and large-scale farmers).[158]

Regardless of the good objectives of the Agricultural Policy in promoting agriculture-focused adaptation, some gaps exist in the Policy. For example, a mechanism for monitoring the implementation of the Policy is seemingly missing. Lack of a monitoring mechanism has been identified as a factor that generally affects the stability of all policies in Nigeria, including the Agricultural Policy.[159] Similarly, changes in government as well as changes in the personnel entrusted with the task of implementing policies always affect the stability of policies (including the Agricultural Policy).[160]

Based on the aforementioned gaps identified in the Agricultural Policy, there is every need to at least review the Policy, or to formulate another agricultural policy. So far, apart from the Agricultural Policy of 2001, the only policy which seems to support adaptation in the context of agriculture in Nigeria is the NARF.

4.3.1.2 National Agricultural Resilience Framework of 2014

The NARF was formulated in 2014 as a framework for enhancing the resilience of agricultural practices against the impacts of climate change pursuant to the "principles of adaptive management and participatory engagement".[161] These principles are reflected in the objectives of the NARF which include:

> [s]trengthening the overall policy/institutional framework for improved resilience and adaptation to climate variability and change in the agricultural sector, including planning and implementation, systems for resource mobilization, and effective project monitoring and evaluation.[162]

The objectives of the NARF appear to have captured the essential requirements that are relevant to adaptation in the context of agriculture. More elaborately, strengthening the overall policy and institutional framework for planning and implementing climate change adaptation could be interpreted as the adoption of measures that tend to improve the knowledge base of agricultural practices (crops and livestock production), especially capacity building. In the context of agricultural adaptation, capacity building could be implemented through briefings, training, and short courses, as well as information sharing for staff and partners working in different sub-sectors of agriculture on (for example) how to deploy land and water management techniques (water harvesting and irrigation methods, erosion control, and soil fertility enhancement).[163]

Likewise, the concept of resource mobilisation as expressed by the objectives of the NARF also takes into consideration transfer mechanisms and risk management, which in the context of agricultural adaptation connotes using insurance services such as weather-index crop insurance schemes to cushion the effects of climate-related risks on agriculture. Resource mobilisation in accordance with the objectives of the NARF also means the use of seasonal and real-time weather forecast technology to determine changes in climate systems, which typically include heat waves, droughts, threats to water availability, and crop failures.[164] Resource mobilisation could be validated through monitoring and evaluation.[165] The form of monitoring and evaluation as articulated in the objectives of the NARF seems to suggest "testing and refining agricultural tools, to improve understanding of climate change adaptation issues".[166]

Given the above objectives of the NARF, climate change adaptation and agriculture were integrated into the Vision 20:2020, which is the most recent national development planning policy of Nigeria.[167] Vision 20:2020 makes provision for the development and integration of flood or drought-tolerant crops into the research activities of the Department of Climate Change at the Federal Ministry of Environment.[168] This integration can be beneficial for climate adaptation and agriculture, especially for areas in Nigeria affected by climate change-induced droughts.[169]

To summarise, the NARF supports adaptation in the context of agriculture by incorporating some of the most cost-efficient and effective methods of adaptation.[170] These cost-efficient and effective methods of adaptation include investments in grazing reserves, switching to sedentary (permanently resident) livestock production systems (as an alternative to pastoralist livestock systems), increasing crop storage facilities, and promoting agro-processing as a means of reducing post-harvest losses.[171] The NARF also facilitates adaptation in the context of agriculture by promoting crop diversification and animal breeding strategies to withstand climate variability.[172] The NARF promotes sustainable soil practices that can reduce GHG emissions and at the same time enhance (high level of) soil carbon sequestration.[173] Lastly, the NARF seeks to enhance water supply for agricultural and domestic use through irrigation by constructing boreholes, dams, and dykes.[174]

As stated above, the NARF is supported by other policies such as the National Water Policy. Hence, on the enhancement of water supply, the NARF only adopted and complemented the strategies that were initially provided by the National Water Policy.

4.3.1.3 National Water Policy

The impacts of climate change on water resources (water scarcity and water stress) make it necessary to conduct the planning, management, and control of Nigeria's water resources within a framework that also promotes adaptive water management, while at the same time incorporating basic principles such as the principles of equitable and reasonable utilisation and that of prevention of significant harm to water.[175] In this regard, the Federal Government of Nigeria formulated and adopted a National Water Policy in 2004, in order "to improve on the nation's water resources management including the management of hydrological risks and vulnerabilities".[176] The National Water Policy supports the Water Resources Management Reform Program (WRMRP) of 1997.[177] Collectively, the National Water Policy and the WRMRP seek to optimise the use of water resources in accordance with the requirements of sustainable development principles, which entail using environmental (water) resources without degrading them, and in a manner consistent with the demands of both present and future generations.[178]

In addition to the principles of water management above, the Water Policy is based on other guiding principles which are drawn from the CFRN, such as the principle of national accountability and monitoring, and the principle of non-discrimination.[179] These principles are to ensure that water is used in a manner that is consistent with environmental requirements and at the same time to the satisfaction of domestic and industrial needs, including agricultural practices. To further promote the principles of adaptive water management, the Federal Government of Nigeria reviewed the Water Policy in 2016, and what resulted was a National Water Resources Policy (NWRP).[180]

The NWRP is formulated "subject to and consistent" with the provisions of section 20 of the CFRN, 1999.[181] In addition, the NWRP is supported by, and is expected to be implemented by, a number of institutions in Nigeria including the NESREA, the National Water Resources Institute, and the River Basins Development Authority (RBDA). The functions of these institutions in relation to water resources management are explained below. However, before discussing the institutions that implement the NWRP, it is important to highlight the broad and specific objectives of the NWRP. Just like the National Water Policy, the broad objective of the NWRP is:

> [t]o foster the integrated management of water resources for optimum, sustainable, efficient, and equitable water resources development and management in order to meet the current and future water demand use, conserve the water quality and protect the environment. [182]

In specific context, one of the objectives of the NWRP is centred on how to promote:

> [s]takeholder participation (governments, communities, Civil Societies and Private Sector) in the water sector development to meet rapidly growing demand for domestic and industrial water supply, sanitation, irrigation and drainage, food and erosion control, hydropower generation, inland transportation, inland fishery, livestock farming and other uses.[183]

In the same regard, the NWRP is intended to support rainwater management techniques in conjunction with sustainable drainage systems as a part method for boosting household water supply, drainage, and flood control, while at the same time aiming to prevent the over-exploitation and adulteration of groundwater quality.[184] In other words, the rationale behind these objectives of the NWRP is to conserve the quality of both surface and groundwater resources, while also promoting the protection of the environment and associated aquatic ecosystems to ensure long term sustainability.[185]

In unequivocal terms, this specific objective of the NWRP supports the essence of adaptation in the context of agriculture. In addition, the NWRP outlines some strategies including integrated water resources management, which can promote all the specific objectives of the Policy.[186] It should be noted that the NWRP directly takes into account the impacts of flood, drought, and climate change when implementing integrated water resources management strategies, as articulated in a key policy statement which reads as follows:

> The management of water resources shall be responsive to climate change resilience measures by adopting environmentally friendly practices and innovative solutions towards maintaining the balance between the environment, water resources and ecosystems.[187]

For this reason, the NWRP outlines some strategies that, among others, include the continuous gathering of information on climatic changes, followed by public awareness and sensitisation programmes, as well as the development of small-scale earth dams for expanded storage of raw water.[188]

Hence, in the context of agricultural adaptation, both the Water Policy and the NWRP appear to have provided sufficient direction on what adaptive water management strategies are, and how they can be implemented. In addition, the guidelines in the two policies were integrated and further elaborated upon in the NARF, which has been explained above as an agricultural policy that also supports adaptive water-management strategies. Notwithstanding the integration of the Water Policy and NWRP into the NARF, the Federal Government still formulated another policy, known as the National Drought Plan, to facilitate and enhance the adaptive capacity of water supply in Nigeria.

4.3.1.4 National Drought Plan 2018

As discussed previously, some parts of Nigeria are prone to droughts and desertification as a result of climate change and variability.[189] In response, the Federal Government of Nigeria formulated a number of policies to manage the impacts of droughts on the environment. The most relevant policy in this context is the National Drought Plan of 2018.[190] The National Drought Plan was formulated in line with the goals of the United Nations Convention to Combat Desertification (UNCCD),[191] which is also premised on the objectives of addressing the effects of the drought that usually manifest as environmental losses, economic stress and social hardship.[192]

Therefore, the National Drought Plan is a framework that recognises droughts and desertification as perennial environmental problems that are capable of undermining agricultural practices, and accordingly promotes a cost-effective and holistic but long-term approach to averting the impacts of droughts and desertification on, among others, water supply, agriculture, energy production, and recreation.[193] The National Drought Plan is set to achieve some broad and specific objectives. The broad objectives of the National Drought Plan are to support:

> comprehensive drought monitoring and early warning systems; complete vulnerability assessments for sectors, populations and regions vulnerable to drought; and implement drought mitigation measures that limit the adverse impacts of drought and provide appropriate response measures when drought occurs.[194]

As for the specific objectives of the National Drought Plan, they are premised on reducing the vulnerability of Nigerian communities to drought and on enhancing adaptive capacities to droughts. The specific objectives of the National Drought Plan are expected to be achieved through drought-monitoring and implementation measures.[195]

The broad objectives of the National Drought Plan revolve around mitigation measures, while the specific objectives focus on adaptation measures. However, since the focus of this book is on adaptation, particularly in the context of agriculture, its point of reference will be the specific objectives of the National Drought Plan.

Therefore, in accordance with the specific objectives of the National Drought Plan, stakeholders in adaptation governance (agencies, institutions, and citizens) are assigned with responsibilities, and they are supported with relevant equipment for the collection, analysis, and utilisation of data in order to address drought and desertification in a systematic manner.[196] Examples of relevant equipment that stakeholders can use to combat drought and desertification include meteorological and hydrological information tools and early warning system facilities. Further, stakeholders are supported through awareness programmes that comprise both formal and informal education on how to use some of the complicated and advanced meteorological and hydrological information tools for addressing droughts and desertification.[197] Additionally, relevant equipment for addressing drought and desertification include awareness campaigns for stakeholders on appropriate land-use techniques such as changing the topography of land, cover crops, and conservation tillage that are not only effective in boosting carbon sequestration but can also boost crop yield.[198]

The National Drought Plan also recommends measures that could promote the rehabilitation of agricultural lands in Nigeria affected by soil erosion and floods.[199] These measures are also adequately described in a later policy (the NPEFC), which is discussed below.

4.3.1.5 National Policy on Erosion and Flood Control of 2006

Erosion and floods are among the impacts of climate change with profound effects on the environment. A regulatory framework is required to ensure that coordinated and systematic measures are implemented in managing and controlling the impacts of erosion and floods on the environment. Consequently, the Federal Government of Nigeria formulated a NPEFC in 2006.[200] As the name suggests, the NPEFC is mainly designed to facilitate "the management and control of erosion and floods" resulting from climate change, through systematic and coordinated strategies.[201] These systematic and coordinated strategies among others include a forecasting, monitoring, and control mechanism for erosion and floods,[202] and they are among the forms of adaptation discussed in Chapter 1.[203] To support systematic and coordinated adaptation strategies, the NPEFC organises symposia or teaching for land users and developers (including farmers) on how to use forecasting, monitoring, and control mechanisms to adjust or reduce the immediate impacts of floods and erosion from escalating.[204]

In addition, the NPEFC seeks to "promote training across the country (national, state, and local levels) in erosion and flood prevention, management and control" as part of systematic and coordinated strategies for erosion and flood control.[205] Some of the trainings embarked on by the NPEFC include creating the required public awareness campaign in order to encourage public participation in erosion and flood control and exercises on how to limit the utilisation and carrying capacity of marginal lands in order to protect them against erosion and floods.

Accordingly, the NPEFC has a role to play in promoting the promulgation of other land-use regulations that could boost soil resilience against erosion and floods at the state level.[206] For example, the Lagos State government of Nigeria promulgated a soil erosion and flood control law in 2011 known as the *Environmental (Soil Erosion and Flood Control) Regulations* of Lagos State, to effect the strategies in the NPEFC.[207] *The Environmental (Soil Erosion and Flood Control) Regulations* of Lagos State seek to protect the environment against losses emanating from flood and erosion by regulating and imposing penalties for unconventional land uses and other land-disturbing activities that could be harmful to health and safety of water.[208]

However, the NPEFC is a policy which, like any other agricultural policy in Nigeria, needs to be translated into a binding law. [209] On this note, relevant agriculture-based laws in Nigeria will be discussed.

4.3.2 *Agriculture-based Laws*

In addition to the agricultural policies discussed above, various laws were enacted to give efficacy to agricultural policies in promoting adaptation in the context of agriculture. These laws include the *Water Resources Act*, the *Land Use Act*, the *NESEFC Regulations*, and the *NAIC Act*. These laws were originally enacted to boost agricultural productivity in Nigeria, but at the same time, the laws provide a legal stimulus for enhancing resilience and adaptive capacity, as well as for reducing the vulnerability of agricultural productivity against risks and disasters such as climate change-induced droughts, floods, and pests. It is worth noting that some of these laws (*Land Use Act, NAIC Act,* and the *Water Resources Act*) were initially promulgated as *Decrees* during military rule when some provisions of the Nigerian Constitution were suspended, especially the provisions mandating the legislative arm to make law, order, and good governance for the Federation.[210] Hence, after the return to civil rule in 1999 and when the CFRN became operational, those Decrees were incorporated in the CFRN by virtue of the provisions of Section 315, which provides that:

> Subject to the provisions of this Constitution, an existing law shall have effect with such modifications as may be necessary to bring it into conformity with the provisions of this Constitution and shall be deemed to be an Act of the National Assembly to the extent that it is a law with respect to any matter on which the National Assembly is empowered by this Constitution to make laws.[211]

Consequently, even though the *Water Resources Act, the RBDA, Land Use,* and the *NAIC Act* (as will be discussed in the following sections) predate the CFRN, they are nonetheless firmly supported by the CRFN.

4.3.2.1 Water Resources Act

The *Water Resources Act* was enacted in 1993[212] to give the Federal Government of Nigeria the sole rights to control shared water bodies in Nigeria and to fulfil certain objectives, which include:

> (a) The promotion of the optimum planning development and use of Nigeria's water resources. (b) The co-ordination of such activities that are likely to influence law, enhance the quality, distribution, use and management of water. (c) The application of appropriate standards and technologies for the investigation, use, control, protection, management and administration of water resources. (d) The facilitation of technical assistance and rehabilitation for water supplies.[213]

The impacts of climate change on water resources were not directly considered by the Federal Government of Nigeria at the time of enacting the *Water Resources Act,* and when it came into force in 1993. In order to specifically promote the resilience of water against the impacts of climate change, and to conform to emerging trends in water-resources governance, the *Water Resources Act* was amended in 2004.[214] The objectives of the 2004 Act are to:

> (a) protect, conserve and control water resources for equitable and sustainable social and economic development and to maintain environmental integrity; (b) ensure allocation, supply and distribution of water resources for all uses to promote equitable, sustainable and sufficient best practices and conduct ... [215]

In protecting, conserving, and controlling the quality and quantity of water resources, the amended *Water Resources Act* also considers natural disasters and risks such as climate change as factors that could undermine the quality and quantity of water resources.[216] The *Water Resources Act* identifies processes that could facilitate adaptive water-management techniques, which are presented as technical, social, and commercial standards that must be adhered to by water developers and water providers.[217] Technical standards for water are exemplified by a set of requirements that among others relate to water acidity, colour, and turbidness. These requirements are necessary in preserving and enhancing the quality of water for designated uses (recreation, domestic, agricultural, and industrial).[218] The technical standards for water that are provided by the *Water Resources Act* are the same as those set by the provisions of sections 23 and 27 of the *NESREA Act*, and they are intended to protect the public health or welfare and enhance the quality of water for agricultural, domestic, industrial and other uses. The provisions of the *NESREA Act* on water standards are already discussed above, as such they do not require further analysis.[219]

4.3.2.2 River Basins Development Authority Act

Another law that promotes adaptive water-management strategies in Nigeria is the *RBDA Act*.[220] River Basins Authorities are the sole government institutions in Nigeria authorised with the function of providing large-scale irrigation services for dry-season farming and for supplementing water requirements for agricultural, industrial use, and domestic uses where there is lack of or poor precipitation.[221] The functions of the RBDA are properly elaborated in section 4 of the *RBDA Act*, and in summary, these functions consist of the development of surface and groundwater through the construction of dams, dykes, polders, wells, and boreholes in a bid to support irrigation services.[222] More precisely, the functions performed by the RBDA are among the adaptation strategies in the context of agriculture that have been discussed previously.[223] Likewise, the RBDA supports adaptation in the context of agriculture by making provision for development of water-storage schemes, for supply to all users (domestic, agricultural, industrial, and recreational), though based on specified fees.[224]

In addition to the provision of large-scale irrigation services and supplementing water requirements for agricultural, industrial, and domestic uses, the *RBDA Act* is a law that has been enacted to partly support the control of soil erosion and floods in Nigeria.[225] As explained above, both soil erosion and floods are consequences of unconventional land uses and land-disturbing activities that could be aggravated by climate change. On the basis of this, it could be concluded that the *RBDA Act* facilitates adaptation in the context of agriculture. It should be noted, however, that land-use activities are mainly regulated by the *Land Use Act*.

4.3.2.3 Land Use Act of 1978

The earliest efforts made by the Government of Nigeria in regulating land use in the country started with the enactment of the *Land Use Act* of 1978.[226] The *Land Use Act* is a law that:

> vest all land comprised in the territory of each State (except land vested in the Federal Government or its agencies) solely in the Governor of the State, who would hold such land in trust for the people and would henceforth be responsible for allocation of land in all urban areas to individuals resident in the State and to organisations for residential, agricultural, commercial and other purposes while similar powers with respect to non-urban areas are conferred on Local Governments.[227]

Based on the above description, it is widely believed that the *Land Use Act* gave all Nigerians the right to use and enjoy all lands in Nigeria including its appurtenances, subject to overriding public interest.[228] For this purpose, the *Land Use Act* recognises and incorporates certain basic international land-use principles such as land tenure rights and communal rights that are necessary in promoting sustainable land-management practices, already explained in Chapter 2.[229]

In addition to basic international land use principles, the *Land Use Act* makes provision for the public trust doctrine (PTD),[230] which advocates for fair and effective management of natural resources including land by the government as a trust for all the people in Nigeria, in order to promote sustainable and equitable land use in Nigeria.[231] In the context of adaptation and agriculture, the promotion of sustainable and equitable use of land through the PTD could also advance adaptive land management by controlling unconventional land uses and land-disturbing activities, such as over-cultivation and overgrazing, which were previously identified as factors that contribute to more GHG emission.[232]

Although the *Land Use Act* serves as the foundation for land-related laws in Nigeria, and has some provisions that prohibit and penalise unauthorised land-use practices,[233] the *Land Use Act* has been criticised for being an outdated law that ignored other consequences of unconventional land uses and land-disturbing activities, such as erosion and flooding, which could be exacerbated by climate change.[234] For this reason, the NPEFC and the *NESEFC Regulations* were developed to complement the *Land Use Act*. The NPEFC has been analysed above,[235] while the *NESEFC Regulations* will be analysed below.

4.3.2.4 NESEFC Regulations of 2011

It has been previously stated that policies are translated into laws, and laws serve as tools for the implementation of policies.[236] Based on that statement, the implementation of the NPEFC is supported by the *NESEFC Regulations 2011*.[237] The *NESEFC Regulations* were promulgated to:

> [p]rotect human life and the environment; minimize losses due to flood and erosion and their effects on vulnerable areas by regulating land-disturbing activities; and to control accelerated soil erosion, flooding and sediment deposition in water bodies and water courses in order to prevent pollution of these water resources.[238]

Flooding, soil erosion, and pollution were identified as environmental problems associated with climate change which, if not adapted to, could have dire consequences for agricultural practices.[239] In order to help agricultural practices in Nigeria to adapt to the emerging risks posed by erosion and floods, the *NESEFC Regulations* make provision for Erosion Control Plan Measures such as the installation of rock beams, sand bags, and the construction of slopes and the planting of cover crops.[240] In addition, the *NESEFC Regulations* facilitate adaptation measures against soil erosion and floods through the deployment of "surface and subsurface drainage facilities, dams, flood walls, high flow diversions and planting of trees, shrubs and grasses".[241] These were previously described as agriculture-based adaptation strategies using technological solutions.[242] Accordingly, these Erosion Control Plan Measures are believed to "provide effective erosion and sedimentation control for rainfall events for 24-hours daily, and up to a period of 25 years when properly designed, installed and maintained".[243]

Additionally, the *NESEFC Regulations* make provision for practices that could help agricultural systems in adapting to climate change-induced erosion. These provisions are otherwise contained in the specific objectives of the *NESEFC Regulations*.[244] The specific objectives of the *NESEFC Regulations* are to control or prevent activities which are harmful to land, natural floodplains, stream channels, and natural protective barriers, and other related activities that constitute threats to the health and safety of living things.[245] For these reasons, the *NESEFC Regulations* seek to restrict or prohibit the:

> [f]illing, grading, dredging and other development projects which may increase soil erosion or flooding incidence and damage; prevent or

> regulate the construction of flood barriers and other structures capable of diverting or escalating flood waters or which may increase flood hazards on human settlements and related infrastructure; protect drainage infrastructure...[246]

It should be emphasised that the above prohibitions were previously covered by other Regulations related to climate change adaptation, which were incorporated into the *NESEFC Regulations*. These are the *Watershed, Mountainous, Hilly and Catchment Areas Regulations,* 2009 and the *Wetlands, River Banks and Lake Shores Regulations, 2009*. The *Watershed, Mountainous, Hilly and Catchment Areas Regulations,* 2009 were aimed at controlling some land activities on mountainous, hilly, and catchment areas, which are capable of causing landslides, floods, drought, and desertification, and are compounded by climate change.[247] On the other hand, the *Wetlands, River Banks, and Lake Shores Regulations,* 2009 were formulated to achieve some objectives relevant for climate change adaptation, which are to:

> (a) provide for the conservation and wise use of wetlands and their resources in Nigeria; (b) ensure water catchments conservation and flood control; (c) ensure the sustainable use of wetlands for ecological and tourism purposes for the common good of the citizens; (d) ensure that the country's wetlands are protected as habitats for species of fauna and flora; and (e) minimize and control pollution. [248]

In practical terms, most of the provisions of the *NESEFC Regulations* support adaptation in the context of agriculture by compelling farmers to alter agricultural practices in areas which could damage the environment – in other words, aggravate climate change. However, as is the case with most Regulations, the *NESEFC Regulations* are not expected to cover all matters relevant to soil erosion and flood control. For example, the provisions of section 8(1) of the *NESEFC Regulations* exempt certain land-disturbing agricultural practices, especially livestock grazing, which is also considered as a contributing factor to soil erosion and climate change. Consequently, the *NESEFC Regulations* appear to contain a gap as far as adaptation in the context of agriculture is concerned.[249] Therefore, farmers who suffer crop or livestock losses caused by erosion or flooding, despite having adhered to the *NESEFC Regulations,* require insurance cover as an alternative solution. In this regard, there is a dedicated insurance corporation for the agriculture sector in Nigeria.

4.3.2.5 Nigerian Agricultural Insurance Corporation Act

The Federal Government of Nigeria established the NAIC in 1993 by virtue of a law known as the *NAIC Act.*[250] The objective of the NAIC is "to protect the Nigerian farmer from the effects of natural hazards by introducing measures which shall keep the farmer in business..."[251] In specific terms, the NAIC was primarily established to provide insurance cover against risks to Nigerian farmers. The forms of risk covered by the NAIC also include risks resulting from the impacts of climate change such as droughts, erosion, floods, pests, and invasive species. To help farmers cope with climate change-based risk, the NAIC offers financial support (credits) and subsidised cover for livestock and crop losses to farmers in the form of index-based and indemnity-based insurance services.[252] These forms of insurance were explained previously.[253] The process for the implementation of insurance services under the NAIC is the same as the process that has been discussed under the Nigerian Agricultural Policy of 2001 and the NARF of 2014;[254] therefore, they do not deserve further analysis in this section. Accordingly, adaptation in the context of agriculture is facilitated by the NAIC to some extent.

However, the main challenge of the *NAIC Act* is the exclusion of non-commercial farmers from the insurance scheme.[255] This means that the majority of farmers in Nigeria who were identified as poor and small-scale rural farmers with weak adaptive capacity[256] could not benefit from the cover provided by the NAIC scheme, which remains the best indemnity against risks emanating from the impacts of climate change. This observation prompts further discussion on the deficiencies in policy and legal frameworks for adaptation and agriculture.

4.4 Deficiencies in the Policy and legal frameworks for adaptation and agriculture in Nigeria

So far, this chapter has analysed the impacts of climate change on agriculture and how agricultural practices also contribute to climate change. The chapter underscored the need for agriculture to adapt to the impacts of climate change, as well as the need to safeguard the environment from agriculture-induced climate change. In light of this, the chapter argued that adaptation strategies can only be effectively implemented in the context of Nigeria's agriculture when they are supported by policies and laws. As a result, the policy and legal framework for climate change adaptation and agriculture in Nigeria were analysed. However, as discussed above, most of the laws and policies for climate change adaptation and agriculture in Nigeria have shortcomings in facilitating agriculture-focused adaptation. This section will summarise the shortcomings.

Firstly, some climate change policies, such as the NAP, NPEC, and laws, especially the *NESREA Act* and *the Climate Change Act*, only minimally provide for adaptation in the context of agriculture because they have been noted to be too broad and do not "provide the country with the required focused response to climate change concerns".[257] Specifically, these laws and policies merely refer to measures which could have satisfied the requirements of adaptation in the context of agriculture, without providing sufficient direction and backing on those measures. With reference to the NAP, it has been formulated as a national policy that sets broad goals, which include the facilitation of low-carbon and climate-resilient development pathways in the country. At the same time, the NAP is supposed to be a monitoring and implementation mechanism for adaptation strategies in Nigeria. For these reasons, it has been noted that the NAP did not explicitly provide for agriculture-focused adaptation. Instead, agriculture-focused adaptation could only be inferred from the goals of the NAP. Similarly, agriculture-focused adaptation could be inferred from the NAP's alignment with other policies that support agriculture such as the National Water Policy. The same observation has been made in respect of some agriculture-related laws such as the *Land Use Act* and the *NESEFC Regulations*. These laws only partly facilitate adaptation measures for agriculture, and the effectiveness or potential of these agriculture-related laws in supporting adaptation measures in the context of agriculture could not be fully ascertained.[258]

Secondly, it has been observed that Nigeria's NAP process failed to precisely elaborate how institutional arrangements for adaptation in the context of agriculture can be implemented.[259] Similarly, as opposed to most NAP processes, which are conceived as part of long-term strategies against climate change impacts, Nigeria's NAP process is built as a short-term to medium-term strategy.[260] A NAP that is being conceived as long-term strategy is cost-effective and could promote socio-economic development, while a short-term or medium-term NAP could lead to mal-adaptation.[261]

Thirdly, Nigeria's adaptation governance framework has been generally criticised because it did not stimulate the required collaboration among stakeholders; it lacks a mechanism for target setting and for monitoring and evaluation.[262] These shortcomings could have fed into other deficiencies in Nigeria's adaptation governance framework, such as duplication of mandates and a lopsided spread of financial burden among climate change agencies. A prominent example of both these deficiencies can be found in the RBDA and the WRMRP.

On a final note, the recently promulgated Climate *Change Act* [263] is a generic law, which could not be expected to provide the required legal stimulus for agriculture-focused adaptation in explicit terms. Therefore, it is too early to ascertain whether the Act can, for instance, promote the successful mainstreaming of climate change issues with agricultural development planning in Nigeria.[264]

4.5 Conclusion

Agriculture is an important sector of Nigeria's economy, second to the petroleum sector in providing employment opportunities, generating revenue and foreign exchange and boosting the GDP. However, climate change and variability that results in droughts and flooding have significant implications for agricultural practices such as livestock and crop productions, and by extension Nigeria's GDP. Consequently, adaptation is essential in enhancing the resilience of agricultural systems against the impacts of climate change and also safeguarding the environment against agriculture-induced climate change in Nigeria. In cognisance of this, Nigeria (as a party to the UNFCCC, Kyoto Protocol, and Paris Agreement) formulated and implemented a host of climate change adaptation policies (such as the NASPA-CCN, NCCPRS, and NAP) as well as climate change adaptation laws enacted pursuant to the provisions of the CFRN (such as the *NESREA Act*, and the *Climate Change Act*). Apart from these policies and laws, adaptation in the context of agriculture is also facilitated by agriculture-related policies and laws such as the *Water Resources Act, RBDA Act, Land Use Act,* and the *NESEFC Regulations* that were initially formulated to address water issues, desertification, and erosion.

A review of the above-mentioned climate change adaptation and agriculture-related policies and laws in Nigeria confirmed that they are too broad and contain gaps, in terms of enabling agriculture to adapt to the impacts of climate change. Given that this book is set to evaluate the development of agriculture-focused adaptation in Kenya, Nigeria, and South Africa, the next chapter will evaluate the South African regulatory framework for climate change adaptation and agriculture in order to determine whether they have sufficient provisions for agriculture-focused adaptation.

Notes

1 *Constitution of the Federal Republic of Nigeria, 1999* (as amended) Cap C23 Laws of the Federation of Nigeria 2014.
2 Federal Ministry of Environment Climate Change Department 2011 National Adaptation Strategy And Plan of Action on Climate Change for Nigeria (NASPA-CCN). https://faolex.fao.org/docs/pdf/nig211219.pdf, 1–101.
3 Climate Policy Database 2012 National Policy on Climate Change and Climate Change Policy Response and Strategy. https://climate-laws.org/document/national-policy-on-climate-change-and-climate-change-policy-response-and-strategy_95ff.
4 FAOLEX 2016 National Policy on the Environment (Revised 2016) https://faolex.fao.org/docs/pdf/nig176320.pdf, 12; UNFCCC 2010 https://unfccc.int/files/adaptation/application/pdf/nigerianeeds.pdf, 9.
5 Government of Nigeria, 2020 Nigeria's National Adaptation Plan Framework https://napglobalnetwork.org/wp-content/uploads/2021/06/napgn-en-2020-Nigeria-National-Adaptation-Plan-NAP-Framework.pdf.
6 The *NESREA Act* 2007 is an Act which generally provides for the establishment and enforcement of environmental standards in Nigeria.
7 The *Climate Change Act* was adopted on 18 November 2021. See https://natlex.ilo.org/dyn/natlex2/natlex2/files/download/112597/NGA112597.pdf.
8 Federal Ministry of Agriculture, Water Resources and Rural Development, Abuja 2001 https://faolex.fao.org/docs/pdf/nig149296.pdf.
9 J Adegoke, A Araba and C Ibe 2014 Agricultural research and technology development for climate resilient agriculture. National Agricultural Resilience Framework. *Federal Ministry of Agriculture and Rural Development, Abuja*. https://boris.unibe.ch/62564/1/Nigeria%27s%20Changing%20Cliamte.pdf, 10–14.
10 Federal Republic of Nigeria 2004 National Water Policy
Federal Republic of Nigeria https://faolex.fao.org/docs/pdf/nig158231.pdf, 1–28.
11 Federal Government of Nigeria 2018 National Drought Plan https://www.unccd.int/sites/default/files/country_profile_documents/1%2520FINAL_NDP_Nigeria.pdf 35.

12 Government of Nigeria 2020 National Policy on Climate Change https://faolex.fao.org/docs/pdf/NIG209876.pdf, 24.
13 *Water Resources Act* No, 101 of 1993.
14 *River Basin Development Authority Act* CAP R9, Laws of the Federation of Nigeria 2004.
15 *Land Use Act* CAP L5, Laws of the Federation of Nigeria 2004.
16 *National Environmental (Soil Erosion and Flood Control) Regulations* No. 39, vol. 98.
17 *Nigerian Agricultural Insurance Corporation (NAIC) Act* Cap. N89 LFN 2004.
18 Specifically, Section 20, CFRN 1999 (as amended) makes provision for the protection of the environment through some measures which also include adaptation. The provisions of Section 20 are explained below.
19 See 2050 Long-Term Vision for Nigeria (LTV-2050), Department of Climate Change, Federal Ministry of Environment, Nigeria 2021 https://unfccc.int/sites/default/files/resource/Nigeria_LTS1.pdf, 1–40.
20 See Section 4.1.
21 UNFCCC 2010 National Environmental, Economic and Development Study (Needs) for Climate Change In Nigeria https://unfccc.int/files/adaptation/application/pdf/nigerianeeds.pdf, 37.
22 See generally, Chapter 1, section.
23 Before the return to civil rule in 1999, Nigeria was operating the Presidential Constitution of 1979. Save for few amendments on provisions dealing with, for instance, citizenship and acquisition of moveable properties in Nigeria, etc., most of the wording and provision of the 1979 Constitution were retained in the 1999 Constitution, though with few modifications that are not within the scope of this chapter.
24 Section 1 CFRN, 1999 (as amended). This means that any law or policy that is inconsistent with the constitution is a nullity.
25 Section 20 CFRN, 1999. The provision of Section 20 of the CFRN should be jointly read with that of Section 13 of the CFRN, 1999 which is to the effect "It shall be the duty and responsibility of all organs of government, and of all authorities and persons, exercising legislative, executive or judicial powers, to conform to, observe and apply the provisions of this Chapter of this Constitution".
26 AB Abdulkadir 2014 The right to a healthful environment in Nigeria: A review of alternative pathways to environmental justice in Nigeria. *Journal of Sustainable Development Law and Policy,* 3(1), 118–131.
27 The provisions of 20 are found under Chapter 2 dealing with fundamental objectives and directive principles of state policy. Unlike article 42, under the Bill of Rights of the Constitution of Kenya, 2010 which makes environmental protection a fundamental right that could be pursued by legal means (see Chapter 4, paragraph *4.3.3*), rights contained under Chapter 2 of the CFRN are non-justiciable rights. See R Ako, N Stewart and EO Ekhator 2016 Overcoming the (non) justiciable Conundrum: The doctrine of Harmonious construction and the interpretation of the right to a healthy environment in Nigeria. In *Justiciability of Human Rights Law in Domestic Jurisdictions*, 123–141.
28 See Chapter 1, Section 1.2.
29 Given that Nigeria is a Dualist State as opposed to a Monist one, section 12 (1) of the CFRN provides that treaties can only bind Nigeria where it has been domesticated through a law-making process of the National Assembly. The *NESREA Act* of 2007 and the *Climate Change Act* of 2021 are few examples of the domestication of the UNFCCC, Kyoto Protocol, and the Paris Agreement. On Monist and Dualist States, see J Laws 2000 Monism and dualism. *La Revue Administrative,* 53(2), 18–22.

30 Nigeria signed the UNFCCC on 13 June 1992 and ratified it on 29 August 1994. However, adaptation governance did not receive considerable attention in Nigeria until 2010, when a series of efforts such as the NASPA-CCN, NCCRS, and NPEC were made by the government in order to cope with climate change impacts and vulnerability in the country. See United Nations 2022 United Nations Treaty Collections. https://treaties.un.org/pages/ViewDetailsIII.aspx?src=TREATY&mtdsg_no=XXVII-7&chapter=27&Temp=mtdsg3&clang=_en.
31 Moreover, by virtue of the powers granted to the Federal Government of Nigeria by the CFRN (through committees and sub-committees of the National Assembly), environmental protection policies and laws are periodically reviewed in order to strengthen the regulatory framework for environmental protection. See for example, section 62 of the CFRN which makes provision for Committees of the National Assembly to form and appoint Committee members.
32 The MDGs were an eight time-bound development agenda and their monitorable indicators of the United Nations that were aimed at promoting sustainable development in developing countries. Nigeria and some other developing 188 countries from across the world endorsed the MDGs in New York in September 2000. The MDGs were replaced by the SDGs in 2015. See Nigeria 2015 Millennium Development Goals End-Point Report, https://www.undp.org/sites/g/files/zskgke326/files/publications/Nigeria_MDGs_Abridged_Sept30.pdf, 1–28.
33 O Afinowi 2018 *An outline and critical assessment of the role of planning laws in the regulatory framework of climate change adaptation in South Africa and Nigeria*. University of Cape Town, 115.
34 See Section 4.1.
35 The legal status of these policies has been previously discussed in Chapter 2, Section 2.2.
36 See Building Nigeria's Response to Climate Change 2011 https://csdevnet.org/wp-content/uploads/NATIONAL-ADAPTATION-STRATEGY-AND-PLAN-OF-ACTION.pdf, 4 and 28.
37 Federal Ministry of Environment Climate Change Department 2011 Federal Ministry of Environment Climate Change Department 2011 National Adaptation Strategy And Plan of Action on Climate Change for Nigeria (NASPA-CCN). https://faolex.fao.org/docs/pdf/nig211219.pdf, 1–101.
38 Federal Ministry of Environment Climate Change Department 2011 Federal Ministry of Environment Climate Change Department 2011 National Adaptation Strategy And Plan of Action on Climate Change for Nigeria (NASPA-CCN). https://faolex.fao.org/docs/pdf/nig211219.pdf, 1–101.
39 CU Ogbonna and E Albrecht 2015 Strategic environmental assessment as a tool to integrate climate change adaptation: A perspective for Nigeria. In FW Leal (Ed.) *Handbook of climate change adaptation*. Springer, 1239–1260.
40 Building Nigeria's Response to Climate Change 2011 http://csdevnet.org/wp-content/uploads/NATIONAL-ADAPTATION-STRATEGY-AND-PLAN-OF-ACTION.pdf, 32.
41 Building Nigeria's Response to Climate Change 2011 http://csdevnet.org/wp-content/uploads/NATIONAL-ADAPTATION-STRATEGY-AND-PLAN-OF-ACTION.pdf, 36–38.
42 See Chapter 1, Sections 1.5.1.1 and 1.5.1.2. These strategies include for example building coastal defences to protect coastal farms against sea level rise and salinised sea water, and cloning or the use of genetic engineering technology to develop new hybrid crop varieties and animal breeds that can withstand unsuitable temperature, droughts, or lack of soil moisture due to climate change.
43 E Woodley 2011 Building Nigeria's response to climate change: Pilot projects for community-based adaptation in Nigeria. In *Experiences of Climate Change Adaptation in Africa*. Springer, 297–315.

44 See Building Nigeria's response to climate change (BNRCC) Project 2011. http://csdevnet.org/wp-content/uploads/NATIONAL-ADAPTATION-STRATEGY-AND-PLAN-OF-ACTION.pdf, 31.
45 The relevance and effectiveness of these adaptation strategies were explained previously. See Chapter 1, Section 1.5.1.
46 Building Nigeria's Response to Climate Change (BNRCC) Project 2011. http://csdevnet.org/wp-content/uploads/NATIONAL-ADAPTATION-STRATEGY-AND-PLAN-OF-ACTION.pdf 25.
47 Government of Nigeria Nigeria's National Adaptation Plan Framework 2020 https://napglobalnetwork.org/wp-content/uploads/2021/06/napgn-en-2020-Nigeria-National-Adaptation-Plan-NAP-Framework.pdf, 16.
48 Government of Nigeria 2020 National Policy on Climate Change https://faolex.fao.org/docs/pdf/NIG209876.pdf,12.
49 BE Olajide, MO Quadri and V Ojakorotu 2018 Climate change, human security and good governance in Nigeria. *African Renaissance,* 15(3), 173–196.
50 A multi-sectoral approach is an approach which recognises diverse ways and manners by which each sector is impacted by climate change and how each sector is able to cope with the impacts of climate change. A multi-sectoral approach has the advantage of promoting a coordinated, and coherent national adaptation response. This is further explained in Chapter 7.
51 Government of Nigeria, 2020 Nigeria's National Adaptation Plan Framework https://napglobalnetwork.org/wp-content/uploads/2021/06/napgn-en-2020-Nigeria-National-Adaptation-Plan-NAP-Framework.pdf, 4.
52 See Chapter 1, Section 1.4.4.
53 See C Abazu and A Alabi 2019 Fostering urban environment adaptation to climate change in Nigeria. *Coou African Journal of Environmental Research,* 2(1),43–55; EG Orie 2021 Climate change adaptation mechanism for sustainable development Goal 1 in Nigeria: Legal Imperative. In *African Handbook of Climate Change Adaptation*. Springer International Publishing, 1897–1917.
54 See K Shehu, Auwal, and AA Dahuwa 2018 A review of climate change: Its impacts and governance for enhancing Nigeria's environmental resources. *Dutse Journal of Pure and Applied Sciences (DUJOPAS),* 4(2), 1–10.
55 See Chapter 2, Section 2.3.3.
56 The reviewed NCCPRS is expected to operate till 2030, the same time when the United Nation Sustainable Development Goals will be ended. Government of Nigeria 2021 National Climate Change Policy for Nigeria https://climatechange.gov.ng/wp-content/uploads/2021/08/NCCP_NIGERIA_REVISED_2-JUNE-2021.pdf, 8.
57 Government of Nigeria 2021 National Climate Change Policy for Nigeria https://climatechange.gov.ng/wp-https://climatechange.gov.ng/wp-content/uploads/2021/08/NCCP_NIGERIA_REVISED_2-JUNE-2021.pdf, 7.
58 Government of Nigeria 2021 National Climate Change Policy for Nigeria https://climatechange.gov.ng/wp-https://climatechange.gov.ng/wp-content/uploads/2021/08/NCCP_NIGERIA_REVISED_2-JUNE-2021.pdf, 7.
59 The first Nigerian national policy on environmental conservation was formulated in 1989 in order to address major environmental problems in the country, which include but are not limited to erosion, deforestation, desertification, pollution, hazardous waste, drought, floods, etc. The policy was revised in 1999 to accommodate new and emerging environmental concerns, especially climate change. In addition, the revised environmental policy of 1999 was intended to promote sustainable development in Nigeria. Adesina and Odekunle 2011 Climate change and adaptation in Nigeria: Some background to Nigeria's response 1 International Conference on Environmental and Agricultural Engineering, 15(2), 146–154.

60 FAOLEX 2016 National Policy on the Environment (Revised 2016) https://faolex.fao.org/docs/pdf/nig176320.pdf, 12; UNFCCC 2010 National Environmental, Economic and Development Study (Needs) for Climate Change In Nigeria https://unfccc.int/files/adaptation/application/pdf/nigerianeeds.pdf; 9.
61 FAOLEX 2016 National Policy on the Environment (Revised 2016) https://faolex.fao.org/docs/pdf/nig176320.pdf, 12; UNFCCC 2010 National Environmental, Economic and Development Study (Needs) for Climate Change In Nigeria https://unfccc.int/files/adaptation/application/pdf/nigerianeeds.pdf, 9.
62 This has been previously emphasised. See Chapter 1, Section 1.4. See also H Mongi, AE Majule, and JG Lyimo 2010 Vulnerability and adaptation of rain fed agriculture to climate change and variability in semi-arid Tanzania. *African Journal of Environmental Science and Technology*, 4 (6), 371–381.
63 FAOLEX 2016 National Policy on the Environment (Revised 2016) https://faolex.fao.org/docs/pdf/nig176320.pdf, 15, 19 and 27.
64 These policies are further discussed below under agricultural policies.
65 FAOLEX 2016 National Policy on the Environment (Revised 2016) https://faolex.fao.org/docs/pdf/nig176320.pdf, 16.
66 FAOLEX 2016 National Policy on the Environment (Revised 2016) https://faolex.fao.org/docs/pdf/nig176320.pdf, 26.
67 See Chapter, Section 2.3.1.
68 UNFCCC 2013 The State of Adaptation under the United Nations Framework Convention on Climate Change 2013 Thematic Report. https://unfccc.int/files/adaptation/cancun_adaptation_framework/adaptation_committee/application/pdf/ac_2013_report_high_res.pdf, 13.
69 A Möhner 2018 The evolution of adaptation metrics under the UNFCCC and its Paris Agreement. In L Christiansen, G Martinez and P Naswa (eds.) *Adaptation metrics: Perspectives on measuring, aggregating and comparing adaptation results*, 1–166.
70 UNFCCC 2016 Assessment of progress on NAPs in developing countries. https://unfccc.int/sites/default/files/resource/201608_questionnaire_nigeria.pdf, 1–14.
71 It is expected that Nigeria's NAP framework, once fully implemented, will foster the adaptation component of Nigeria under the INDC as submitted to the UNFCCC secretariat. Government of Nigeria, 2020 Nigeria's National Adaptation Plan Framework. https://napglobalnetwork.org/wp-content/uploads/2021/06/napgn-en-2020-Nigeria-National-Adaptation-Plan-NAP-Framework.pdf, 1–56.
72 Dazé Angie 2019 International Institute for Sustainable development 2019 why gender matters in climate change adaptation https://www.iisd.org/blog/gender-climate-change; Government of Nigeria, 2020 Nigeria's National Adaptation Plan Framework. https://napglobalnetwork.org/wp-content/uploads/2021/06/napgn-en-2020-Nigeria-National-Adaptation-Plan-NAP-Framework.pdf, 1–56.
73 Government of Nigeria 2020 Nigeria's National Adaptation Plan Framework. https://napglobalnetwork.org/wp-content/uploads/2021/06/napgn-en-2020-Nigeria-National-Adaptation-Plan-NAP-Framework.pdf, 8.
74 See UNFCCC, 2011 Decision 5/CP.17: National adaptation plans https://unfccc.int/sites/default/files/decision_5_cp_17.pdf, 1–7.
75 Government of Nigeria 2020 Nigeria's National Adaptation Plan Framework. https://napglobalnetwork.org/wp-content/uploads/2021/06/napgn-en-2020-Nigeria-National-Adaptation-Plan-NAP-Framework.pdf, 8, 9.
76 Government of Nigeria 2020 Nigeria's National Adaptation Plan Framework. https://napglobalnetwork.org/wp-content/uploads/2021/06/napgn-en-2020-Nigeria-National-Adaptation-Plan-NAP-Framework.pdf, 4.
77 Nigeria NAP Framework 2020 Nigeria's National Adaptation Plan Framework. https://napglobalnetwork.org/wp-content/uploads/2021/06/napgn-en-2020-Nigeria-National-Adaptation-Plan-NAP-Framework.pdf, viii.

78 Adesina and Odekunle 2011 Climate change and adaptation in Nigeria: Some background to Nigeria's response 1 International Conference on Environmental and Agricultural Engineering, 15(2),146–154.
79 Article, 4(2), Paris Agreement; Government of Nigeria 2020 Nigeria's National Adaptation Plan Framework. https://napglobalnetwork.org/wp-content/uploads/2021/06/napgn-en-2020-Nigeria-National-Adaptation-Plan-NAP-Framework.pdf, 10.
80 Article 4.2 of the Paris Agreement.
81 Though Nigeria signed the Paris Agreement in 2015, actual ratification of the Paris Agreement by the government of Nigeria only took place in March 2017. See *Thisday News* December 28, 2017, https://www.thisdaylive.com/index.php/2017/12/28/paris-climate-change-agreement-how-has-nigeria-fared/.
82 Article 4.2 of the Paris Agreement.
The NDC should not be confused with National Communications, in which Non-Annex I Parties are expected to report NAP progress at the COP of the UNFCCC within three years of entering the Convention, and every four years thereafter. See Decision 17/CP.8 of the UNFCCC https://unfccc.int/sites/default/files/17_cp.8.pdf, 1–11; Chapter 1, Section 1.3.1. So far, Nigerian has submitted its first National Communication to the UNFCCC secretariat in November 2003; followed by a Second National Communication that was submitted in February 2014 and a third National Communication that was submitted in April 2020. See National Communication Submissions from Non-Annex I Parties https://unfccc.int/non-annex-I-NCs.
83 See Federal Government of Nigeria 2021 NDC update https://climatechange.gov.ng/wp-content/uploads/2021/08/NDC_File-Amended-_11222.pdf, 1–60.
84 See G I Malumfashi 2024 Nigeria's *Climate Change Act*, 2021 and the Global Climate Commitments: Legal Implications, Issues and Prospects. *Nigerian Journal of Policy and Strategy*, 21(1), 19–38.
85 Government of Nigeria 2021 National Climate Change Policy for Nigeria https://climatechange.gov.ng/wp-content/uploads/2021/08/NCCP_NIGERIA_REVISED_2-JUNE-2021.pdf, 12.
86 UNFCCC 2015 Nigeria's Intended Nationally Determined Contribution. https://unfccc.int/sites/default/files/NDC/2022-06/Approved%20Nigeria%27s%20INDC_271115.pdf, 2. See also Chapter 1, Sections 1.5.1.1 and 1.5.1.2 for an in-depth discussion of these strategies.
87 DFID 2009 Response to the *International Development* (Reporting and Transparency) *Act* 2006 https://assets.publishing.service.gov.uk/media/5a78cc83ed915d04220656e2/dfid-in-2009-10-revised-6-sept-2010.pdf, 34.
88 UNFCCC 2015 Nigeria's Intended Nationally Determined Contribution. https://unfccc.int/sites/default/files/NDC/2022-06/Approved%20Nigeria%27s%20INDC_271115.pdf,4.
89 UNFCCC 2015 Nigeria's Intended Nationally Determined Contribution. https://unfccc.int/sites/default/files/NDC/2022-06/Approved%20Nigeria%27s%20INDC_271115.pdf, 4.
90 Government of Nigeria, 2020 Nigeria's National Adaptation Plan Framework. https://napglobalnetwork.org/wp-content/uploads/2021/06/napgn-en-2020-Nigeria-National-Adaptation-Plan-NAP-Framework.pdf, 7.
91 Government of Nigeria, 2020 Nigeria's National Adaptation Plan Framework. https://napglobalnetwork.org/wp-content/uploads/2021/06/napgn-en-2020-Nigeria-National-Adaptation-Plan-NAP-Framework.pdf, 7. It is worth noting that the implementation of Nigeria's NDC falls under the mandate of the NCCPRS. The NDC is expected to reaffirm the strategic goals of the NCCPRS, which among others include the promotion of low-carbon and climate-resilient society. See Federal Government of Nigeria 2015. Nigeria's Intended Nationally Determined

Contribution. https://unfccc.int/sites/default/files/NDC/2022-06/Approved%20 Nigeria%27s%20INDC_271115.pdf, 18.

92 UNFCCC 2015 Nigeria's Intended Nationally Determined Contribution. https:// unfccc.int/sites/default/files/NDC/2022-06/Approved%20Nigeria%27s%20 INDC_271115.pdf, 18.

93 Government of Nigeria 2021 National Climate Change Policy for Nigeria https://climatechange.gov.ng/wp-content/uploads/2021/08/NCCP_NIGERIA_ REVISED_2-JUNE-2021.pdf, 11.

94 UNFCCC 2015 Nigeria's Intended Nationally Determined Contribution. https:// unfccc.int/sites/default/files/NDC/2022-06/Approved%20Nigeria%27s%20 INDC_271115.pdf, 18.

95 This alignment is exemplified by the NAP, NCCRPS, and the NASPA-CCN documents which encapsulate adaptation strategies and programmes in 13 sectors similar to those sectors reflected in Nigeria's NDC. These sectors include agriculture, energy, water, land use, transport, tourism, oil, and gas. See D Okoh AI Mulumo and O Orokpo 2020 Nigeria's intended Nationally Determined Contributions implementation and Transition to Net Zero carbon emission in post Paris era. *Dutse Journal of Economics and Development Studies,* 4(2),1–20.8–21.

96 Vanguard 21 November 2021 https://www.vanguardngr.com/2021/11/opinion-significance-of-nigerias-new-climate-change-law/.

97 See especially Section 20, CFRN, 1999.

98 Section 1, *NESREA (Establishment) Act* 25 of 2007.

99 Section 2, *NESREA Act.*

100 See Section 7 (c), *NESREA Act.*

101 See Chapter 2, Sections 2.3.1, 2.3.2 and 2.3.3.

102 Section 20, *NESREA Act.*

103 Section 23, *NESREA Act.*

104 Section 21, *NESREA Act.*

105 Section 7 *NESREA Act* provides that "in establishing such standards, the Agency shall take into consideration the use and value for public water supplies, propagation of fish and wildlife, recreational purposes, agricultural, industrial and other legitimate uses".

106 The *Revised African Convention on Conservation of Nature and Natural Resources* is one of such water conventions which Nigeria signed and ratified. The objectives of the Convention and how it facilitates adaptation in the context of agriculture at the regional level have been previously explained. See Chapter 2, Section 2.4.2.1.

107 Integrated water resources management techniques were explained as the process of making up for the shortfall in water through the installation or augmentation of power generation equipment for water purification plants, construction of integrated drainages, reservoirs, and recharge areas and the desalinisation of seawater. See Chapter 1, Section 1.5.

108 See paragraph, 3 (3) (a) of National Environmental Policy 1988, (as amended in 2009).

109 From 31 October 2021 to 12 November 2021, country parties to the UNFCCC met in Glasgow Scotland under the auspices of COP 26 to agree on net zero Carbon emission. The Government of Nigeria finally passed the Climate Change Bill 2021 to a law on 18 November 2021. It is believed that in "signing the law, the President of Nigeria has included the country into an elite group of countries that have enacted emissions-target legislation aiming to eliminate carbon emissions." See Vanguard 21 November 2021 https://www.vanguardngr.com/2021/11/ opinion-significance-of-nigerias-new-climate-change-law/; Premium Times 15 December 2021 https://www.premiumtimesng.com/news/more-news/501032-climate-change-act-life-changing-legislation-gbajabiamila.html.

110 Accordingly, a Bill in Nigeria becomes a law when it has passed through both houses of the National Assembly (House of Representatives and Senate) as required by sections 58 and 59 of the CFRN,1999; the Bill is then sent to the President of the Federal Republic of Nigeria for assent. Where the President withholds assent, the Bill becomes a law (without the need for a Presidential assent) through a passage for the second time by both Houses of the National Assembly. Section 58(4) and (5) of the CFRN, 1999.

111 Section 1, *Climate Change Act*.

112 Section 2, *Climate Change Act*.
The benefit of including public and private entities in the administration of climate change governance by the *Climate Change Act* is likely to lead to the facilitation of effective adaptation measures in Nigeria. At the same time, the active participation of both public and private entities in climate change governance could promote Nigeria's quest for achieving its climate-resilient development targets, through the mainstreaming of climate change concerns in national planning and development agenda. An explanation of how mainstreaming takes place in Nigeria is already provided above. See Section 4.2.2.2 above.

113 Already, there are climate desks or climate change units in these ministries. A ministry, department, or agency without a climate change desk is required by the *Climate Change Act*, to establish one. See section 22 of the *Climate Change Act* https://climatechange.gov.ng/structure/.

114 See section 1, *Climate Change Act*.

115 Section 3, *Climate Change Act*. Just as Kenya's *Climate Change Act*, the Nigerian *Climate Change Act* also makes the President of the Federal Republic of Nigeria the chairman and head of the National Climate Council. Membership of the NCC also includes the Minister for Agriculture and Rural Development. Meaning that agriculture is considered and represented in the NCC. See generally, section 5 of the *Climate Change Act*.

116 Section 19, *Climate Change Act*.

117 Section 20, *Climate Change Act*.

118 A National Climate Change Action Plan is accommodated by the *Climate Change Act*, and it is expected that an Action Plan must be formulated every five years. The Climate Action Plan (which is not different from the NCCPRS and NASPA-CCN explained above) is intended to serve as a mechanism "for establishing national goals, objectives and priorities on climate adaptation." See section 20 (4) (a) (ii) *Climate Change Act*.

119 According to section 27, "the Council shall promote and adopt nature-based solutions to reducing GHG emissions and mitigating climate change issues in Nigeria."

120 These were also explained as forms of adaptation in the context of agriculture. See Chapter 1, Section 1.5.1.

121 Chapter 1, Section 1.5.1.4.

122 See Chapter 3, Section 3.3.1.

123 OS Adesina and AM Loboguerrero 2021 Enhancing food security through climate-smart agriculture and sustainable policy in Nigeria. In JM Luetz and D Ayal (Eds.) *Handbook of climate change management: research, leadership, transformation*. Springer, 657–673.

124 Examples of these animal breeds include the *Bos indicus* (Cow) breeds, Boer goats, Myotonic goats, and the Karakul sheep. The hemolytic complement activity in these animals was observed to be higher as such; the animals have high resistance to climate change-induced diseases such as brucellosis, tuberculosis, and salmonellosis. See A Pal and AK Chakravarty 2020 Disease resistance for different livestock species. *Genetics and Breeding for Disease Resistance of Livestock*. Academic press,271.

125 See Chapter 1, Section 1.2.
126 J Pretty 2008 Agricultural sustainability: Concepts, principles and evidence. *Philosophical Transactions of the Royal Society B: Biological Sciences,* 363(1491), 447–465; CR Moreno and AE Racelis 2015 Attraction, repellence, and predation: Role of companion plants in regulating Myzus persicae (Sulzer) (Hemiptera: Aphidae) in organic kale systems of South Texas. *Southwestern Entomologist,* 40(1),1–14.
127 The PROACT CSA projects were launched in Nigeria in 2016 by Oxfam Nigeria in collaboration with the EU with the aim of enhancing food security by helping farmers boost agricultural productivity notwithstanding ensuing challenges and risks such as climate change. See PROACT 2016 Pro Resilience Action (PROACT) project, Nigeria. https://oxfamilibrary.openrepository.com/bitstream/handle/10546/621156/cs-improving-agricultural-practices-proact-120221-en.pdf;jsessionid=0CEA92FB9469A41984B6ACEA2824CA1B?sequence=1
128 Adamawa, Borno, and Yobe States have faced serious security challenges since 2009, when Boko Haram group started launching attacks on the inhabitants of these states. There were rampant cases of killing, maiming, and kidnappings among others. As a result, farming activities have been drastically affected. See I Emmanuelar 2015 Insurgency and humanitarian crises in Northern Nigeria: The case of Boko Haram. *African Journal of Political Science and International Relations,* 9(7), 284–296.
129 See FAO and ICRISAT 2019a *Climate-smart agriculture in Adamawa state of Nigeria* https://cgspace.cgiar.org/handle/10568/106101; FAO and ICRISAT 2019b *Climate-smart agriculture in Borno state of Nigeria* https://www.fao.org/3/ca5416en/CA5416EN.pdf,22.
130 See FAO and ICRISAT 2019a *Climate-smart agriculture in Adamawa state of Nigeria* https://cgspace.cgiar.org/handle/10568/106101.
131 See FAO, ICRISAT 2019b *Climate-smart agriculture in Borno state of Nigeria* https://www.fao.org/3/ca5416en/CA5416EN.pdf, 22.
132 I A Adedeji, N E Tiku, P R Waziri-Ugwu and S O Sanusi 2017 The effect of climate change on rice production in Adamawa State, Nigeria. *Agroeconomia Croatica,* 7(1),1–13.
133 H Richard and S Olajide 2020 Insight into agricultural education in Nigeria: A scoping study to identify potential areas of Netherlands support. *Netherlands Enterprise Agency, Ministry of foreign Affairs, Amsterdam,* 1–20.
134 OS Adesina and AM Loboguerrero 2021 Enhancing food security through climate-smart agriculture and sustainable policy in Nigeria. In JM Luetz and D Ayal (Eds.) *Handbook of climate change management: research, leadership, transformation.* Springer, 657–673.
135 OS Adesina and AM Loboguerrero 2021 Enhancing food security through climate-smart agriculture and sustainable policy in Nigeria. In JM Luetz and D Ayal (Eds.) *Handbook of climate change management: research, leadership, transformation.* Springer, 657–673.
136 I A Adedeji, N E Tiku, P R Waziri-Ugwu and S O Sanusi 2017 The effect of climate change on rice production in Adamawa State, Nigeria. *Agroeconomia Croatica,* 7(1), 1–13; FAO, ICRISAT 2019a Climate-smart agriculture in Adamawa state of Nigeria https://cgspace.cgiar.org/handle/10568/106101.
137 Government of Nigeria, 2020 https://napglobalnetwork.org/wp-content/uploads/2021/06/napgn-en-2020-Nigeria-National-Adaptation-Plan-NAP-Framework.pdf, 23.
138 RU Onyeneke, CA Nwajiuba, CC Emenekwe, A Nwajiuba, CJ Onyeneke, P Ohalete and UI Uwazie 2019 Climate change adaptation in Nigerian agricultural sector: A systematic review and resilience check of adaptation measures. *AIMS Agriculture and Food,* 4(4),967–1006. See also AA Enete and TA Amusa 2010 Challenges of

agricultural adaptation to climate change in Nigeria: A synthesis from the literature. Field Actions Science Reports. *The Journal of Field Actions*, 4.
139 See Chapter 1, Section 1.5.1.1.
140 See Chapter 1, Section 1.3.2.
141 UNFCCC 2010 National Environmental, Economic and Development Study (Needs) for Climate Change In Nigeria https://unfccc.int/files/adaptation/application/pdf/nigerianeeds.pdf, 11.
142 Federal Ministry of Agriculture Water resources and rural development, Abuja 2001 Agricultural Policy for Nigeria https://faolex.fao.org/docs/pdf/nig149296.pdf, 7.
143 See Federal Ministry of Agriculture Water resources and rural development, Abuja 2001 Agricultural Policy for Nigeria https://faolex.fao.org/docs/pdf/nig149296.pdf, 29.
144 Historically, River Basin Development Authorities were established in 1973 as an agency of the Federal Ministry of Water Resources, following the drought that ravaged the Sahel region of Africa. The aim of creating the River Basin Development Authorities was to create large water bodies (dams) that would assist Nigeria to cope with water scarcity and to prevent the reoccurrence of droughts. See for example, AT Ogundele 2019 River Basin development authorities in Nigeria: The neglected tools for national development. *International Journal of Advanced Academic Research Sciences, Technology and Engineering*, 5(3), 39–49.
145 See Federal Ministry of Agriculture Water resources and rural development, Abuja 2001 Agricultural Policy for Nigeria https://faolex.fao.org/docs/pdf/nig149296.pdf, 30.
146 The *Water Resources Act* of 1993 and the amended *Water Resources Act of* 2004 are aimed at developing and improving the quality and quantity of water resources in Nigeria, and they are discussed below.
147 See Chapter 1, Section 1.5.1.1.
148 On the expansion of livestock grazing reserves in Nigeria, the Agricultural Policy recommends the use of legal means (such as lease) to allocate more lands to grazers. Another means of expanding grazing lands which the Agricultural policy recommends is the zoning of agricultural lands to provide a quota for livestock production. For the purpose of ensuring availability of grazing lands, each state in Nigeria is expected to designate a minimum of 10% of total land in the state as grazing area. See Federal Ministry of Agriculture Water resources and rural development, Abuja 2001 Agricultural Policy for Nigeria https://faolex.fao.org/docs/pdf/nig149296.pdf, 10.
149 See Chapter 1, Section 1.3.2.
150 See Federal Ministry of Agriculture Water resources and rural development, Abuja 2001 Agricultural Policy for Nigeria https://faolex.fao.org/docs/pdf/nig149296.pdf, 10.
151 See Federal Ministry of Agriculture Water resources and rural development, Abuja 2001 Agricultural Policy for Nigeria https://faolex.fao.org/docs/pdf/nig149296.pdf, 10.
152 See Chapter 1, Section 1.2.
153 Federal Ministry of Agriculture Water resources and rural development, Abuja 2001 Agricultural Policy for Nigeria https://faolex.fao.org/docs/pdf/nig149296.pdf, 7.
154 See Federal Ministry of Agriculture Water resources and rural development, Abuja 2001 Agricultural Policy for Nigeria https://faolex.fao.org/docs/pdf/nig149296.pdf, 25.
155 Federal Ministry of Agriculture Water resources and rural development, Abuja 2001 Agricultural Policy for Nigeria https://faolex.fao.org/docs/pdf/nig149296.pdf, 27.

156 See Chapter 1, Section 1.5.1.3.
157 Federal Ministry of Agriculture Water resources and rural development, Abuja 2001 Agricultural Policy for Nigeria https://faolex.fao.org/docs/pdf/nig149296.pdf, 21.
158 Federal Ministry of Agriculture Water resources and rural development, Abuja 2001 Agricultural Policy for Nigeria https://faolex.fao.org/docs/pdf/nig149296.pdf, 21.
159 OJ Offiong and R Dibie 2014 Environmental policy and issues in Nigeria. In *Comparative Perspectives on environmental policies and issues,* Routledge, 268–300.
160 PA Okuneye and IA Ayinde 2011 The Nigerian agricultural policy inadequacy: The way forward. *Nigerian Journal of Agricultural Economics,* 2(1), 1–16.
161 Government of Nigeria 2020 Nigeria's National Adaptation Plan Framework https://napglobalnetwork.org/wp-content/uploads/2021/06/napgn-en-2020-Nigeria-National-Adaptation-Plan-NAP-Framework.pdf, 3. The NARF also adopts a mainstreaming approach to support the principles of adaptive management and participatory engagement. In this regard, the NARF dedicates a whole chapter to the concept of mainstreaming. See Chapter 5, NARF 2014.
162 Adegoke et al. 2014 New framework for building resilience in Nigeria's agriculture value-chains through climate smart policies, adaptation programs and actions. http://uest.ntua.gr/adapttoclimate/proceedings/full_paper/Adegoke_et_al.pdf, 6
163 J Adegoke, A Araba and C Ibe 2014 Agricultural research and technology development for climate resilient agriculture. National Agricultural Resilience Framework. *Federal Ministry of Agriculture and Rural Development, Abuja*. https://boris.unibe.ch/62564/1/Nigeria%27s%20Changing%20Cliamte.pdf, 10–14.
164 J Adegoke, A Araba and C Ibe 2014 Agricultural research and technology development for climate resilient agriculture. National Agricultural Resilience Framework. *Federal Ministry of Agriculture and Rural Development, Abuja*.https://boris.unibe.ch/62564/1/Nigeria%27s%20Changing%20Cliamte.pdf, 10–14.
165 D Bours, C McGinn and P Pringle 2014 *Monitoring and evaluation for climate change adaptation and resilience: A synthesis of tools, frameworks and approaches.* SEA Change Community of Practice, and UK Climate Impacts Programme, 15.
166 J Adegoke, A Araba and C Ibe 2014 Agricultural research and technology development for climate resilient agriculture. National Agricultural Resilience Framework. *Federal Ministry of Agriculture and Rural Development, Abuja*. https://boris.unibe.ch/62564/1/Nigeria%27s%20Changing%20Cliamte.pdf, 10–14.
167 Vision 20:2020 was aimed at "promoting efficient use of human and natural resources to achieve rapid economic growth; and to translate the economic growth into equitable social development for all citizens." *The Guardian* 10 September 2020 https://guardian.ng/news/nigeria-abandons-vision-202020-dreams-agenda-2050/.
168 Nigeria Long-term vision https://unfccc.int/sites/default/files/resource/Nigeria_LTS1.pdf, 19.
169 States located in northern part of Nigeria especially Sokoto, Zamfara, Katsina, Jigawa, Kano Yobe, and Borno are areas that are mostly susceptible to droughts. See D Bashir 1976 Sustainable management of drought and desertification in Nigeria. *Change*, 78 (1993/95).
170 J Adegoke, A Araba and C Ibe 2014 Agricultural research and technology development for climate resilient agriculture. National Agricultural Resilience Framework. *Federal Ministry of Agriculture and Rural Development, Abuja*. https://boris.unibe.ch/62564/1/Nigeria%27s%20Changing%20Cliamte.pdf, 10–14.
171 J Adegoke, A Araba and C Ibe 2014 Agricultural research and technology development for climate resilient agriculture. National Agricultural Resilience

Framework. *Federal Ministry of Agriculture and Rural Development, Abuja.* https://boris.unibe.ch/62564/1/Nigeria%27s%20Changing%20Cliamte.pdf, 10–14.

172 J Adegoke, A Araba and C Ibe 2014 Agricultural research and technology development for climate resilient agriculture. National Agricultural Resilience Framework. *Federal Ministry of Agriculture and Rural Development, Abuja.* https://boris.unibe.ch/62564/1/Nigeria%27s%20Changing%20Cliamte.pdf, 10–14.

173 J Adegoke, A Araba and C Ibe 2014 Agricultural research and technology development for climate resilient agriculture. National Agricultural Resilience Framework. *Federal Ministry of Agriculture and Rural Development, Abuja.*https://boris.unibe.ch/62564/1/Nigeria%27s%20Changing%20Cliamte.pdf, 10–14.

174 J Adegoke, A Araba and C Ibe 2014 Agricultural research and technology development for climate resilient agriculture. National Agricultural Resilience Framework. *Federal Ministry of Agriculture and Rural Development, Abuja.* https://boris.unibe.ch/62564/1/Nigeria%27s%20Changing%20Cliamte.pdf, 10–14.

175 These principles were previously explained. See Chapter 2, Section 2.4.2.1.

176 Federal Republic of Nigeria 2004 National Water Policy https://faolex.fao.org/docs/pdf/nig158231.pdf, 10.

177 Federal Republic of Nigeria 2004 National Water Policy https://faolex.fao.org/docs/pdf/nig158231.pdf, 6.

178 Federal Ministry of Environment 2011 National Environmental, Economic and Development Study (NEEDS) for Climate Change in Nigeria https://unfccc.int/files/adaptation/application/pdf/nigerianeeds.pdf, 11.

179 See item 64 of the exclusive legislative list, part 1 of the second schedule to the CFRN, 1999 (as amended).

180 Federal Ministry of Water Resources 2016 National Water Resources Policy https://faolex.fao.org/docs/pdf/nig181288.pdf, 1–45.

181 Federal Ministry of Water Resources 2016 National Water Resources Policy https://faolex.fao.org/docs/pdf/nig181288.pdf, 12.

182 Federal Ministry of Water Resources 2016 National Water Resources Policy https://faolex.fao.org/docs/pdf/nig181288.pdf, 14.

183 Federal Ministry of Water Resources 2016 National Water Resources Policy https://faolex.fao.org/docs/pdf/nig181288.pdf, 14.

184 Federal Ministry of Water Resources 2016 National Water Resources Policy https://faolex.fao.org/docs/pdf/nig181288.pdf, 15.

185 Federal Ministry of Water Resources 2016 National Water Resources Policy https://faolex.fao.org/docs/pdf/nig181288.pdf, 15.

186 Federal Ministry of Water Resources 2016 National Water Resources Policy https://faolex.fao.org/docs/pdf/nig181288.pdf, 16. Integrated water resources management has been explained previously in Chapter 1, Section 1.5.1.2, and revisited in Section 4.2.3.1 above.

187 See Key Policy Statement 10 NWRP, Federal Ministry of Water Resources 2016 National Water Resources Policy https://faolex.fao.org/docs/pdf/nig181288.pdf, 26.

188 Federal Ministry of Water Resources 2016 National Water Resources Policy https://faolex.fao.org/docs/pdf/nig181288.pdf, 26.

189 See Chapter 1, Section 1.3.2.

190 See Federal Government of Nigeria 2018 National Drought Plan https://www.unccd.int/sites/default/files/country_profile_documents/1%2520FINAL_NDP_Nigeria.pdf, 35.

191 The UNCCD was adopted in 1994. Most African and developing countries are parties to the Convention. The UNCCD acknowledges desertification and drought as environmental problems with global implication on sustainable development given that they serve as vectors for social problems such as poverty,

poor health and nutrition, lack of food security, and displacement of persons. As such the UNCCD recommends for "a joint action by the international community to combat desertification and/or mitigate the effects of drought, Mindful of the role that National Governments can play on the local implementation of action programmes in affected areas." See preamble to the UNCCD 1994 https://catalogue.unccd.int/936_UNCCD_Convention_ENG.pdf, 1–3.

192 See article 10 UNCDD.

193 Federal Government of Nigeria 2018 National Drought Plan. https://www.unccd.int/sites/default/files/country_profile_documents/1%2520FINAL_NDP_Nigeria.pdf, 97.

194 Federal Government of Nigeria 2018 National Drought Plan. https://www.unccd.int/sites/default/files/country_profile_documents/1%2520FINAL_NDP_Nigeria.pdf, 65.

195 Government of Nigeria 2018 National Drought Plan https://www.unccd.int/sites/default/files/country_profile_documents/1%2520FINAL_NDP_Nigeria.pdf, 21.

196 Government of Nigeria 2021 National Climate Change Policy for Nigeria https://climatechange.gov.ng/wp-https://climatechange.gov.ng/wp-content/uploads/2021/08/NCCP_NIGERIA_REVISED_2-JUNE-2021.pdf, 23.

197 It is expected that when these strategies are implemented, sustainability of the agricultural sector will be guaranteed, as there will be accuracy in predicting droughts, as well less livestock loss, minimal migration as a result of drought, etc. Federal Government of Nigeria 2018 National Drought Plan. https://www.unccd.int/sites/default/files/country_profile_documents/1%2520FINAL_NDP_Nigeria.pdf, 91.

198 These methods were practiced in 1995 for instance when droughts led to massive crop failure and loss of livestock in Nigeria. See S J Chang 2013 Solving the problem of carbon dioxide emissions. *Forest policy and economics,* 35, 92–97.

199 See Federal Ministry of Agriculture Water resources and rural development, Abuja 2001 Agricultural Policy for Nigeria https://faolex.fao.org/docs/pdf/nig149296.pdf, 27.

200 Government of Nigeria 2021 National Climate Change Policy for Nigeria https://climatechange.gov.ng/wp-https://climatechange.gov.ng/wp-content/uploads/2021/08/NCCP_NIGERIA_REVISED_2-JUNE-2021.pdf, 24.

201 Government of Nigeria 2021 National Climate Change Policy for Nigeria https://climatechange.gov.ng/wp-https://climatechange.gov.ng/wp-content/uploads/2021/08/NCCP_NIGERIA_REVISED_2-JUNE-2021.pdf, 24.

202 Government of Nigeria 2021 National Climate Change Policy for Nigeria https://climatechange.gov.ng/wp-https://climatechange.gov.ng/wp-content/uploads/2021/08/NCCP_NIGERIA_REVISED_2-JUNE-2021.pdf, 24. See also S I Omofonmwan and G I Osa-Edoh 2008 The challenges of environmental problems in Nigeria. *Journal of Human Ecology,* 23(1), 53–57.

203 See Chapter 1, Section 1.5.1.2.

204 Government of Nigeria 2021 National Climate Change Policy for Nigeria https://climatechange.gov.ng/wp-https://climatechange.gov.ng/wp-content/uploads/2021/08/NCCP_NIGERIA_REVISED_2-JUNE-2021.pdf, 24.

205 Government of Nigeria 2021 National Climate Change Policy for Nigeria https://climatechange.gov.ng/wp-https://climatechange.gov.ng/wp-content/uploads/2021/08/NCCP_NIGERIA_REVISED_2-JUNE-2021.pdf, 24.

206 UNFCCC 2010 National Environmental, Economic And Development Study (Needs) For Climate Change In Nigeria https://unfccc.int/files/adaptation/application/pdf/nigerianeeds.pdf, 1–45.

207 See *National Environmental (Soil Erosion and Flood Control) Regulations,* No. 39 Lagos, 2011.

208 Examples of these types of unconventional land uses and land-disturbing activities were previously given in Chapter 1, Section 1.2.
209 See Chapter 2, Section 2.2.
210 Section 4(2) of the CFRN for instance provides that: "The National Assembly shall have power to make laws for the peace, order and good government of the Federation or any part thereof with respect to any matter included in the Exclusive Legislative List set out in Part I of the Second Schedule to this Constitution."
211 Section 315(1) (a) CFRN, 1999.
212 No. 101 of 1993.
213 Section 4.
214 The amendment of the *Water Resources Act* was done pursuant to the provisions of Sections 4 and 5 CFRN, 1999, which gives the legislative arm of government and executive arm of government the powers to make laws and execute laws, respectively.
215 Section 1.
216 Section 1(1) (b).
217 Section 1(1) (c).
218 Section 5 (ii) (h).
219 See Section 4.2.3.1 above.
220 CAP R9, Laws of the Federation of Nigeria 2004.
221 There is a total of eight river basins authorities in Nigeria, which include the Benue, Delta and Cross Rivers, the Imo-Anambra, Hadejia-Chad, Sokoto-Rima, Niger, Owena and Ogun, and Osun Basins. See E O Ita 1994 Aquatic plants and wetland wildlife resources of Nigeria. CIFA Occasional Paper. *No. 21. Rome, FAO* 52.
222 Section 4 *RBDA Act*.
223 See Chapter 1, Section 1.5.1.
224 Section 4 *RBDA Act*.
225 Section 4 *RBDA Act*.
226 Cap L5, Laws of the Federation of Nigeria 2004.
227 See Long title to the *Land Use Act*.
228 Overriding public interest could mean the interest of the State Government to construct schools, park, or to expand a local road network. In this instance, the right to own a land is not absolute. In other words, the State Governor (as the custodian of all lands in the State) can revoke the right to own and occupy a land under section 28 of the *Land Use Act*. However, the provisions of section 28 of the *Land Use Act* seem to be in conflict with the provisions of sections 43 and 44 of the CFRN, which guarantees the right acquires and own immovable property (land) in any part of Nigeria, and the right to be compensated where there is compulsory acquisition of an immovable property. Therefore, the provisions of section 28 of the *Land Use Act* were challenged in a court of law. See for example, the cases of *Alhaji Tsoho Dan Amale v Sokoto Local Government and Others* (unreported) suit no SC290/2001 of 20 January 2012 para, 4–6; *Goldmark Nigeria Limited and Others v Ibafon Company Limited and Others* (unreported) suit no SC421/2001 of 23 March 2012 para, 17–18; *Ohochukwu v Attorney General of Rivers State and Others* (unreported) suit no SC207/2004 of 17 February 2012 para, 1–3.
229 See Chapter 2, Section 2.4.2.2.
230 The PTD is provided by section 1 of the *Land Use Act* which vest all lands in the State in the hands of the Governor, who "holds and administers all lands in trust for the use and common benefit of all Nigerians in accordance with the provisions of this Act."
231 It is believed that when land is held in trust by the government (on behalf of the citizen) for instance, there could be equitable use and adequate protection

by and on behalf of present and future citizens. See JS Alao and RB Shuaibu 2013 Agroforestry practices and concepts in sustainable land use systems in Nigeria. *Journal of Horticulture and Forestry,* 5(10), 156–159. The PTD was initially applied to a few sets of natural resources which consist of submerged lands and shellfish beds. The doctrine was later expanded to include wildlife, oceans, and ecosystem services. See RJ Lazarus 1985 Changing conceptions of property and sovereignty in natural resources: Questioning the public trust doctrine. *Iowa Law Review,* 71, 631; MC Blumm and RD Guthrie 2011 Internationalizing the public trust doctrine: Natural law and constitutional and statutory approaches to fulfilling the Saxion vision. *UCDL Review,* 45(3), 741.

232 See Chapter 1, Section 1.3.2.

233 See section 43 of the *Land Use Act* which prohibits the erection of any building, wall, fence, or other structure or the cultivation of crops or trees, which obstruct land in urban areas, and makes an offender liable to a term of imprisonment or a fine.

234 A Otubu 2018 The *land use Act* and land administration in 21st century Nigeria: Need for reforms. *Journal of Sustainable Development Law and Policy,* 9(1),100–108.

235 See Section 4.3.1.5 above.

236 See Chapter 2, Section 2.2.

237 *National Environmental (Soil Erosion and Flood Control) Regulations* No 39, vol.98.

238 Section 2(1).

239 See Chapter 1, Section 1.5 generally.

240 Section 20(1), *National Environmental (Soil Erosion and Flood Control) Regulations.*

241 Regulation 14 (1).

242 See Chapter 1, Section 1.5.1.2. See also F Alcoforado 2018 Flood control and its management. *Journal of Atmospheric & Earth Sciences,* 1(1), 1–13.

243 Section 20(2) (c), *National Environmental (Soil Erosion and Flood Control) Regulations.*

244 Section 2(2), *National Environmental (Soil Erosion and Flood Control) Regulations.*

245 Section 2(2), *National Environmental (Soil Erosion and Flood Control) Regulations.*

246 Section 2(2), *National Environmental (Soil Erosion and Flood Control) Regulations.*

247 Section 1, *National Environmental (watershed, Mountainous, Hilly, and Catchment Areas) Regulations* 2009 Government Notice No. 280.

248 Section 2, *National Environmental (Wetlands River Banks and Lake Shores) Regulations,* 2009, Government Notice No. 279.

249 MA Ibrahim, BY Abubakar and ML Balarabe 2019 Sequestrated carbon content in tree species and diurnal temperature influence for adaptive climate change resilience in Nigeria. In *Handbook of climate change resilience*. Springer, 45.

250 Section 1, *NAIC Act,* Cap N89.

251 Section 6 of the *NAIC Act.*

252 Section 7, *NAIC Act.*

253 See Chapter 1, Section 1.5.1.3.

254 See Sections 4.3.1.1 and 4.3.1.2 above.

255 According to Section 7 of the *NAIC Act,* the Scheme shall apply to crops, livestock, and other agricultural items and enterprises, on a commercial basis only.

256 See Section 4.2.4 above.

257 UNFCCC 2010 National Environmental, Economic And Development Study (Needs) For Climate Change In Nigeria https://unfccc.int/files/adaptation/application/pdf/nigerianeeds.pdf, 19.

258 E Oladipo 2021 Analysis of the adaptation components that could be included in Nigeria's Nationally Determined Contributions (NDCs) Revision. In *Center*

for Climate and Development. Alex Ekwueme Federal University Ndufu-Alike Nigeria Discussion Paper, 6–10; EG Orie 2021 Climate change adaptation mechanism for sustainable development Goal 1 in Nigeria: Legal Imperative. In *African handbook of climate change adaptation*. Springer International Publishing, 1897–1917.

259 F A Adesina, T O Odekunle, O Ajayi, O Eludoyin, O Babatimehin, A Dami and O T Adetiloye2008 Adaptation strategies of action for Nigeria. *A working document prepared for the special climate change unit of the Federal Ministry of environment,* Housing and Urban Development, Heinrich Boll Stiftung, Nigeria.

260 RU Onyeneke, CU Nwajiuba, B Tegler and CA Nwajiuba 2020 Evidence-based policy development: National adaptation strategy and plan of action on climate change for Nigeria (NASPA-CCN). In *African handbook of climate change adaptation*, Cham: Springer International Publishing, 2547–2564.

261 Government of Nigeria, 2020 Nigeria's National Adaptation Plan Framework: https://napglobalnetwork.org/wp-content/uploads/2021/06/napgn-en-2020-Nigeria-National-Adaptation-Plan-NAP-Framework.pdf, 14.

262 K Shehu, Auwal, and AA Dahuwa 2018 A review of climate change: Its impacts and governance for enhancing Nigeria's environmental resources. *Dutse Journal of Pure and Applied Sciences (DUJOPAS)*, 4(2), 1–10; Y Tarfa and F Adesina 2020 Making Adaptation More Effective in Nigeria: Nigeria's NAP process. https://napglobalnetwork.org/2020/06/making-adaptation-more-effective-in-nigeria-nigerias-nap-process

263 The process of enacting the *Climate Change Act* was initiated at the Nigerian National Assembly since 2007. See S Matemilola, OH Adedeji, I Elegbede and F Kies 2019 Mainstreaming climate change into the EIA process in Nigeria: Perspectives from projects in the Niger Delta Region. *Climate*, 7(2), 29; S Ojo, H Mensah, E Albrecht and B Ibrahim 2020 Adaptation to climate change effects on water resources: Understanding institutional barriers in Nigeria. *Climate*, 8(11), 134.

264 EM Okon, BM Falana, SO Solaja, SO Yakubu, OO Alabi, BT Okikiola and AB Edeme 2021 Systematic review of climate change impact research in Nigeria: Implication for sustainable development. *Heliyon*, 7(9), e07941.

References

Abazu C and Alabi A 2019 Fostering urban environment adaptation to climate change in Nigeria. *COOU African Journal of Environmental Research*, 2(1), 43–55.

Abdulkadir AB 2014 The right to a healthful environment in Nigeria: A review of alternative pathways to environmental justice in Nigeria. *Journal of Sustainable Development Law and Policy*, 3(1), 118–131.

Adedeji et al. 2017 The effect of climate change on rice production in Adamawa State, Nigeria. *Agroeconomia Croatica*, 7(1), 1–13.

Adegoke Ibe and Araba 2014 New framework for building resilience in Nigeria's agriculture value-Chains through climate smart policies, adaptation programs and actions https://uest.ntua.gr/adapttoclimate/proceedings/full_paper/Adegoke_et_al.pdf (accessed 12 June 2019).

Adesina et al. 2008 *Working document prepared for the Special climate change unit of the Federal Ministry of environment*. Housing and Urban Development, Heinrich Boll Stiftung, Nigeria.

Adesina and Odekunle 2011 Climate change and adaptation in Nigeria: Some background to Nigeria's response 1 International Conference on Environmental and Agricultural Engineering, 1146–1154.

Adesina OS and Loboguerrero AM 2021 Enhancing food security through climate-smart agriculture and sustainable policy in Nigeria. In JM Luetz and D Ayal (Eds.) *Handbook of climate change management*. Springer, 657–673.

Afinowi O 2018 *An outline and critical assessment of the role of planning laws in the regulatory framework of climate change adaptation in South Africa and Nigeria*. PhD thesis University of Cape Town.

Ako R, Stewart N and Ekhator EO 2016 Overcoming the (non) justiciable Conundrum: The doctrine of Harmonious construction and the interpretation of the right to a healthy environment in Nigeria. In Justiciability of Human Rights Law in Domestic Jurisdictions, 123–141.

Alao JS and Shuaibu RB 2013 Agroforestry practices and concepts in sustainable land use systems in Nigeria. *Journal of Horticulture and Forestry*, 5(10), 156–159.

Alcoforado FAG 2018 Flood control and its management. *Journal of Atmospheric and Earth Sciences*, 1(1), 1–13.

Angie Dazé 2019 International Institute for sustainable development 2019 why gender matters in climate change adaptation https://www.iisd.org/blog/gender-climate-change (accessed 10 November 2020).

Bashir D 1976 *Sustainable management of drought and desertification in Nigeria1*. Change, 1993–1995.

Blumm MC and Guthrie RD 2011 Internationalizing the public trust doctrine: Natural law and constitutional and statutory approaches to fulfilling the Saxion vision. *UCDL Review*, 45(3), 741.

Bours D, McGinn C and Pringle P 2014 *Monitoring and evaluation for climate change adaptation and resilience: A synthesis of tools, frameworks and approaches*. SEA Change Community of Practice, and UK Climate Impacts Programme, 1–87.

Enete AA and Amusa TA 2010 Challenges of agricultural adaptation to climate change in Nigeria: A synthesis from the literature. *Field Actions Science Reports. The Journal of Field Actions*, 4.

Ibe et al. 2014 *Agricultural research and technology development for climate resilient agriculture. National Agricultural Resilience Framework*. Ministry of Agriculture, Government of Nigeria. https://boris.unibe.ch/62564/1/Nigeria%27s%20Changing%20Cliamte.pdf (accessed 14 July 2022).

Ibrahim MA, Abubakar BY and Balarabe ML 2019 Sequestrated carbon content in tree species and diurnal temperature influence for adaptive climate change resilience in Nigeria. In *Handbook of climate change resilience*. Springer, 3–27.

Laws J 2000 Monism and dualism. *La Revue Administrative*, 53(2), 18–22.

Lazarus RJ 1985 Changing conceptions of property and sovereignty in natural resources: Questioning the public trust doctrine. *Iowa Law Review*, 71, 631.

Malumfashi SL 2021 Wealth out of wastes: Sociological examinations of life styles of waste Dealers in Northwestern Nigeria. *Al-Qalam University Journal of Sociology and Anthropological Studies*, 1.

Matemilola S, Adedeji OH, Elegbede I and Kies F 2019 Mainstreaming climate change into the EIA process in Nigeria: Perspectives from projects in the Niger Delta Region. *Climate*, 7(2), 29.

Möhner A 2018 The evolution of adaptation metrics under the UNFCCC and its Paris Agreement. In L Christiansen, G Martinez, and P Naswa (Eds.), *Adaptation metrics: Perspectives on measuring, aggregating and comparing adaptation results*, 15.

Mongi H, Majule AE and Lyimo JG 2010 Vulnerability and adaptation of rain fed agriculture to climate change and variability in semi-arid Tanzania. *African Journal of Environmental Science and Technology*, 4(6), 371–381.

Moreno CR and Racelis AE 2015 Attraction, repellence, and predation: Role of companion plants in regulating Myzus persicae (Sulzer) (Hemiptera: Aphidae) in organic kale systems of south Texas. Southwestern Entomologist, 40(1), 1–14.

Offiong OJ and Dibie R 2014 Environmental policy and issues in Nigeria. In *Comparative Perspectives on environmental policies and issues*. Routledge, 4–326.

Ogbonna CU and Albrecht E 2015 Strategic environmental assessment as a tool to integrate climate change adaptation: A perspective for Nigeria. In FW Leal (Ed.) Handbook of climate change adaptation. Springer, 1239–1260.

Ogundele AT 2019 River Basin development authorities in Nigeria: The neglected tools for national development. *International Journal of Advanced Academic Research Sciences, Technology and Engineering*, 5(3), 39–49.

Ojo S, Mensah H, Albrecht E and Ibrahim B 2020 Adaptation to climate change effects on water resources: Understanding institutional barriers in Nigeria. *Climate*, 8(11), 134.

Okoh AI Mulumo D and Orokpo O 2020 Nigeria's intended Nationally Determined Contributions implementation and Transition to Net Zero carbon emission in post Paris era. *Dutse Journal of Economics and Development Studies*, 4(2), 1–21.

Okon EM, Falana BM, Solaja SO, Yakubu SO, Alabi OO, Okikiola BT and Edeme AB 2021 Systematic review of climate change impact research in Nigeria: Implication for sustainable development. *Heliyon*, 7(9), e07941.

Okuneye PA and Ayinde IA 2011 The Nigerian agricultural policy inadequacy: The way forward. *Nigerian Journal of Agricultural Economics*, 2, 1–16.

Oladipo E 2021 Analysis of the adaptation components that could be included in Nigeria's Nationally Determined Contributions (NDCs) Revision. In *Center for Climate and Development*. Alex Ekwueme Federal University Ndufu-Alike Nigeria Discussion Paper, 1–14.

Olajide BE, Quadri MO and Ojakorotu V 2018 Climate change, human security and good governance in Nigeria. *African Renaissance*, 15(3), 173–196.

Onyeneke RU, Nwajiuba CA, Emenekwe CC, Nwajiuba A, Onyeneke CJ, Ohalete P and Uwazie UI 2019 Climate change adaptation in Nigerian agricultural sector: A systematic review and resilience check of adaptation measures. *AIMS Agriculture and Food*, 4(4), 967–1006.

Onyeneke RU, Nwajiuba CU, Tegler B and Nwajiuba CA 2020 Evidence-based policy development: National adaptation strategy and plan of action on climate change for Nigeria (NASPA-CCN). In *African handbook of climate change adaptation*, 1–18.

Orie EG 2021 Climate change adaptation mechanism for sustainable development Goal 1 in Nigeria: Legal Imperative. In *African handbook of climate change adaptation*. Springer International Publishing, 1897–1917.

Otubu 2018 The land use act and land administration in 21st century Nigeria: Need for reforms. *Journal of Sustainable Development Law and Policy*, 9(1), 80–108.

Pal A and Chakravarty AK 2020 Disease resistance for different livestock species. *Genetics and Breeding for Disease Resistance of Livestock*, Academic press, 271.

Pretty J 2008 Agricultural sustainability: Concepts, principles and evidence. *Philosophical Transactions of the Royal Society B: Biological Sciences*, 363(1491), 447–465.

Richard H and Olajide S 2020 Insight into agricultural education in Nigeria: A scoping study to identify potential areas of NL support Netherlands Enterprise Agency https://www.rvo.nl/sites/default/files/2020/10/Insight-into-agricultural-education-in-Nigeria.pdf 20.

Shehu K, Auwal IN and Dahuwa AA 2018 A review of climate change: Its impacts and governance for enhancing Nigeria's environmental resources. *Dutse Journal of Pure and Applied Sciences (DUJOPAS)*, 4(2), 1–10.

Tarfa Y and Adesina F 2020 Making adaptation more effective in Nigeria: Nigeria's NAP process. https://napglobalnetwork.org/2020/06/making-adaptation-more-effective-in-nigeria-nigerias-nap-process/(accessed 16 January 2022).

Woodley E 2011 Building Nigeria's response to climate change: Pilot projects for community-based adaptation in Nigeria. In *Experiences of climate change adaptation in Africa 297–315*. Springer.

5 South Africa

The policy and legal framework for adaptation and agriculture

5.1 Introduction

As has been explained in Chapter 2, effective adaptation measures whether in a broad context or in the context of agriculture hinges on an effective and well-established regulatory framework.[1] Therefore, as was done in Chapters 3 and 4, this chapter will also investigate key climate change and agriculture-related policies and laws in South Africa – to determine whether they contain the type of provisions that could enable and promote adaptation in the context of agriculture.

Adaptation in South Africa is broadly regulated by climate change-related policies that were formulated based on the guiding principles of the UNFCCC, the Kyoto Protocol, and the Paris Agreement (which the country signed and ratified), as well as the provisions of the *Constitution of the Republic of South Africa*, 1996 (hereinafter referred to as the *Constitution)*.[2] These policies are exemplified by the National Climate Change Response Strategy (NCCRS),[3] National Climate Change Response Policy White Paper (NCCRP),[4] National Climate Change Adaptation Strategy (NCCAS),[5] the National Development Plan (NDP),[6] and a Nationally Determined Contributions (NDCs).[7] In addition, South Africa's commitment to adapting to the impacts of climate change is demonstrated by the enactment of laws that are intended to protect the environment from various hazards, including climate change. These laws have been enacted pursuant to the country's *Constitution*, and they include the *National Environmental Management Act* (*NEMA*),[8] the *Disaster Management Act* (*DMA*),[9] the *Carbon Tax Act* (*CTA*),[10] and the *Climate Change Act*.[11]

Because agriculture is heavily impacted by climate change, the national government formulated several agriculture-based policies and enacted some agriculture-related laws in order to assist agricultural practices in coping with the impacts of climate change.[12] Agricultural policies which seek to support adaptation include the draft Climate-Smart Agriculture Strategic Framework (CSASF),[13] Climate Change Sector Plan for Agriculture, Forestry, and Fisheries (CCSPAFF),[14] and the Second National Action Programme for South Africa to Combat Desertification, Land Degradation and the effects of Drought (NAP-CDLDD).[15] Agriculture-related laws in South Africa consist of the *Water*

DOI: 10.4324/9781003536437-5

Services Act (*WSA*),[16] the *National Water Act* (*NWA*),[17] the *Conservation of Agricultural Resources Act* (*CARA*),[18] and the Draft *Conservation of Agricultural Resources Amendment Act*[19] as well as the *Spatial Planning and Land Use Management Act* (*SPLUMA*).[20]

Consequently, it is relevant to also ask: how do these South African climate change and agriculture-related policies and laws provide for adaptation measures in the context of agriculture, and are there observable gaps in this regulatory framework?

5.2 Regulatory framework for adaptation in South Africa

South Africa is a non-Annex 1 country of the UNFCCC, and as previously discussed, non-Annex 1 countries are not bound by any specific emission targets.[21] Nevertheless, the country is prepared to implement climate change measures (mitigation and adaptation) as informed by the country's ratification of the UNFCCC, the Kyoto Protocol, and the Paris Agreement.[22] As a signatory to these international legal instruments, South Africa is among others expected to:

> [p]repare and periodically update a national inventory of greenhouse gas emissions and sinks; formulate and implement national and, where appropriate, regional programmes to mitigate climate change and facilitate adequate adaptation to climate change....[23]

For example, through its NDC, South Africa indicated its readiness to lower greenhouse gas (GHG) emissions in the country to net zero emissions by 2050.[24] South Africa's commitment to minimising, as well as coping with the impacts of climate change, is also demonstrated by the provisions of the country's *Constitution*, which led to the formulation of a series of climate change policies and the enactment of climate change-related laws. These climate change policies and laws which are already mentioned above are implemented by different, but interrelated levels of government in South Africa. Because South Africa operates a semi-federal structure of governance, each of the spheres of government (national, provincial, and local government) is saddled with the role of governing specific aspects contained in the *Constitution*.[25] In the words of Taylor et al.[26]: "All three spheres of government are noted as having key roles to play in adaptation, involving both sectoral interventions as well as measures that require co-ordination between sectors and departments".

At the national level, it is the Department of Forestry, Fisheries, and the Environment that implements provisions of the UNFCCC, the *NEMA*, and the *CTA*.[27] As well, it is at the national level that institutional structures for mainstreaming climate change concerns into development planning exist. In this regard, the Presidential Climate Commission (PCC) advises the national government on how to achieve "a just and equitable transition towards a low-emissions and climate-resilient economy"[28] in the country through mainstreaming climate change response strategies across government departments in South Africa.[29]

At the provincial level, there is a Department for Environmental Affairs, or a related department, which is responsible for integrating climate change responses into other sectors within a province. For example, in the North West province, the Department of Economic Development, Environment, Conservation, and Tourism is saddled with the task of promoting inclusive economic growth for the people of the North West province.[30] Inclusive economic growth is also achieved through agriculture. However, as explained above, agricultural-based inclusive economic growth appears to be hampered by some barriers, especially those created by climate change.[31] Therefore, adaptation strategies could be considered as one of the solutions adopted by the North-West Department of Economic Development, Environment, Conservation, and Tourism in addressing climate change-related challenges to inclusive economic growth in the province.

Given that the realisation of the right to a safe and healthy environment (as enshrined in the *Constitution*) is threatened by climate change, it is the function of local governments to oversee the integration of climate change responses (adaptation) into all economic sectors at the local level.[32] In other words, local governments are integral facilitators of adaptation due to their unique position to take action in this sphere.[33] There is a plethora of legal opinions on the role of local governments in promoting a safe and healthy environment in South Africa through adaptation. Du Plessis and Kotzé for instance argue that "the *Constitution* grants almost limitless scope for specifically local government's intervention in responding to climate change".[34] Taylor et al.[35] further assert that a purposive interpretation of the *Constitution* in the context of climate change governance could provide the legal basis:

> [o]n which municipalities can and should build and implement a comprehensive set of climate change adaptation measures in order to ensure the safety and sustainability of the communities and local environment within its boundaries.

In this regard, local governments use some established mechanisms and a variety of tools at their disposal, especially integrated development plans (IDPs) to mount a flexible response to the impacts of climate change.[36] Through IDPs, local governments will be able to find the most suitable long-term solutions to climate change that appear to have negative impacts on development.[37] This is because an IDP is considered a strategic planning instrument that also considers all existing problems (social, economic, and environmental) against resources that are relevant to development, such as water, land, and agricultural products.[38] In addition, IDPs outline the types of infrastructure and services that are required for development and give insights into how infrastructure and services are to be used without compromising environmental standards.[39] For this reason, IDPs recognise adaptation strategies that could sustain environmental standards, such as disaster risk reduction and management (early warning systems) and adaptive water interventions (planning and conservation methods).[40]

Even though environmental protection is a concurrent matter that is jointly regulated by all three spheres of government in South Africa,[41] the enactment of climate change laws remains the sole responsibility of the national government.[42] Likewise, it is the specific function of the national government to formulate climate change policies and to establish an institutional forum for climate change governance in South Africa.[43] The Climate Change and Air Quality Management (CCAQM) Agency[44] and the Inter-Ministerial Committee on Climate Change (IMCCC)[45] serve as examples of institutional fora for climate change governance established by the national government. The functions of the CCAQM include the gathering, analysis, and distribution of climate change data and information in order to facilitate a well-informed climate response decision-making process. In addition, the CCAQM is also saddled with the tasked of providing support and monitoring of climate change mitigation and adaptation responses at the provincial and local levels in South Africa. While the purpose of the IMCCC is to coordinate and oversee the alignment, integration and implementation of climate change responses across and within ministries and agencies with national climate change policies and legislation.[46]

However, it must be emphasised that these institutions also have some challenges in performing their roles. Masters and Duff[47] for instance identified a lack of alignment between the CCAQM Agency and the IMCCC with other departments in integrating specific adaptation measures. The Department of Environment, Forestry and Fisheries (DEFF); the Department of Water and Sanitation; the Department of Agriculture, Land Reform and Rural Development (DALRRD)[48]; as well as the Departments for Energy and Transport are examples of the departments, which are unaligned with the CCAQM Agency and the IMCCC while integrating specific adaptation measures.[49]

In order to offer a better understanding of the operationalisation of the institutional framework for adaptation in South Africa, an extensive analysis of key policies for adaptation in the country is required. Hence, relying on the assertion that all policies are formulated in accordance with a country's *Constitution* and then translated into laws,[50] it is essential to first discuss the *Constitution* as the basis for further analysis of adaptation policies in the country.

5.2.1 The Constitution

Broadly speaking, constitutions serve as the supreme law in any country,[51] and they define the responsibilities of the government towards its citizens.[52] For example, in South Africa, the *Constitution* makes environmental protection part of the responsibilities of the national government towards its citizens. Specifically, section 24 of the *Constitution* provides that:

> [e]veryone has the right: (a) to an environment that is not harmful to their health or wellbeing and (b) to the environment protected, for the benefit of present and future generations, through reasonable legislative and other measures that – (i) prevent pollution and ecological degradation; (ii) promote conservation; and (iii) secure ecologically sustainable development and use of natural resources while promoting justifiable economic and social development.[53]

The phrase, "reasonable legislative and other measures" for the protection of the environment, has been interpreted to also include adaptive measures supported by both legal and non-legal mechanisms.[54] Legal mechanisms for adaptive measures in this context also include some international legal instruments which South Africa signed and ratified in accordance with the provisions of section 231 of the *Constitution*.[55] For example, South Africa ratified the UNFCCC in 1997 and the Kyoto Protocol in 2002, and signed the Paris Agreement in 2016.[56] These international legal instruments are part of the reasonable legislative measures that are intended to achieve a dual purpose of enabling the environment (including agriculture) to cope with hazards; and promote "ecologically sustainable development and justifiable economic and social development"[57] in the country.[58] It is on the basis of the ratification of these international instruments that South Africa formulated climate change-related policies and enacted some laws, which are discussed in subsequent sections of this chapter.

There is a range of views further elaborating the meaning of the provisions of section 24 of the *Constitution* in the context of adaptation. For example, Van der Berg, Du Plessis, and Murphree[59] believe that section 24 provides for measures which seek to adjust activities that are capable of inducing pollution and ecological degradation while promoting the sustainable use of natural resources and justifiable economic and social development. Along the same line, it has been argued that "section 24 compels municipalities and other authorities to intervene for the protection of people's health and safety, inter alia in the event of and in preparation for potential disasters triggered by climate change".[60]

In addition to the provisions of section 24, section 27 of the *Constitution* could also be relied upon in promoting adaptation, especially in the context of agriculture. Section 27 provides the right of access to food and water.[61] As noted in Chapter 1, access to sufficient food and water in most African countries, including South Africa, is impaired by the impacts of climate change.[62] Section 27 of the *Constitution* obliges the State (national government of South Africa) to apply appropriate measures that can aid in the realisation of the right of access to food and water.[63] Adaptation in the context of agriculture, as already noted, is one of the appropriate measures that could aid in the realisation of access to sufficient food and water.

Section 27 of the *Constitution* also partly provides for the right to social security in South Africa, including insurance services.[64] Jegede and Mokoena[65] argue that insurance against climate change-related risks is a *Constitutional* right that is derived from the right to social security and the right to social assistance. The analysis in preceding chapters confirmed that insurance is a means for supporting human needs (food and water) as well as an adaptation strategy against climate change-induced risks in all sectors of the economy, including agriculture.[66] In sum then, the *Constitutional* provisions highlighted above provide the foundation for environmental and climate change policies in South Africa, from which adaptation can be inferred. The major climate change adaptation policies in South Africa will now be discussed.

5.2.2 Adaptation-related Policies

As mentioned in the introduction, South Africa's climate change policies include the NCCRS, NCCRP, draft NCCAS, and the NDP. Just like Kenya and Nigeria, South Africa's climate change policies also provide a set of foundational principles that are important in guiding climate change responses (especially adaptation) in the country.[67] Therefore, these climate change policies are evaluated with a view to determining whether they contain the type of provisions that could enable and promote adaptation in the context of agriculture.

5.2.2.1 National Climate Change Response Strategy

While ratifying the UNFCCC in 1997, the national government of South Africa saw the need and urgency of preparing a national climate change response policy.[68] It was felt that:

> [a] national climate change response policy can potentially act as a significant factor in boosting sustainable economic and social development, a national strategy specifically designed to bring this about is clearly in the national interest, supporting the major objectives of the government including poverty alleviation and the creation of jobs.[69]

It was also envisaged that a national climate change response policy would stimulate the mainstreaming of climate change concerns within various government departments and different sectors of the economy (including agriculture) in a coordinated manner.[70] It is believed that the mainstreaming of climate change concerns into government departments could efficiently harness climate change response strategies in South Africa.[71] It was on this basis that the national government formulated the NCCRS in 2004.[72] The NCCRS recognises the susceptibility of human and natural systems in South Africa to climate change impacts, and as such also recommends the development of adaptation measures to cope with the impacts of climate change.[73]

In order to support adaptation, the NCCRS provides the required complementary support to other sectoral policies such as energy, water, and agriculture.[74] With regard to agriculture, the NCCRS facilitates adaptation through capacity building and awareness of how to use appropriate technology and technological skills to enhance the resilience of agricultural practices.[75] This set of technological skills is illustrated by rainwater-harvesting technologies and other water conservation techniques.[76] Rainwater-harvesting techniques and water conservation, including their benefits, were already explained in Chapter 1,[77] and are further supported by the NWA (introduced in Section 5.1 and to be discussed below). In addition, the NCCRS facilitates capacity building and awareness of agricultural management practices such as "changes in planting dates, row spacing, planting density and cultivar choice, and other measures".[78]

Based on the above objectives of the NCCRS, it is reasonably clear that adaptation in the context of agriculture has been catered for. However, just like any national policy which requires constant review and updating in order to conform to prevailing circumstances, the national government decided to improve the NCCRS with the NCCRP.

5.2.2.2 National Climate Change Response Policy White Paper

The national government developed an NCCRP in 2011.[79] Cognisant of the risk posed by climate change on critical sectors of the economy such as water, agriculture, and forestry, the NCCRP outlines the responsibilities of the government of South Africa in mitigating and adapting to the impacts of climate change.[80] For these purposes, the NCCRP provides for both mitigation and adaptation programmes and strategies that were supposed to be implemented in 2016 as short-term measures; as medium-term measures by 2031; and as long-term measures by 2050.[81] It is believed that these short-term to long-term measures could facilitate the required climate-resilient development in South Africa.[82] However, it should be noted that as of September 2022, those short-term climate change measures of 2016 were not integrated into all spheres and levels of government in South Africa.[83]

In connection with adaptation, the NCCRP aims to promote strategies such as early warning systems and integrated resource planning (particularly IDPs), which are mostly required in coping with the impacts of climate change on vulnerable sectors identified as agriculture, biodiversity, health, water, and human settlements.[84] As a result of this, the NCCRP makes provision for interventions that can enhance the resilience and adaptive capacity of vulnerable sectors (including agriculture) against risks and emergencies, especially those emanating from the impacts of climate change.[85] In this context, the NCCRP recognises and affirms insurance as a viable option for responding to the impacts of climate change. More specifically, the NCCRP "identifies insurers as important development partners for the government including in the development of innovative climate change-related products".[86]

Additionally, the NCCRP makes provision for a micro-insurance mechanism that allows both highly developed commercial-based farmers and poor small-scale subsistence farmers in South Africa to recover from climate-induced disasters.[87]

Consequently, the NCCRP aims to enhance the resilience and adaptive capacity of social, economic, and environmental systems through an integrated process (such as IDPs).[88] Water, health, agriculture, and human settlements are among the sectors identified by the NCCRP as being more susceptible to the impacts of climate change as well as deserving of an integrated adaptation response.[89] Examples of integrated processes are given by Biazin et al.,[90] and they include strengthening biodiversity management, empowering local communities, and developing climate-smart agriculture (CSA) strategies such as the promotion of catchment and water management practices. For the proper integration of adaptation measures, the NCCRP "adopts a strongly sectoral approach, but also identifies the need for coordination of responses between sectors".[91] For example in the water sector, the NCCRP seeks to promote strategies developed by National Water Resource Strategy (NWRS), which is further discussed below.[92] To further support climate change responses including adaptation through a sectoral approach, "the NCCRP directs all government departments to review policies, strategies, legislation, regulations and plans in their jurisdiction to fully align them with the NCCRP within 2 years of publication [by 2013]".[93]

In addition to the sectoral approach of integrating adaptation measures, the NCCRP advocates for adaptation responses with stronger local content, which could equally foster green-jobs opportunities and sustainable development in South Africa.[94] Based on this, the NCCRP recognises the role of local governments in South Africa in responding to the impacts of climate change.[95] Specifically, the NCCRP promotes strategies for reducing the vulnerability of water, human settlements, and energy against the impacts of climate change at the local government level.[96] Van der Berg, Du Plessis, and Murphree[97] illustrate these strategies as short-term adaptation measures in the form of disaster risk-reduction responses against the impacts of climate change such as heat waves, floods, and droughts.[98]

However, the NCCRP has some limitations. Lehman[99] observes that adaptation received limited attention in the NCCRP when compared to mitigation, which appears to be the main focus of the NCCRP. This could be due to the fact that when the NCCRP was formulated in 2011, mitigation (in contrast to adaptation) dominated the international and domestic climate change governance landscape.[100] Similarly, it has been noted that short-term adaptation strategies intended for implementation from 2016 to 2030 are yet to be mainstreamed into all spheres and levels of government in South Africa.[101] In addition, the necessary cross-sectoral linkages required for mainstreaming adaptation strategies at the national level are lacking. This gap in the NCCRP could, for example, be attributed to institutional challenges, including lack of synergy and duplication in the roles and functions of government departments such as the Department of Forestry, Fisheries and the Environment (DFFE) and the DALRRD.[102] Arguably this shortcoming has been noted by the national government of South Africa, and in response a national climate change adaptation strategy was prepared by the DFFE.[103]

5.2.2.3 National Climate Change Adaptation Strategy 2019

The NCCAS is informed by other South African climate change policies and documents such as the NCCRP, NDP, and the NDC, as well as sub-regional climate change policies, particularly the draft South African Development Commission (SADC) Climate Change Strategy and Action Plan.[104] The main goal of the NCCAS is to develop a NAP for South Africa as part of efforts in operationalising the NCCRS from 2020 to 2030.[105] In the same context, the NCCAS is aimed at developing a framework for vulnerability assessment, as well as a framework that could support South Africa's adaptation needs as of 2020.[106] More importantly, the NCCAS serves as "the cornerstone for climate change adaptation in the country and [is intended] to reflect a unified, coherent, cross-sectoral, economy-wide approach to climate change adaptation".[107]

This means that all organs and levels of government as well as the private sector in South Africa are supposed to be guided by the NCCAS in the planning and implementation of adaptation measures and actions.[108] Simply put, the purpose of the NCCAS is to have a common vision of adaptation in South Africa, particularly in critical sectors of the country's economy that appear susceptible to climate change, and these include agriculture, forestry, water, coastal zones, health, and terrestrial ecosystems.[109]

In facilitating adaptation, the NCCAS provides for strategic intervention in mainstreaming adaptation responses into other sectoral planning and implementation efforts. For example, it is possible to have agricultural adaptation strategies such as irrigation services embedded in the activities of the National Planning Commission.[110] In the agricultural sector, South Africa's adaptation needs would be coping with water stress and land use changes and, in this regard, the NCCAS aims to provide strategic intervention guidance that could promote the development of South Africa's water resources infrastructure, including surface storage capacity, conservation efficiency, and integrated catchment management of water resources.[111] In addition, the NCCAS seeks to facilitate the development of "a risk, early warning, vulnerability and adaptation monitoring system for crops and livestock through agricultural technologies".[112]

However, it must be noted that the NCCAS is not expected to provide in-depth direction for the implementation of adaptation strategies in most of the critical sectors of the economy, including agriculture, that are vulnerable to the impacts of climate change. A detailed explanation of how adaptation should take place may instead be found in the NDP or a related policy document such as the NDC. Thus, the NDP and the NDC are analysed below.

5.2.2.4 National Development Plan 2030

Having acknowledged climate change as a challenge to South Africa's economic growth and development,[113] the national government formulated the NDP in 2011.[114] The NDP (also known as Vision 2030) seeks to eliminate poverty through the promotion of economic development without compromising the quality of the environment.[115] One of the measures proposed by the NDP in protecting the environment is to reaffirm and support the implementation of an existing environmental management framework that supports the protection of natural resources (through adaptation) such as the NEMA[116] (discussed below).

Adaptation is included in Chapter 5 of the NDP as part of the environmental sustainability and resilience objectives of South Africa.[117] Concerning agriculture for instance (while recognising the sector as the economic backbone of South Africa with the potential of adding about one million new jobs by 2030[118]), the sustainability and resilience objectives of the NDP are to invest in irrigation, water storage, and distribution infrastructures for a ten-year period (2012–2022).[119] According to the NDP, irrigation has the potential to expand South Africa's field crops from 1.5 million hectares to two million hectares within the projected period of ten years.[120] It is believed that investing in water infrastructure including irrigation services would allow for the management and monitoring of South Africa's scarce water resources in a manner that will not compromise economic growth in the country.[121] In addition, investing in water infrastructure could promote safe wastewater reusage and could allow water users in South Africa to understand and respond to emerging constraints in various aspects of water uses such as supply planning, development, and disposal of wastewater operations.[122]

Although the sustainability and resilience objectives of the NDP appear to have satisfied some of the requirements of adaptation in the context of agriculture, there are minor concerns about the policy. For example, the assertion that irrigation services under the NDP could possibly expand South Africa's field crops from 1.5 million hectares to two million hectares in ten years has been criticised as being unrealistic, because the projection did not factor in underused lands in communal areas.[123] Similarly, the promotion of economic development in South Africa, while enhancing the resilience of critical sectors of the economy against the impacts of climate change, could only be successful when all budgetary processes and finances earmarked for other development needs are properly mainstreamed. This seems to be missing in the NDP.

5.2.2.5 National Determined Contribution

South Africa submitted its first NDC to the UNFCCC secretariat in 2016 pursuant to the requirement of article 4.9 of the Paris Agreement, and in compliance with the provisions of section 24 of the *Constitution*.[124] The 2016 NDC is part of South Africa's short-term response to climate change challenges.[125] The 2016 NDC outlines both mitigation and adaptation interventions for South Africa.[126] Regarding adaptation, the 2016 NDC adopts a cross-sectoral approach, which in the opinion of Mantlana et al.,[127] is intended to bolster an all-inclusive governance that consolidates national planning with climate change concerns. In a bid to promote adaptation in the context of agriculture, the NDC identifies programmes and measures that could support crops and livestock production, especially processes that tend to promote resilience against droughts and floods (such as the rehabilitation of vulnerable catchment areas and landscapes, water conservation techniques, and land restoration methods).[128]

The DFFE communicated an update to the 2016 NDC in September 2021.[129] Accordingly, the updated NDC is believed to be a demonstration of South Africa's progressiveness in addressing the impacts of climate change relative to the country's national circumstances.[130] In the context of adaptation, the updated NDC outlines South Africa's planned adaptation efforts (from 2021 until the next NDC communication in 2025) in promoting the global goal of adaptation as required by article 7 of the Paris Agreement.[131] Towards this end, the updated NDC describes agriculture, water, transport, energy, biodiversity, and human settlements as some of the critical sectors that would require more adaptation interventions.[132]

The NDP and the NDC are policies which, like any other policy discussed above, need to be translated into a binding law. Hence, this prompts further discussion of adaptation laws in South Africa.

5.2.3 Adaptation-related Laws

As previously discussed, law facilitates adaptation by guiding stakeholders, and in giving legitimacy to the set of action plans and appropriate measures to be taken.[133] Climate change adaptation in South Africa is broadly facilitated by some environment-focused laws that were enacted pursuant to the *Constitution*, such as the *NEMA*, which is a framework law for environmental protection in South Africa.[134] Other environmental laws also providing for adaptation include the *DMA, CTA*, as well as *Climate Change Act*. An in-depth analysis of all provisions of these laws is not within the scope of this section. Instead, only the provisions of these laws which appear directly relevant to adaptation and agriculture will be examined.

5.2.3.1 National Environmental Management Act

The primary purpose of enacting the *NEMA* was to regulate the exploitation of natural resources, promote the conservation of natural resources, and control all forms of pollution and waste management as well as the regulation of land uses and management (a tool for agricultural practice).[135] Accordingly, the *NEMA* serves as the foundation from which most of South Africa's environmental governance principles are articulated,[136] and "provides generic or umbrella principles and obligations in relation to all activities that affect or may affect the environment".[137] The environmental principles in the *NEMA* are intended to also guide South Africa's contribution to the protection of the global climate system in accordance with the country's national circumstances, responsibilities, capability, and development goals (described by the Paris Agreement as the principle of Common – but differentiated responsibility and respective capabilities – CBDR and RC).[138] Section 2 of *NEMA* contains the main environmental principles which, among others, require the environment to be protected for the benefit of present and future generations of humankind (otherwise known as the principles of inter-generational equity and intra-generational equity).[139] These principles partly reflect the provisions of section 24(1) (b) of the *Constitution*, which has been explained as part of the legal foundation for adaptation. The principles also confirm previous discussions about the *Constitution* as the centrepiece for all legislation in the country.[140]

Strydom et al.[141] observe that in addition to the above principles, the *NEMA* also adopts other principles that are relevant to adaptation in a broad context, and these include "environmental management must place people and their needs at the forefront of its concerns, and serve their physical, psychological, developmental, cultural and social interest equitably".[142] In a strict sense, this principle is provided by section 2 of *NEMA*, and it requires environmental management actions such as adaptation to be conducted anthropocentrically (i.e., equitably in consideration of the physical, psychological, developmental, and social needs and in the interest of people) rather than bio-centrically.[143] Other principles adopted by the *NEMA* are the sustainable development principle,[144] integration principle,[145] environmental justice principle,[146] equitable access principle,[147] public trust principle,[148] precautionary principle,[149] and the polluter pays principle.[150] These environmental management principles are expected to offer guidance as well as serve as a reference point for each organ of the state responsible for the protection of the environment.[151] The DALRRD, for instance, relies on these environmental management principles in developing strategies that could enhance the adaptive capacity of water, which is essential for crops and livestock production.[152]

The *NEMA* also employs a range of environmental protection tools, especially Environmental Impact Assessments (EIAs), in order to satisfy the objectives of the Act and to operationalise the above-stated principles.[153] According to section 24 of the *NEMA*, certain development activities (such as farming near wetlands) can only be carried out when their potential consequences and impacts on the environment are assessed.[154] An EIA report is expected to include either mitigation or adaptation measures as solutions to envisaged or existing environmental hazards.[155] In the words of Humby,[156] an EIA report must take into account the correlation between vulnerability, impact, and potential disaster risk.[157]

However, it has been observed that the provisions of the *NEMA*, especially those dealing with EIAs, overlook the impacts of climate change as issues that could arise due to a poorly conducted EIA.[158] In the same vein, as a framework law, the *NEMA* is not expected to contain provisions that expressly outline the types of adaptation that are required in sectors that support social and economic needs such as agriculture, and how they can be implemented. These might rather be included in other sectoral laws such as the *NWA*, the *WSA*, *CARA*, and the *SPLUMA* discussed below under agriculture-related laws.

5.2.3.2 Disaster Management Act

Sadoff and Muller[159] believe that effective adaptation strategies depend on a perfect understanding of the risks and impacts of climate change and how they apply to any given circumstance. Hence, due to South Africa's susceptibility to drought (as an example of a potential climate change-induced disaster), the national government earmarked different measures, including legal measures, to tackle drought in eight of the nine provinces of the country (excluding Limpopo) considered as drought disaster zones.[160] The *DMA* of 2002 is an example of a legal measure embarked upon by the national government in tackling droughts in South Africa. As noted by Van der Berg, Du Plessis, and Murphree,[161] the provisions of the *DMA* apply to all disaster situations in South Africa. Disaster situations within the contemplation of the *DMA* could be natural or human-induced disasters, and these could manifest as climate change and floods.[162]

Therefore, in addition to tackling droughts, the *DMA* was also enacted to address severe flooding in the country.[163] In sum, the aim of the *DMA* is to:

> [p]rovide for an integrated and coordinated disaster management policy that focuses on preventing or reducing the risk of disasters, mitigating the severity of disasters, emergency preparedness, rapid and effective response to disasters and post-disaster recovery; the establishment of national, provincial and municipal disaster management centres; disaster management volunteers; and matters incidental thereto.[164]

Based on the above aim, the *DMA* conceptualises disasters and risks such as climate change-induced droughts and floods as disasters that require a multi-sectoral and multi-disciplinary approach.[165] The *DMA* establishes an institutional structure which supports and facilitates effective disaster management in South Africa.[166] The principal institutional structure for disaster management is the Intergovernmental Committee on Disaster Management (ICDM).[167] The main function of the ICDM is to advise the Cabinet on issues concerning disaster management which, among others, include emergency preparedness, rapid and effective response, as well as post-disaster recovery measures.[168] The ICDM consists of Cabinet Members who are involved in disaster management (the Department of Cooperative Governance, and disaster management centres at the national, provincial, and municipal levels),[169] and Members of the Executive of each province involved in disaster management, as well as members of municipal councils.[170] In a bid to support and facilitate disaster management, especially at the municipal level, the *DMA* requires municipalities (local government) to integrate disaster management plans into their IDP projects and programmes.[171]

However, a critical review of the *DMA* indicates that the provisions of the law did not expressly include adaptation among the strategies of adjusting to the impacts of risk and natural disasters such as climate change-induced droughts and floods.[172] This could be considered a major gap. Therefore, in order to reflect current realities about risks and natural disasters in South Africa, especially those risks that are compounded by climate change, the law was amended in 2015.[173]

Consequently, adaptation was added to the 2015 amendment. In addition to adaptation, the 2015 amendment also included a comprehensive definition of 'disaster risk reduction' in the context of climate change adaptation, which seems to be missing in the 2002 version of the *DMA*.[174] Adaptation and disaster risk reduction are explained by the 2015 amendment as ecosystem and community-based measures against risk and disasters.[175]

More importantly, ecosystem and community-based measures were outlined in the 2015 amendment to accommodate a symbiotic relationship between plants and animals. For present purposes, plants and animals could also be interpreted as crops and livestock, both of which were previously identified as vulnerable to risk and natural disasters including climate change.[176] In the same manner, vulnerability has been redefined in the 2015 amendment to reflect the definition of the Intergovernmental Panel on Climate Change (IPCC),[177] while 'risk assessment' was added in the 2015 amendment to mean:

> [a] methodology to determine the nature and extent of risk by analysing potential hazards and evaluating existing conditions of vulnerability that together could potentially harm exposed people, property, services, livelihoods and the environment on which they depend.[178]

Based on the above definition of risk assessment, as given by the *DMA* of 2015, it could be stated that the aim of risk assessment is similar to the aims of the Strategic Environment Assessment (SEA) and EIA under article 4 of the UNFCCC,[179] the *EMCA* of Kenya,[180] and section 24 of the *NEMA* above, all of which are intended to support adaptation measures directly or indirectly. Thus, the *DMA* facilitates adaptation in the context of agriculture, though with insufficient contents and focus, because the Act does not elaborate on the required adaptation measures and how they could be implemented.

5.2.3.3 Carbon Tax Act

The national government adopted the *CTA* in 2019 as part of measures for regulating carbon emissions in the country.[181] Pursuant to sections 7 and 8, the *CTA* regulates carbon emissions in the country through the imposition of levies.[182] The levies for carbon emissions are calculated based on the percentage and on the basis of the gravity of the emission.[183] The rationale behind the imposition of levies is to compel those responsible for carbon emissions to alter or adjust their activities by conforming to standards provided by the *CTA*. As explained previously, adaptation relies on altering or adjusting carbon-intensive activities, among other measures.[184]

The *CTA* is expected to operate in two phases. The first phase will run from 2019 to 2022, while the second phase will run from 2023 to 2030.[185] The classes of carbon that are regulated under the first phase of the *CTA* include carbon generated from fuel combustion, electricity generation (especially from coal) as well as industrial processes.[186] However, agricultural practices, especially commercialised crops and livestock production, are excluded from the first phase of the *CTA*.[187] It is anticipated that farmers will be liable for direct and indirect carbon tax in the second phase of the administration of the *CTA*.[188]

The *CTA* is guided by environmental principles derived from the *NEMA* as highlighted above, such as the environment should be protected for the benefit of present and future generations. In addition, the *CTA* also seeks to advance the objectives of the *NEMA*, which include the protection of the global climate system in accordance with South Africa's national circumstances, responsibilities, capability, and development goals.[189]

Judging by the provisions of sections 7 and 8 of the *CTA*, it is obvious that the law placed much emphasis on mitigation and not on adaptation. This is because the *CTA* is a generic law and as such its broad provisions seemingly facilitate adaptation in a broad context. Likewise, the *NEMA* and *DMA* are generic environmental-based laws from which agriculture-focused adaptation could be only inferred. Hence, just like Kenya and Nigeria, South Africa also probably needs a specific climate change law that supports adaptation in the context of agriculture. According to Taylor et al.,[190] having a specific climate change law would make the implementation of climate change adaptation measures legally binding. In support of this assertion, the *Climate Change Act* is discussed below.

5.2.3.4 Climate Change Act

Unlike Kenya and Nigeria, South Africa's specific climate change law is yet to fully take effect. Although the President of the Republic of South Africa signed and assented to the long-awaited *Climate Change Act* on 23 July 2024,[191] the Act will only become operational when the President fix a date for commencement by a proclamation in a Government Gazette.[192] What this means is that the *Climate Change Act* must be declared, and this will be done after the DFFE is able to finalise all the relevant regulations under the Act.[193]

The *Climate Change Act* has been prepared pursuant to the *Constitution* and is guided by environmental management principles derived from the *NEMA*.[194] In other words, the *Climate Change Act* also seeks to advance constitutionally guaranteed environmental rights by protecting the environment.[195] It should, however, be noted that the *Climate Change Act* is not intended to abrogate or contradict the provisions of existing environmental legislation such as the *NEMA* from which climate change governance could be inferred, but is meant to complement them.[196] Likewise, the *Climate Change Act* incorporates and builds on the general provisions of the *CTA*, which regulates carbon emissions in South Africa (see above).[197]

Therefore, relying on the regulatory foundations provided by other laws such as the *NEMA* and *CTA*, as well as in cognisance of the existing threat posed by climate change on "human health, access to food and water, biodiversity, coastal infrastructure and human settlements",[198] the *Climate Change Act* seek to give legal effect to a long-term climate change response and coping strategies that are in conformity with a climate-resilient and low-carbon development pathway for South Africa.[199] In addition, the purpose of the *Climate Change Act* is to design and implement effective climate change mitigation and adaptation actions in South Africa that are "consistent to the country's contribution to the global climate change response".[200] To this end, the *Climate Change Act* has been acknowledged as the first "comprehensive South African legal framework for the regulation of the impacts of climate change, with the ultimate overall goal of achieving net-zero emissions by 2050".[201] Specifically, the objectives of the *Climate Change Act* are to:

> provide for the effective management of inevitable climate change impacts by enhancing adaptive capacity, strengthening resilience and reducing vulnerability to climate change, with a view to building social, economic, and environmental resilience and an adequate national adaptation response in the context of the global climate change response...[202]

As demonstrated by the specific objectives above, it is apparent that most of the provisions of the *Climate Change Act* are tilted towards adaptation actions rather than mitigation.[203] These adaptation actions are also found in other existing climate change-related policies such as the NCCRS, NCCRP, the NCCAS, and the NDCs of South Africa discussed above. In the same context, the *Climate Change Act* seeks to mainstream adaptation actions into "all spheres of governance in accordance with the principles of cooperative governance".[204] To support this principle, the *Climate Change Act* makes provision for the continuity of a PCC that was already formed in September 2020.[205] The PCC is expected to coordinate the operationalisation of the *Climate Change Act* when it becomes enforceable.[206] The general functions and membership composition of the PCC seem to be more or less similar to those of the National Climate Change Council (NCCC) under the Kenyan *Climate Change Act* and the National Council on Climate Change (NCC) of the Nigerian *Climate Change Act*.[207]

More importantly, the *Climate Change Act* outlines the modalities for managing adaptation measures in South Africa. It is required that "climate change adaptation within the Republic must be managed in a coherent and coordinated manner and in accordance with a National Adaptation Strategy and Plan".[208]

The *Climate Change Act* also elaborately provides that adaptation options must be managed in accordance with South Africa's international commitments and obligations on climate change adaptation; without compromising monitoring and evaluation results; and in line with technological, scientific, or informational standards.[209] Moreover, the *Climate Change Act* specifies the guidelines for attaining the national adaptation objectives of South Africa, such as a deadline for gazetting the national adaptation objectives, benchmarks for measuring the progress of national adaptation objectives, a timeframe for mainstreaming adaptation options into development planning, and a periodic interval for reviewing and amending the NCCAS.[210]

Having acknowledged the impacts of climate change on critical sectors of the economy (such as agriculture, forestry, fisheries, and energy), the *Climate Change Act* also provides for a sectoral adaptation strategy plan. Section 19 of the *Climate Change Act* requires the ministers of various ministries to:

> [c]onduct an assessment of the functions under the Minister's operational control which—(i) identifies and spatially maps risks and vulnerabilities, areas, ecosystems and communities that will arise and that are vulnerable to the impacts of climate change; and (ii) determines measures and mechanisms to manage and implement the required adaptation response.[211]

In addition to the provision of section 19, Schedule 2 of the *Climate Change Act* identifies and lists the departments that are responsible for managing agriculture,[212] forestry, fisheries, disaster risk reduction, energy, and environment as some of the national departments that are required to develop a sectoral adaptation strategy and plan.[213]

Thus, when it becomes enforceable, the *Climate Change Act* could probably derive the required level of adaptation in critical sectors of the South African economy such as agriculture, forestry, energy, and water that are susceptible to climate change, notwithstanding that the sectors are already regulated by various laws. The next part of this chapter will examine the regulatory framework for agriculture in South Africa.

5.3 Regulatory framework for agriculture

Agricultural practices such as crop and livestock production are partly regulated in order to protect the environment from agriculture-induced climate change and to support agricultural practices in adapting to a changing climate.[214] As noted above, the regulatory framework for agriculture as an important sector of the economy is also derived from the *Constitution*.[215] The *Constitution* places agriculture under concurrent matters which are to be jointly regulated by the national and provincial governments.[216] In line with the approach followed while analysing the Kenyan and Nigerian agriculture-related policies and laws, and in reaffirmation of arguments that policies precede laws,[217] the next section of this chapter will first analyse agriculture-related policies before analysing agriculture-based laws.

5.3.1 Agricultural Policies

South Africa formulated agricultural-based policies to provide direction on how to boost productivity and growth of agricultural products (crops and livestock) severely affected by the impacts of climate change – which manifest as droughts, water scarcity, and flooding in some parts of the country.[218] The following sections will discuss these agricultural policies to establish their relevance in facilitating adaptation in the context of agriculture.

5.3.1.1 Draft Climate-Smart Agriculture Strategic Framework 2018

South Africa is in the process of finalising a national CSA framework, already introduced above as the draft CSASF.[219] When approved, the draft CSASF is supposed to operate as a short-term measure (ten years from the date of commencement), and a broad strategic guideline that provides direction on how South Africa's climate change policies, laws, and action plans could be applied in addressing, and in coping with, the impacts of climate change on the agricultural sector.[220]

The goal of the draft CSASF is to enhance resilience and adaptive capacity in the agriculture, forestry, and fisheries sectors through effective adaptation responses.[221] Adaptation responses identified by the draft CSASF include, among others, the:

> [a]doption of efficient environmental resources management practices such as the planting of early maturing crops, adoption of hardy varieties of crops and selective keeping of livestock in areas where rainfall has declined.[222]

In addition to the adaptation responses above, the draft CSASF seeks to integrate other adaptation strategies such as disaster risk management, irrigation and water resource management, conservation agriculture, watershed and land management, and livestock management,[223] which were also provided by key South African climate change policies (discussed above) such as the NCCRS, NCCRP, and the NCCAS.

The core objectives of the draft CSASF are to guide the government (at all levels), investors, and development partners in mainstreaming CSA into their plans, programmes, and projects in order to boost productivity and growth of agricultural, forestry- and fisheries-related value chains.[224] By connecting adaptation strategies with critical and vulnerable sectors such as agriculture, energy, and water, the draft CSASF could help transform policies for these sectors into efficient and productive development plans. In addition, the objective of the draft CSASF is to enhance the resilience of the social, environmental, and economic aspects of agriculture, forestry, and fisheries systems against climatic and weather shocks, while also strengthening governance and institutional aspects of CSA programmes at the national, provincial, and local levels.[225]

Therefore, the draft CSASF adopts an all-inclusive approach in its formulation and governance structure, which consists of stakeholders such as representatives of the FAO, government departments, farmers, NGOs, scientists, and other stakeholders.[226] Consequently, the draft CSASF symbolises a balance of scientific and political ideas as well as the interests, demands, and realities of farmers (including foresters and fishers) in a bid to manage agricultural resources in a sustainable manner.[227] Just like the NCCRS which focuses on the linkages between climate change, agriculture, energy, and water, the draft CSASF also seeks to focus on the same linkages through its nexus approach.[228]

As can be concluded from the foregoing discussion, the draft NCSASF has captured many of the requirements of adaptation in the context of agriculture. Other agriculture-related policies – especially the draft Climate Change Sector Plan for Agriculture, Forestry, and Fisheries (CCSPAFF) and the Second NAP-CDLDD are also relevant.

5.3.1.2 Climate Change Sector Plan for Agriculture, Forestry, and Fisheries

The national government developed a CCSPAFF in 2013 to address the impacts of climate change on agriculture, forestry, and fisheries in South Africa.[229] The CCSPAFF seeks to promote comprehensive adaptation strategies against the impacts of climate change on the agriculture, forestry, and fisheries sectors through a vulnerability assessment and vulnerability audit processes.[230] In formulating the CCSPAFF, consideration was given to the requirements of the NCCRP and the provisions of the *DMA* of 2002.[231] The CCSPAFF conceptualised the need for climate-resilient agriculture in South Africa, while also recognising the negative impact of agriculture on the environment.[232]

Consequently, one of the goals of the CCSPAFF is to increase resilience against climate change impacts on the agriculture, forestry, and fisheries sectors through the promotion of public awareness and knowledge on the implications of some agricultural practices that lead to climate change, such as organic agriculture and agroecology and the application of industrial fertilisers (NH3 and NH4).[233] Another goal of the CCSPAFF is to evaluate and cope with the potential impacts of climate change on agriculture, forestry, and fisheries through various adaptation responses in the short-term (ten years), medium-term (20–30 years), and long-term (50 years and above).[234] Within these given periods, the CCSPAFF is expected to "identify adaptation gaps and opportunities that require research, development and capacity building to increase resilience to climate change".[235] Likewise, within the three terms, the CCSPAFF aims to provide relevant approaches for integrating adaptation-response policies and planning in sectors such as energy and water with those in the agriculture, forestry, and fisheries sectors.[236]

Therefore, it is apparent that the CCSPAFF contains the type of provisions that could enable and promote adaptation in the context of agriculture. However, a major problem is that, since the CCSPAFF was formulated (in 2013, before the Paris Agreement),[237] the CCSPAFF cannot be expected to reflect all agricultural-adaptation norms contained in the Paris Agreement, especially those norms that were elaborated by the Koronivia Joint Work on Agriculture (KJWA).[238]

5.3.1.3 Second National Action Programme for South Africa to Combat Desertification, Land Degradation, and the Effects of Droughts

South Africa has suffered many years of droughts, and the country was during the period of this study under the lingering effects of droughts.[239] Agricultural practices, especially crop and livestock production, are mostly affected by droughts.[240] Therefore, South Africa's most recent attempts and commitments in managing the effects of droughts on the environment and its components are confirmed by the Second NAP-CDLDD (2018–2030), simply described as the Action Programme.[241] The Action Programme is designed to promote measures that can combat the effects of drought on the environment (and associated components, especially land and water).[242] In other words, part of the strategic objective of the Action Programme is "to mitigate, adapt to, and manage the effects of drought in order to enhance the resilience of vulnerable populations and ecosystems".[243]

The measures for adapting to drought under the Action Programme are guided by international drought-control initiatives, especially the United Nations Convention to Combat Desertification (UNCCD).[244] Examples of these measures include the use of monitoring and early warning systems and vulnerability/impact assessment to combat droughts.[245] The relevance of these adaptation measures to agriculture and their advantages were previously discussed.[246] In addition, the Action Programme has provisions for the development of CSA practices that are intended to preserve and restore critical habitats, protect watersheds, and improve the health of soil and water.[247] Also, the Action Programme promotes CSA practices that can "protect land, water runoff, soil nutrients and genetic resources and also contribute to carbon stocks".[248] These measures are part of on-farm adaptation strategies that were discussed initially.[249]

Hence, the Action Programme seems to facilitate adaptation in the context of agriculture. However, the Action Programme and other agricultural policies such as the draft CSASF and the CCSPAFF are simply policies that need to be translated into binding laws.[250] For this reason, the next section of this chapter will discuss a few agriculture-related laws in South Africa with a view to determining if the laws contain the type of provisions that could enable and promote adaptation in the context of agriculture.

5.3.2 Agriculture-related Laws

As stated earlier, all South African laws are enacted pursuant to the *Constitution*. Since agriculture and resources that support agriculture (for instance, water and soil) are heavily impacted by climate change, adaptive measures which are supported by effective legal provisions are necessary in order to help water and soil cope with a changing climate.[251] Section 27 of the *Constitution* serves as the legal basis for access to water, upon which the State is compelled to take reasonable measures to provide access to water.[252] By inference, "reasonable measures" could also mean adaptation measures which can enhance water availability and quality against hazards such as climate change. The provisions of section 27 of the *Constitution* on access to water led to the enactment of specific water laws such as the *WSA* and *NWA*. These laws will be analysed with a view to determining their relevance to adaptation in the context of agriculture. However, it is noteworthy to state that the *WSA* and *NWA* are broad laws with varying interpretations, which cannot be exhaustively discussed in this chapter.[253]

5.3.2.1 Water Services Act

Water security is a priority in South Africa, given the perennial water scarcity challenges, which are further compounded by climate change.[254] Therefore, the *WSA* was enacted in 1997,[255] though not with the explicit intent to deal with the impacts of climate change, but rather to achieve certain objectives which are in accordance with the provisions of sections 24 and 27 of the *Constitution*.[256] The main objectives of the *WSA*, as provided by section 2 of the Act, are to support the right of access to clean water in sufficient quantity, provide a regulatory framework for water services, and promote effective water resource management and conservation in South Africa.[257] Effective water resource management and conservation could be promoted through adaptive water management strategies such as the protection of aquifers, rainwater-harvesting methods and the construction of strategic boreholes to boost water availability in dry seasons. As previously explained, these water resource management and conservation techniques are required to boost crop yields and animal productivity, especially in the event of climate change-induced water scarcity and droughts.[258]

The *WSA* also partly regulates the manner in which water is used in South Africa. Based on the provision of section 7 of *WSA*, any industrial activity (including commercialised agriculture) capable of introducing industrial effluents into water services is prohibited.[259] By implication, the provisions of section 7 seek to compel water users in South Africa to preserve water quality, while also altering their (agricultural) practices. In practical terms, the alteration of agricultural practices could be likened to the process of modification or adjustment, which is what adaptation is also about.[260] In other words, it could be stated that adaptation in the context of agriculture is provided by section 7 of the *WSA*, though not explicitly.

Aside from the provisions of sections 2 and 7 of the *WSA*, the Act has less significance to adaptation, especially in the context of agriculture. Therefore, in addition to the *WSA*, another law which seeks to facilitate adaptive water management is the *NWA*.

5.3.2.2 National Water Act

The NWA[261] was enacted to provide a comprehensive approach to water management strategies in South Africa, pursuant to the provisions of sections 24 and 27 of the *Constitution*.[262] More specifically, section 2 of the *NWA* outlines the purpose of the law, which is to:

> [e]nsure that the nation's water resources are protected, used, developed, conserved, managed and controlled in ways which take into account amongst other factors… meeting the basic human needs of present and future generations; promoting equitable access to water; promoting the efficient sustainable and beneficial use of water in the public interest, protecting aquatic and associated ecosystems and their biological diversity; reducing and preventing pollution and degradation of water resources; managing floods and droughts.[263]

As a legal mechanism which also seeks to promote equitable allocation and distribution of water, the *NWA* is anchored based on the principles of the Public Trust Doctrine (PTD), which vests custody of water resources in the State on behalf of the people.[264] The rationale of including the PTD in the *NWA* is to make the national government responsible for ensuring that water is:

> [p]rotected, used, developed, conserved, managed and controlled in a sustainable and equitable manner, for the benefit of all persons and in accordance with its *Constitutional* mandate.[265]

To elaborate, the constitutional mandate of the National Water Affairs Minister has been interpreted in many ways, which also includes the mandate to implement the provisions of section 24 of the *Constitution* on the promotion of environmental values.[266]

In addition to the reasons given by section 3 of the *NWA*, the purpose of the PTD is also elaborated by a White Paper on National Water Policy for South Africa, which emphasises the need for the PTD, because it:

> [g]uarantees access to sufficient water for basic domestic needs; makes sure that the requirements of the environment are met; takes into account the interconnected nature of the water cycle—a process

> on which the sustainability and renewability of the resource depends; makes provision for the transfer of water between catchments; respects South Africa's obligations to its neighbours; and fulfils its commitment as custodian of the nation's water.[267]

The PTD is considered to be flexible and more effective in climate-sensitive and water-constrained situations such as in South Africa when compared to the former Riparian Rule system, because the PTD promotes the allocation of water among users while reflecting hydrological realities and differing priorities.[268] In the view of Alder,[269] a Riparian system creates uncertainty and scarcity concerning the quantity of water to be used – resulting in losses for services that depend on water, especially agriculture, and it makes it difficult for competing users to make choices.

According to Rumble,[270] the *NWA* was drafted when the State recognised climate change as a threat to water resources, and at a time when a regulatory mechanism was required to allow for state intervention regarding water uses. Accordingly, the *NWA* provides the necessary tools (such as licences) for state intervention in terms of water allocation and redistribution. Pursuant to the provisions of section 43(1) of the *NWA*, a licence is required for identified water uses[271] to satisfy the needs of fair and equitable allocation of water in situations of demand due to water stress, or to facilitate water use, which is beneficial for public needs and interest, to promote the efficient management of water and to preserve the quality of water.[272]

The provisions of section 43(1) could be applied in promoting adaptation in the context of agriculture. Given that agricultural practices require the right amount of water supply to flourish, irrigation becomes necessary in cases where there is water stress and water scarcity. For this reason, the *NWA* makes it obligatory for farmers who seek to irrigate their farms using water in catchment areas to obtain a licence.[273] It is believed that licences could force water users to apply efficient adaptive water use measures, especially where there is water stress or scarcity as a result of climate change.[274]

In addition, Bank and Karsten[275] observe that most water challenges in South Africa, such as limited and uneven distribution resulting from climatic change and variation, could be overcome through application of water management processes facilitated by section 37 of the *NWA*, otherwise described as "controlled activities". These water management processes include the irrigation of farmlands with wastewater, modification of atmospheric precipitation, the generation of power for the alteration of the flow of water, and the recharging of aquifers by wastewater.[276] These activities were previously explained as forms of adaptation in the context of agriculture.[277]

Furthermore, it is believed that controlled activities under section 37 of the *NWA* could also facilitate:

> [s]trategic resource management, flexibility in water use allocations, water demand management and water conservation measures, contingency

> planning for extreme events such as floods and droughts, communication, optimising the operation of existing infrastructure and constructing new infrastructure.[278]

Rumble[279] also argues that the types of controlled activities provided by the *NWA*, which are also regulated through general authorisations under section 39, could promote climate-resilient development (especially high crop yield and animal productivity).[280]

Also of relevance to agriculture and adaptation is section 145 of the *NWA*. The provisions of section 145 mandate water management institutions, especially the Department of Water and Sanitation, to promptly inform the public about existing or impending floods, droughts, and other associated risks to water quality.[281] In other words, the provisions of section 145 could be described as anticipatory adaptation, which has been initially identified as the most suitable type of adaptation for agriculture.[282]

In order to achieve its purpose as stated in section 2, the *NWA* classified water resources in South Africa for industrial, agricultural, and recreational uses.[283] The type of water management processes that are relevant for agriculture are facilitated by section 37 of the *NWA* (discussed above). The implementation of adaptive water management strategies that are provided by section 37 of the *NWA* is done through the NWRS. The NWRS has been established pursuant to section 5 of the *NWA*, and it is expected to provide strategic direction on how South Africa's water resources should be managed in 20 years from the date of formulation (2013–2033). In 2013 for instance, the NWRS formulated and steered the implementation of the Impala Irrigation Scheme Water Management Plan as part of its functions.[284] Accordingly, the NWRS has been described as:

> [a]n institutional framework for the protection, use, development, conservation, management and control of water resources, as well as the framework within which water must be managed at catchment level, in defined water management areas.[285]

For connected purposes, Rumble[286] believes that the NWRS, as facilitated by the *NWA*:

> [h]as achieved its purpose especially in translating the theory of integrated water resources management into legislation, allowing for a holistic approach, at least on paper, to water resource management and flexible response mechanisms.

Consequently, it could be stated that the *NWA* appears to contain the type of provisions that enhance the resilience of agricultural practices against climate change-induced water stress and scarcity. However, water uses (in addition to land uses) are also facilitated by the *CARA*.

5.3.2.3 Conservation of Agricultural Resources Act

Previous discussion suggested that law is required to facilitate adaptive land management practices (such as conservation tillage) which are necessary for enhancing the properties, composition, and nutrients of agricultural lands.[287] On the basis of this, South Africa has laws that support adaptive land management. A prominent example of these laws is the *CARA*.[288] Apart from supporting adaptive land management, the *CARA*[289] also served as the basis for the promulgation of other water resources and water-induced erosion regulations.[290] Even though the *CARA* was enacted at a time when the impacts of climate change on agriculture were not so glaringly obvious, the law is still considered to have facilitated several adaptation measures that are currently considered relevant for agriculture.[291] This assertion is illuminated by the objectives of the *CARA*, which are to:

> [p]rovide for the conservation of the natural agricultural resources of the Republic through the maintenance of the production potential of land by the combating and prevention of erosion and weakening or destruction of the water sources, and by the protection of the vegetation and the combating of weeds and invader plants.[292]

In order to achieve the above objectives, section 6 of the *CARA* requires the implementation of strategies such as the cultivation of virgin soils[293] in order to allow previously cultivated soils to replenish their nutrients; the irrigation of lands to provide more soil moisture; the control of soil erosion through the prevention of waterlogs on lands; and the control of water run-off flow pattern.[294] The strategies which section 6 seeks to facilitate are the same types of on-farm adaptation strategies that were previously explained as necessary for agriculture. [295] In addition, section 6(2) of the *CARA* provides for other on-farm adaptation strategies such as control of water run-off flow pattern, protection of water sources against pollution on account of farming practices, and land restoration measures.[296]

Section 9 of the *CARA* is yet another provision which appears relevant to adaptation. Section 9 makes provision for financial support in the form of subsidies and grant schemes to conservation-minded farmers at the local and regional levels.[297] The subsidies and grants are meant to fund farmers who seek to undertake disaster management or disaster risk measures such as the repairing of agricultural lands destroyed by floods, or the restoration of eroded land, as well as for:

> [t]he reduction of the number of animals being kept on land in order to restrict the detrimental effect of drought on that land, and the planting and cultivation of particular crops which improve soil fertility or counteract the vulnerability of soil to erosion.[298]

As an example, these types of subsidies and grants were disbursed by an organisation known as *FundsforNGOs* in 2021, to support smallholder and emerging farmers in Kouga Local Municipality in the Eastern Cape province from 2022 to 2023.[299] Additional examples of subsidies and grants include those that were offered to emerging farmers by other organisations such as the National Development Agency and the Department of Social Development.[300] According to Humby,[301] the provisions of section 9 of *CARA* have (in principle) provided the legal stimulus for insurance against risks and disasters in the agricultural sector, especially climate change-induced droughts and flooding. The relevance of financial assistance, especially subsidies, grants, and insurance to adaptation in the context of agriculture, was discussed in previous chapters of this book.[302]

In spite of the provisions of the *CARA*, which seem to have provided for adaptation in the context of agriculture, some gaps have been noted. For example, a lack of skilled personnel (agricultural technicians and investigators) to advise farmers on soil-conservation strategies and to enforce provisions of the Act is considered a major gap in the *CARA*.[303] This gap forms part of the reasons that necessitated an amendment of the *CARA*; and as a result, a draft Bill on the conservation of agricultural resources has been before parliament since 2013.[304]

As a means of improving the *CARA*, the draft *Conservation of Agricultural* Resources Amendment Bill, 2013 adopts the PTD and makes the national government (through the Minister of Agriculture, Forestry, and Fisheries) the trustee of agricultural resources (including agricultural lands) in South Africa on behalf of the people of South Africa.[305] The rationale for adopting the PTD is to ensure that agricultural resources (particularly lands) are accessed and used in an equitable manner, without compromising environmental quality or economic and social development in the country.[306] In other words, equitable allocation of land, for instance through the PTD, would allow for enough land to accommodate grazing and crop production with less GHG emissions. The PTD is required to support adaptive land management in a country like South Africa where agricultural practices take up much of the land.[307]

Even though the draft *Conservation of Agricultural Resources Amendment Bill,* 2013 seeks to preserve and protect the quality of environmental resources, the draft Bill is considered to have some gaps. For example, the draft Bill lacks reference to important environmental management principles provided by the *NEMA* (discussed above) such as the integration principle, the principle of sustainable development, the equitable access principle, and the polluter pays principle.[308] As previously noted, all environmental policies and laws in South Africa that support adaptation are at least influenced or guided by some of these principles.[309] The draft Bill is supposed to be an improvement of the *CARA*, which was enacted long before the *NEMA*. Without incorporating basic environmental principles, the draft Bill would have failed in achieving its purpose of improving the *CARA*, and it would not likely facilitate the required adaptation processes that it is intended to promote. The gaps in the draft *Conservation of Agricultural Resources Amendment Bill* are just parts of the main challenges for the regulatory framework for agriculture-focused adaptation in South Africa, which are summarised in the next section of this chapter. However, before summarising the main gaps in the South African regulatory framework for agriculture-focused adaptation, it is necessary to review another law in addition to the CARA, which is relevant to adaptive land management – the *Spatial Planning and Land Use Management Act*.

5.3.2.4 Spatial Planning and Land Use Management Act

Driven by the need to redress spatial inequity, and to promote sustainable land use in South Africa, the national government enacted the *SPLUMA* in 2013.[310] The *SPLUMA* is a law that was enacted complementarily to the provisions of the *Constitution*, especially section 24, which guarantees the right to a clean and healthy environment, and section 27, which partly promotes access to sufficient food and water, as well as the right to social security (which in the context of this book also means agricultural insurance).[311] In simplistic terms, part of the objectives of the *SPLUMA* according to section 3 is to "provide for the sustainable and efficient use of land".[312] Taylor et al.[313] argue that spatial planning could lead to sustainable and efficient land use, and sustainable and efficient land use is what informs adaptive land management. Therefore, in order to promote adaptive land management, the *SPLUMA* adopts certain principles of development such as the principle of spatial sustainability and the principle of spatial resilience.[314]

On the one hand, the principle of spatial sustainability is intended to "ensure that special consideration is given to the protection of prime and unique agricultural land".[315] According to Van der Berg, Du Plessis, and Murphree,[316] an example of spatial sustainability is where land use schemes are used in order to improve the usefulness of agricultural lands that depreciated due to drought and flood risk or to control the exploitation (cultivation) of unstable or fragile lands. Taylor et al.[317] rightly point out that the principle of spatial sustainability is not only significant in promoting land use management practices but is also relevant for spatial planning activities which among others is crucial for the protection of agricultural land, food security, and local employment. Furthermore, spatial planning has been noted to be essential in the:

> [t]he provision of infrastructure and social services (for example water provision under lower annual rainfall and higher temperatures); land development in locations that are not at risk to damage from climate events such as flooding from heavy rains and sea storm surges. [318]

Likewise, spatial planning supported by the principle of spatial sustainability could help in limiting certain activities (such as urban influx) which has the tendency of destroying "ecosystems that provide natural buffer capacity from extreme weather and climate events, [which by extension] undermines agricultural productivity".[319]

On the other hand, the principle of spatial resilience has been defined as the use of flexible spatial plans and land use management policies to promote sustainable livelihoods in communities which are susceptible to the impacts of environmental shocks (climate change) and economic losses.[320] The South African Cities Network's Spatial Planning and Land Use Management Programme[321] is an illustration of a spatial plan, while land use management policy can be illustrated by the Spatial Planning and Land Use Management White Paper.[322]

According to Van der Berg, Du Plessis, and Murphree,[323] the principles of "spatial sustainability" and "spatial resilience" in the *SPLUMA* would promote adaptive land management in locations (especially agricultural lands) that were destroyed by, or which appear vulnerable to the impacts of climate change. Consequently, it could be stated that the *SPLUMA* seems to have some provisions that facilitate adaptation in the context of agriculture. However, these notable provisions of the *SPLUMA* do not obviate the need for an overall scrutiny of the South African regulatory framework on climate change adaptation and agriculture.

5.4 Weaknesses of the South African regulatory frameworks for climate change adaptation and agriculture

South Africa's commitment to adapting to the impacts of climate change has been demonstrated through various policies and laws targeting critical sectors of interest such as agriculture. The *Constitution* sets the tone for the formulation of policies and enactment of laws that support adaptation. In this chapter, climate change policies such as the NCCRS, NCCRP, NCCAS, NDP, and the NDC were analysed in addition to environmental laws – the *NEMA*, *DMA*, *CTA*, and the *Climate Change Act*. Aside from these climate change policies and environmental laws, South African agriculture-based policies and laws such as the draft CSASF, CCSPAFF, the Second NAP-CDLDD, as well as the *WSA*, *NWA*, *CARA*, and *SPLUMA* were also analysed. Certain gaps were identified in some of these laws as far as adaptation in the context of agriculture is concerned. This section will summarise these gaps and make further observations where necessary in preparation for solutions, which are discussed in the next chapter of this book.

The NCCRP is among the first set of climate change policies that was formulated in South Africa. However, the policy is faulted for placing more emphasis on mitigation instead of adaptation.[324] In the same context, the mainstreaming of adaptation measures within key government departments such as the DFFE, Department of Water and Sanitation, and the DALRRD are poorly implemented under the NCCRP.[325] Even though the NCCRP recognises the need for adaptation, the policy is not properly aligned with other policies, either vertically or horizontally, in promoting a low-carbon and climate-resilient development in South Africa. For example, the NCCRP is not properly incorporated into the NDC (neither the 2016s nor the 2021 update).[326]

Another climate change policy that facilitates adaptation is the NCCAS. Though the NCCAS acknowledges the susceptibility of climate change impacts on critical sectors of the economy and recommends adaptive measures in those sectors, the policy does not provide direction on how adaptation strategies should be implemented in the sectors.[327] While the NDP underscores the need to enhance the resilience of critical sectors of the economy especially agriculture, and it recommends suitable adaptation strategies with sufficient direction on how the strategies should be implemented, the NDP fails to earmark the budgetary processes and finances for implementing agricultural adaptation strategies.[328] As well, the NDP's projection with regards to the potential of irrigation services in boosting crop production has been noted to be unrealistic, for failing to consider underused lands in communal areas.[329]

Also, it is not possible to have a common reference point for regulating agricultural adaptation (especially at the local level) while relying on the *NEMA*, *DMA*, *CTA*, and the *Climate Change Act*, for they are fragments and duplications of each other.[330] This observation on fragmentation and duplication brings to the fore a similar observation made by Du Plessis and Kotzé[331] that:

> [t]he current law and policy framework for adaptation in South Africa is disjointed and patchy at best and there is still no coherent and explicit outline of what municipalities can and should do in the climate governance context.

Laws (as previously explained) are supposed to coordinate governance actions by guiding the implementation of policy and institutional frameworks.[332] It would appear that the current climate change-related laws somehow breed institutional-based challenges that could impair the effective facilitation of adaptation measures in the context of South Africa's agriculture.[333] For example, institutions that are responsible for enforcing agricultural standards in South Africa, such as the DFFE and the DALRRD, could be hindered from integrating specific adaptation measures that are coordinated by other agencies such as the CCAQM and IMCCC; and by other sectors especially energy and transport.[334]

A final observation with regard to the South African regulatory framework for both climate change adaptation and agriculture is that, despite receiving public comments a long time ago, the draft CSASF and the draft *Conservation Agricultural Resources Amendment Bill* are yet to be finalised.[335] The delay in concluding the formulation of the draft CSASF and promulgating the draft *Conservation Agricultural Resources Amendment Bill* could likely contribute to the general stagnation of adaptation initiatives in the context of agriculture, as well as the inability to mainstream climate change concerns with agricultural development planning in South Africa.

5.5 Conclusion

Climate change in South Africa, just as in most African countries, leads to an increase in the frequency and intensity of extreme weather events such as flooding, water scarcity, and droughts. Given that South Africa is already described as an arid and semi-arid region, climate change impacts such as droughts, water shortages, and rainfall fluctuations in the country could have negative impacts on agricultural practices, especially crops and livestock production.[336] This is because crops and livestock production rely on processes that are dependent on climatic conditions (precipitation, temperature, and fertile soil). Climate change and variability could mean a substantial reduction or loss in crops and animal yield while facing more demand for food and water in South Africa. The impacts of climate change on agriculture in South Africa could also have repercussions on national food security, job creation, and prospects for future economic growth and development in the country.

Therefore, adaptation measures are necessary in responding to the impacts of climate change on the environment and in enabling agricultural practices to adapt to the impacts of climate change. Climate change adaptation in South Africa is supported by many policies such as the NCCRS, NCCRP, draft NCCAS, and laws including the *NEMA, CTA, DMA*, and a *Climate Change Act*; all of which were formulated and enacted in accordance with the UNFCCC, the Kyoto Protocol, and the Paris Agreement and the *Constitution*. These regulatory mechanisms outline the action plans and objectives of South Africa's government in addressing climate change impacts on critical sectors of the economy such as agriculture and water, and in promoting social, economic, and environmental resilience in line with the principles of equity and sustainability. Equally, adaptation in the context of agriculture is facilitated by agriculture-based policies and laws such as the draft NCSASF, CCSPAFF, Second NAP-CDLDD, the *WSA, NWA, CARA*, and the *SPLUMA*.

These policies and laws affirm South Africa's readiness to adapt to the impacts of climate change. However, an analysis of these laws and policies reveals that they contain gaps, as far as adaptation in critical sectors of South Africa's economy including agriculture are concerned.

Notes

1 See Chapter 2, Section 2.2.
2 See The *Constitution of the Republic of South Africa*, 1996. https://www.justice.gov.za/legislation/constitution/SAConstitution-web-eng.pdf.
3 South African National Climate Change Response Strategy 2004 https://unfccc.int/sites/default/files/sem_sup3_south_africa.pdf.
4 National Climate Change Response White Paper GN 757/Government Gazette No. 34695 https://www.gov.za/sites/default/files/gcis_document/201409/nationalclimatechangeresponsewhitepaper0.pdf.
5 National Climate Change Adaptation Strategy 2019 https:// https://www.dffe.gov.za/sites/default/files/docs/nationalclimatechange_adaptationstrategy_ue10november2019.pdf.

6 National Development Plan 2011 https://www.nstf.org.za/wp-content/uploads/2018/04/All-The-NDP.pdf.
7 South Africa First Nationally Determined Contribution under the Paris Agreement https://unfccc.int/sites/default/files/NDC/2022-06/South%20Africa%20updated%20first%20NDC%20September%202021.pdf, 2.
8 *National Environmental Management Act* 107 of 1998.
9 Act 57 of 2002.
10 Act No. 15 of 2019.
11 Apart from making efforts to address the impacts of climate change within the country, the Government of South Africa has also played key role in combatting the effects climate change in a global context, by serving as a host to some international climate change agreement and agenda such as the Johannesburg Plan on Sustainable Development and the Durban COP. For example, the World Summit on Sustainable Development (WSSD) was named the Johannesburg Declaration, because Johannesburg South Africa served as the venue for the WSSD in 2002. The summit while affirming the Rio Declaration of 1992 require world nations "to build a humane, equitable and carry global society, cognizant of the need for human dignity for all." See Johannesburg Declaration on Sustainable Development https://www.un.org/esa/sustdev/documents/WSSD_POI_PD/English/POI_PD.htm.
South Africa also hosted the UNFCCC COP 17 in Durban from 28 November to 11 December 2011. The aim of COP 17 was to negotiate a new treaty for limiting global carbon emissions. UNFCCC 2011 COP17-Documents. https://unfccc.int/process-and-meetings/conferences/past-conferences/durban-climate-change-conference-november-2011/cop-17/cop-17-documents.
12 It must be emphasised that there is no specific agricultural law that directly provide for adaptation in the context of agriculture. There are however agriculture-related laws that contains provisions which promotes adaptation measures for agriculture in South Africa.
13 Draft Climate Smart Agriculture Strategic Framework for Agriculture, Forestry, and Fisheries https://www.gov.za/sites/default/files/gcis_document/201808/41811gen428.pdf.
14 Department of Agriculture, Forestry, and Fisheries, Notice 7 of 2013 https://www.gov.za/sites/default/files/gcis_document/201409/36063gen7.pdf.
15 Department of Forestry, Fisheries, and the Environment 2018 https://www.dffe.gov.za/sites/default/files/docs/nap_desertification_land_degradation_drought effects.pdf.
16 *Water Services Act*, 108 of 1997.
17 *National Water Act*, 36 of 1998.
18 *Conservation of Agricultural Resources Act* 43 of 1983.
19 *Draft Conservation of Agricultural Resources Amendment Bill* 2013 https://www.gov.za/sites/default/files/gcis_document/201409/36605gen673.pdf.
20 *Spatial Planning and Land Use Management Act* 16 of 2013.
21 See Chapter 2, Section 2.3.
22 The signing and ratification of these international legal instruments was done in accordance with Sections 39, 231 and 233 of the *Constitution of South Africa*, 1996 which is discussed below.
23 South African National Climate Change Response Strategy 2004 https://unfccc.int/sites/default/files/sem_sup3_south_africa.pdf5. See also UNFCCC National Reports from non-Annex I Parties. https://unfccc.int/national-reports-from-non-annex-i-parties.
24 South Africa's GHG emission is expected to rise from 398 MtCO2e to 614 MtCO2e between 2025 and 2030. See Carbon Brief 2018 https://www.carbonbrief.org/the-carbon-brief-profile-south-africa/ .

25 See Chapters 5, 6, and 7 of the *Constitution*.
26 A Taylor, H Davies, G Oelofse and S Roux 2016 2016 Urban adaptation, In T Humby, L J Kotzé O Rumble and A Gilder (Eds.) *Climate Change: Law and Governance in South Africa*. Juta, 280.
27 Department of Forestry, Fisheries, and the Environment https://www.dffe.gov.za/.
28 See Climate Commission 2020 https://www.climatecommission.org.za/.
29 Presidential Climate Commission Report 2022 https://pccommissionflo.imgix.net/uploads/documents/A-Just-Transition-Framework-for-South-Africa-with-dedication-FSP-002.pdf, 20.
30 See North West Department: Economic Development, Environment, Conservation and Tourism https://provincialgovernment.co.za/units/view/243/north-west/economic-development-environment-conservation-and-tourism.
31 SeeChapter1,Section1.3.3;UNFCCC2020ClimateChangeIsanIncreasingThreatto Africa. https://unfccc.int/news/climate-change-is-an-increasing-threat-to-africa.
32 The *Constitution* gave local governments broad rights to manage the affairs of its constituents, including the promotion of a safe and healthy environment. See Section 152(1) (b) and (d) of the *Constitution*; A Du Plessis 2010 " Local environmental governance" and the role of local government in realising Section 24 of the South African Constitution. *Stellenbosch Law Review*, 21(2), 265–297.
33 J Eyzaguirre and FJ Warren 2014 Adaptation: Linking research and practice. In FJ Warren and DS Lemmen (Eds) *Canada in a changing climate: Sector perspectives on impacts and adaptation*. Government of Canada, 266–267.
34 A Du Plessis and LJ Kotzé 2014 The heat is on: Local government and climate governance in South Africa. *Journal of African Law,* 58(1), 145–174.
35 A Taylor, H Davies, G Oelofse and S Roux 2016 2016 Urban adaptation, In T Humby, L J Kotzé O Rumble and A Gilder (Eds.) *Climate Change: Law and Governance in South Africa*. Juta, 280.
36 For a comprehensive definition of IDPs, see section 25 of the *Local Government: Municipal Systems Act*, 32 of 2000. See in addition, H Bulkeley and M Betsill 2005 Rethinking sustainable cities: Multilevel governance and the'urban'politics of climate change. *Environmental Politics,* 14(1), 42–63; DJ Gordon and CA Johnson 2017 The orchestration of global urban climate governance: Conducting power in the post-Paris climate regime. *Environmental Politics,* 26(4), 694–714; N Stern and D Zenghelis Climate change and cities: a prime source of problems, yet key to a solution. *The Guardian*, https://www.theguardian.com/cities/2015/nov/17/cities-climate-change-problems-solution.
37 MX Lethoko 2016 Inclusion of climate change strategies in municipal Integrated Development Plans: A case from seven municipalities in Limpopo Province, South Africa. *Jàmbá: Journal of Disaster Risk Studies*, 8(3), a245.
38 A Taylor, H Davies, G Oelofse and S Roux 2016 2016 Urban adaptation, In T Humby, L J Kotzé O Rumble and A Gilder (Eds.) *Climate Change: Law and Governance in South Africa*. Juta, 285.
39 A van der Berg, A Du Plessis and M Murphree 2016 Local disaster risk reduction and management in South Africa's response to climate change In T Humby, LJ Kotzé, O Rumble and A Gilder (Eds.) *Climate change: Law and governance in South Africa*. Juta, 382.
40 A Taylor, H Davies, G Oelofse and S Roux 2016 2016 Urban adaptation, In T Humby, L J Kotzé O Rumble and A Gilder (Eds.) *Climate Change: Law and Governance in South Africa*. Juta, 282. 282.
41 Schedule 4, part A to the *Constitution*.
42 Section 55 of the *Constitution*.
43 Section 84 of the *Constitution*.
44 See Department of Forestry, Fisheries, and the Environment https://www.dffe.gov.za/ccaq_home

45 Climate Change Institutional Arrangement – South Africa https://www.un-gsp.org/sites/default/files/documentos/sa_insitutional_arrangement_case_study.pdf, 3.
46 For more information on the roles of the CCAQM and the IMCCC, see for example, A Averchenkova, KE Gannon and P Curran 2019 Governance of climate change policy: A case study of South Africa. *Grantham Research Institute on Climate Change and the Environment Policy Report,* 1–40.
47 L Masters and L Duff 2012 *Overcoming barriers to climate change adaptation implementation in Southern Africa*. Africa Institute of South Africa, 26.
48 (Previously called Department of Agriculture, Forestry, and Fisheries).
49 L Masters and L Duff 2012 *Overcoming barriers to climate change adaptation implementation in Southern Africa*. Africa Institute of South Africa, 26.
50 This has been stated in Chapter 2, Section 2.2.
51 Section 2 of the *Constitution* provides that: "This Constitution is the supreme law of the Republic; law or conduct inconsistent with it is invalid, and the obligations imposed by it must be fulfilled".
52 J Locke 1689 *Locke: Two treatises of government.* Cambridge University Press, 43–46. This is also described as social contract. An agreement which creates obligation for the government as a common trustee of the community. See also JA Harris 2020 The interpretation of Locke's two treatises in Britain, 1778–1956. *British Journal for the History of Philosophy,* 494.
53 Section 24(1) of the *Constitution*.
54 For example, Du Plessis and Kotzé argue that the measures provided by section 24 "could include anything that answers to the dictates of the environmental right." See A Du Plessis and LJ Kotzé 2014 The heat is on: Local government and climate governance in South Africa. *Journal of African Law,* 58(1),145–174.
55 See the provisions of section 231 (1) to (5) of the *Constitution*, which accommodates international treaties and conventions.
56 See DFFE 2016 Minister Edna Molewa signs Paris Agreement on climate change https://www.gov.za/speeches/south-africa-signs-paris-agreement-climate-change-new-york-22-apr-2016-0000#:~:text=Minister%20Molewa%20signed%20the%20Agreement,change%20regime%20under%20the%20UNFCCC.
57 Section 24(1) (b) (iii) of the *Constitution*.
58 Ecologically sustainable development and justifiable economic and social development is also supported by the UNFCCC, and this has been previously explained in Chapter 2, Section 2.3.1.
59 A van der Berg, A Du Plessis and M Murphree 2016 Local disaster risk reduction and management in South Africa's response to climate change In T Humby, LJ Kotzé, O Rumble and A Gilder (Eds.) *Climate change: Law and governance in South Africa*. Juta, 324.
60 A van der Berg, A Du Plessis and M Murphree 2016 Local disaster risk reduction and management in South Africa's response to climate change In T Humby, LJ Kotzé, O Rumble and A Gilder (Eds.) *Climate change: Law and governance in South Africa*. Juta, 324.
61 Section 27(1) (b) of the Constitution. See also CB Soyapi 2017 Water security and the right to water in Southern Africa: An overview. *Potchefstroom Electronic Law Journal,* 20 (1),14.
62 See Chapter 1, Section 1.3.3.
63 Section 27(2) of the *Constitution*.
64 Section 27(1) (c) of the *Constitution*. See also the preamble to the *Social Assistance Act* 13 of 2004, which recognise insurance services as ways of addressing human security needs.
65 AO Jegede and UC Mokoena 2019 Thinking into the future: Constructing social security law as climate change adaptation strategy in urban South Africa. In P B

Cobbinah and M Addaney (Eds.), *The geography of climate change adaptation in Urban Africa*. Palgrave Macmillan, 329–353.
66 See Chapter 1, Section 1.5.1.3; Chapter 2, Section 2.4.2.3.
67 See also National Climate Change Adaptation Strategy Republic of South Africa https://www.dffe.gov.za/sites/default/files/docs/nationalclimatechange_adaptationstrategy_ue10november2019.pdf, 38.
68 South African National Climate Change Response Strategy 2004 https://unfccc.int/sites/default/files/sem_sup3_south_africa.pdf 2.
69 South African National Climate Change Response Strategy 2004 https://unfccc.int/sites/default/files/sem_sup3_south_africa.pdf, 11.
70 South African National Climate Change Response Strategy 2004 https://unfccc.int/sites/default/files/sem_sup3_south_africa.pdf, 10. See also RM Garland 2014 National policy response to climate change in South Africa. *South African Medical Journal,* 584–584; A Averchenkova, KE Gannon and P Curran 2019 *Governance of climate change policy: A case study of South Africa. Grantham Research Institute on Climate Change and the Environment Policy Report,* 1–40.
71 South African National Climate Change Response Strategy 2004 https://unfccc.int/sites/default/files/sem_sup3_south_africa.pdf, 6.
72 South African National Climate Change Response Strategy 2004 https://unfccc.int/sites/default/files/sem_sup3_south_africa.pdf, 1–48.
73 South African National Climate Change Response Strategy https://unfccc.int/sites/default/files/sem_sup3_south_africa.pdf, 8.
74 M Madzwamuse 2010 "Drowning voices: The climate change discourse in South Africa." *Policy Brief* 5, no. 8. It should be noted that in addition to climate change adaptation, the NCCRS was also formulated to promote poverty alleviation, access to basic amenities including infrastructure development, job creation, rural development, foreign investment, human resource development, and improved health, leading to sustainable economic growth. See South African National Climate Change Response Strategy 2004 https://unfccc.int/sites/default/files/sem_sup3_south_africa.pdf, 2.
75 South African National Climate Change Response Strategy 2004 https://unfccc.int/sites/default/files/sem_sup3_south_africa.pdf, 10.
76 South African National Climate Change Response Strategy 2004 https://unfccc.int/sites/default/files/sem_sup3_south_africa.pdf, 19.
77 See Chapter 1, Section 1.5.1.
78 South African National Climate Change Response Strategy 2004 https://unfccc.int/sites/default/files/sem_sup3_south_africa.pdf, 19.
79 National Climate Change Response White Paper GN 757/Government Gazette No. 34695 https://www.gov.za/sites/default/files/gcis_document/201409/nationalclimatechangeresponsewhitepaper0.pdf, 38. The NCCRP was published shortly before the commencement of the UNFCCC COP 17 in Durban.
80 National Climate Change Response White Paper GN 757/Government Gazette No. 34695 https://www.gov.za/sites/default/files/gcis_document/201409/nationalclimatechangeresponsewhitepaper0.pdf, 11.
81 National Climate Change Response White Paper GN 757/Government Gazette No. 34695 https://www.gov.za/sites/default/files/gcis_document/201409/nationalclimatechangeresponsewhitepaper0.pdf, 16.
82 K Lehman 2016 South Africa's climate change commitments and regulatory response potential. In T Humby, LJ Kotzé, O Rumble and A Gilder (Eds.) *Climate change: Law and governance in South Africa*. Juta, 211.
83 M Madondo and HM Nkwana 2021 Exploring South Africa's National Climate Change Response White Paper, 2011. *Administratio Publica* 29(2),187–206.
84 K Lehman 2016 South Africa's climate change commitments and regulatory response potential. In T Humby, LJ Kotzé, O Rumble and A Gilder (Eds.) *Climate change: Law and governance in South Africa*. Juta, 212.

85 National Climate Change Response White Paper GN 757/Government Gazette No. 34695 https://www.gov.za/sites/default/files/gcis_document/201409/nationalclimatechangeresponsewhitepaper0.pdf, 24. 16–
86 National Climate Change Response White Paper 2011 https://www.dffe.gov.za/sites/default/files/legislations/national_climatechange_response_whitepaper.pdf, 44.
87 National Climate Change Response White Paper GN 757/Government Gazette No. 34695 https://www.gov.za/sites/default/files/gcis_document/201409/nationalclimatechangeresponsewhitepaper0.pdf, 24.
88 Chapter 5 of the NCCRP provides for an overall approach to adaptation including policies, legislation and how adaptation should be integrated into development plans and strategies within and across various sectors such as water agriculture and forestry, health, biodiversity and ecosystems, human settlements disaster risk reduction, and management. See National Climate Change Response White Paper GN 757/Government Gazette No. 34695 https://www.gov.za/sites/default/files/gcis_document/201409/nationalclimatechangeresponsewhitepaper0.pdf, 16–24.
89 National Climate Change Response White Paper GN 757/Government Gazette No. 34695 https://www.gov.za/sites/default/files/gcis_document/201409/nationalclimatechangeresponsewhitepaper0.pdf, 17–29.
90 B Biazin, G Sterk, M Temesgen, A Abdulkedir and L Stroosnijder 2012 Rainwater harvesting and management in rainfed agricultural systems in sub-Saharan Africa – A review. *Physics and Chemistry of the Earth, Parts A/B/C*, 47, 139–151.
91 National Climate Change Response White Paper GN 757/Government Gazette No. 34695 https://www.gov.za/sites/default/files/gcis_document/201409/nationalclimatechangeresponsewhitepaper0.pdf, 17. 17.
92 In the same regard, Rumble also highlighted the role of NWRS in managing the challenges, especially climate change impacts against water resources in South Africa. See O Rumble 2016 Water resources management and climate change. In T Humby, LJ Kotzé, O Rumble and A Gilder (Eds.) *Climate change: Law and governance in South Africa*. Juta, 230.
93 A Taylor, H Davies, G Oelofse and S Roux 2016 2016 Urban adaptation, In T Humby, L J Kotzé O Rumble and A Gilder (Eds.) *Climate Change: Law and Governance in South Africa*. Juta, 281.
94 National Climate Change Response White Paper GN 757/Government Gazette No. 34695 https://www.gov.za/sites/default/files/gcis_document/201409/nationalclimatechangeresponsewhitepaper0.pdf, 16.
95 Most of the people who are vulnerable to the impacts of climate change in African countries live in rural areas and these people are under the direct control of local or municipal governments. See JS Wunsch 2001 Decentralization, local governance and 'recentralization' in Africa. *Public Administration and Development: The International Journal of Management Research and Practice,* 21(4) 277–288; L Mai 2019 Book Reviews: Governing climate change: Global cities and transnational lawmaking. *Transnational Environmental Law,* 8(1),193–197.
96 See Section 10.2.6 of the NCCRP.
97 A van der Berg, A Du Plessis and M Murphree 2016 Local disaster risk reduction and management in South Africa's response to climate change In T Humby, LJ Kotzé, O Rumble and A Gilder (Eds.), *Climate change: Law and governance in South Africa*. Juta, 315.
98 Disaster risk reduction responses are also facilitated by laws such as the *DMA* (discussed below). See for example, Chapter 2 of the *DMA*.
99 K Lehman 2016 South Africa's climate change commitments and regulatory response potential. In T Humby, LJ Kotzé, O Rumble and A Gilder (Eds.) *South Africa's climate change commitments*. Juta, 211.

100 IPCC 2014 Climate Change *Synthesis report.* https://www.ipcc.ch/site/assets/uploads/2018/05/SYR_AR5_FINAL_full_wcover.pdf45.

101 M Madondo and HM Nkwana 2021 Exploring South Africa's National Climate Change Response White Paper, 2011 *Administratio Publica* 29(2), 187–206.

102 G Ziervogel, M New, E Archer van Garderen, G Midgley, A Taylor, R Hamann and M Warburton 2014 Climate change impacts and adaptation in South Africa. *Wiley Interdisciplinary Reviews: Climate Change,* 5(5), 605–620.

103 National Climate Change Adaptation Strategy 2019 https://www.dffe.gov.za/sites/default/files/docs/nationalclimatechange_adaptationstrategy_ue10november2019.pdf, 27.

104 National Climate Change Adaptation Strategy 2019 https://www.dffe.gov.za/sites/default/files/docs/nationalclimatechange_adaptationstrategy_ue10november2019.pdf, 12.

It should be noted that most national adaptation strategies in the SADC region are guided by the [draft] SADC Climate Change Strategy and Action Plan of 2015. Given the susceptibility of Africa to climate change impacts, the [draft] SADC Climate Change Strategy and Action Plan calls for enhanced adaptation responses in the SADC region and Africa in general. The [draft] SADC action plan seek to integrate regional and national policies, strategies and protocols for climate change responses in Africa. See Southern African Development Community 2015 SADC Climate Change Strategy and Action Plan. https://www.sadc.int/sites/default/files/2021-11/SADC_Climate_Change_Strategy_and_Action_Plan-English.pdf.

105 National Climate Change Adaptation Strategy 2019 https://www.dffe.gov.za/sites/default/files/docs/nationalclimatechange_adaptationstrategy_ue10november2019.pdf, 12.

106 National Climate Change Adaptation Strategy 2019 https://www.dffe.gov.za/sites/default/files/docs/nationalclimatechange_adaptationstrategy_ue10november2019.pdf, 12.

107 National Climate Change Adaptation Strategy 2019 https://www.dffe.gov.za/sites/default/files/docs/nationalclimatechange_adaptationstrategy_ue10november2019.pdf, 7.

108 National Climate Change Adaptation Strategy 2019 https://www.dffe.gov.za/sites/default/files/docs/nationalclimatechange_adaptationstrategy_ue10november2019.pdf, 21.

109 National Climate Change Adaptation Strategy 2019 https://www.dffe.gov.za/sites/default/files/docs/nationalclimatechange_adaptationstrategy_ue10november2019.pdf, 19. 19.

110 See for example National Planning Commission Revised Green Paper 2010 https://www.nationalplanningcommission.org.za/assets/Documents/NPC%20Revised%20Green%20Paper%202010.pdf, 1–13.

111 National Climate Change Adaptation Strategy 2019 https://www.dffe.gov.za/sites/default/files/docs/nationalclimatechange_adaptationstrategy_ue10november2019.pdf, 19.

112 National Climate Change Adaptation Strategy 2019 https://www.dffe.gov.za/sites/default/files/docs/nationalclimatechange_adaptationstrategy_ue10november2019.pdf, 7.

113 G Ziervogel, M New, E Archer van Garderen, G Midgley, A Taylor, R Hamann and M Warburton 2014 Climate change impacts and adaptation in South Africa. *Wiley Interdisciplinary Reviews: Climate Change,* 5(5), 605–620.

114 National Development Plan 2011 https://www.nstf.org.za/wp-content/uploads/2018/04/All-The-NDP.pdf.

115 Some economic promoting activities such as mineral exploitation, especially coal mining, which destroy the environment have been taking place in South

Africa since the late 19th century. Therefore, the NDP seeks to protect South Africa's environment while allowing the country to benefit from its mineral deposits. National Development Plan 2011 https://www.nstf.org.za/wp-content/uploads/2018/04/All-The-NDP.pdf.

116 National Development Plan 2011 https://www.nstf.org.za/wp-content/uploads/2018/04/All-The-NDP.pdf. It has been discussed below that the NEMA consider adaptation as one of the strategies for protecting natural resources in South Africa, such as water and land (which are vital for agricultural practices) against the impacts of climate change.

117 National Development Plan 2011 https://www.nstf.org.za/wp-content/uploads/2018/04/All-The-NDP.pdf.

118 National Planning Commission of South Africa, National Development Plan 2030 https://www.nationalplanningcommission.org.za/assets/Documents/ndp-2030-our-future-make-it-work.pdf, 1–148.

119 National Development Plan 2011 2011 https://www.nstf.org.za/wp-content/uploads/2018/04/All-The-NDP.pdf.

120 National Development Plan 2011 2011 https://www.nstf.org.za/wp-content/uploads/2018/04/All-The-NDP.pdf.

121 IP Molobela and P Sinha 2011 Management of water resources in South Africa: A review. *African Journal of Environmental Science and Technology,* 5(12), 993–1002.

122 See for example, OECD 2018 OECD Work on Water. https://www.oecd.org/content/dam/oecd/en/topics/policy-sub-issues/water/brochure-OECD-work-on-water.pdf, 1–36.

123 See RJ Scholes, M Scholes and M Lucas 2015 *Climate change: Briefings from Southern Africa*. Wits University Press, 114.

124 The provisions of article 4.9 have been discussed previously. See Chapter 2, Section 2.3.3. It must be noted that South Africa submitted its first NDC to the UNFCCC secretariat in November 2016 the same time when the country ratified the Paris Agreement. However, the first NDC was updated in 2021. See https://unfccc.int/sites/default/files/NDC/2022-06/South%20Africa%20updated%20first%20NDC%20September%202021.pdf, 1–33.

125 UNFCC 2022 South Africa First Nationally Determined Contribution under the Paris Agreement (Updated September 2021) https://unfccc.int/sites/default/files/NDC/2022-06/South%20Africa%20updated%20first%20NDC%20September%202021.pdf, 2.

126 F Kalame and B Mantlana 2021 NDC implementation in South Africa through green investments by private sector - A Scoping Study https://www.afdb.org/en/documents/ndc-implementation-south-africa-through-green-investments-private-sector-scoping-study.

127 KB Mantlana, MA Maoela and G Nhamo 2021 Mapping South Africa 's nationally determined contributions to the targets of the sustainable development goals. In *Natural Resources Forum*. Oxford, UK: Blackwell Publishing Ltd, 45(1), 3–17.

128 F Kalame and B Mantlana 2021 NDC implementation in South Africa through green investments by private sector - A Scoping Study https://www.afdb.org/en/documents/ndc-implementation-south-africa-through-green-investments-private-sector-scoping-study.

129 By virtue of article 4.9 of the Paris Agreement, parties are expected to submit their NDC to the UNFCCC secretariat after every five years interval. Therefore, the updated NDC will operate till 2025, when the second NDC of South Africa is expected to be communicated. See Presidential Climate Comission 2021 South Africa's NDC Targets for 2025 and 2030. https://pccommissionflow.imgix.net/uploads/images/1eb85a_75d745eb859d43c288f461810b336dd3-compressed.pdf.

130 South Africa's First NDC 2021 Update https://unfccc.int/sites/default/files/NDC/2022-06/South%20Africa%20updated%20first%20NDC%20September%202021.pdf, 3.
131 UNFCC 2022 South Africa First Nationally Determined Contribution under the Paris Agreement (Updated September 2021). https://unfccc.int/sites/default/files/NDC/2022-06/South%20Africa%20updated%20first%20NDC%20September%202021.pdf, 4.
132 UNFCC 2022 South Africa First Nationally Determined Contribution under the Paris Agreement (Updated September 2021). https://unfccc.int/sites/default/files/NDC/2022-06/South%20Africa%20updated%20first%20NDC%20September%202021.pdf, 10.
133 See Chapter 2, Section 2.2.
134 M Marais, FP Retief, LA Sandham and DP Cilliers 2015 Environmental management frameworks: Results and inferences of report quality performance in South Africa. *South African Geographical Journal,* 97(1),83–99; Johan Nel and W du Plessis 2001 An evaluation of NEMA based on a generic framework for environmental framework legislation. *South African Journal of Environmental Law and Policy,* 8(1),1–36.
135 L Du Toit and J Glazewski 2013 International climate change law. In, J Glazewski and L Du Toit (eds.) *Environmental Law in South Africa.*Lexis Nexis. 6.
136 These principles are provided in Section 2(1) (c) NEMA, and they are further listed below.
137 A Du Plessis and LJ Kotzé 2014 The heat is on: Local government and climate governance in South Africa. *Journal of African Law,* 58(1),145–174.
138 Section 23, *NEMA*. The principles of CBDR&RC have been highlighted in Chapter 2, Section 2.3.3.
139 Section 2, *NEMA*.
140 See Section 5.2.1.
141 HA Strydom, ND King and FP Retief 2018 *Environmental management in South Africa* 3rd ed. Juta, 138–145.
142 See also M Kidd 2008 *Environmental Law.* Juta.
143 Section 2(2) of *NEMA*.
144 Section 2(3)–(4), *NEMA*. The sustainable development principle is also entrenched in Section 24 of the Constitution which guarantees the right of everyone in the country to a healthy environment, as well as the protection of natural resources in the pursuit of economic or social development for the present generations which does not compromise the ability of future generation to meet their own needs. See also A Taylor, H Davies, G Oelofse and S Roux 2016 2016 Urban adaptation, In T Humby, L J Kotzé O Rumble and A Gilder (Eds.) *Climate Change: Law and Governance in South Africa*. Juta, 289.
145 Section 2(4) (b), *NEMA*. This environmental management principle advocates for the integration of all elements of the environment (as an entity that is linked and interrelated) in decision making considering its implications on the people of South Africa. Section 23 of the *NEMA* further explains the purpose of integration means and it further provides how integration can be implemented. See the provisions of Section 23 of the *NEMA*. See also LJ Kotzé 2005 *A legal framework for integrated environmental governance in South Africa and the North-West Province*. Doctoral-thesis North-West University, 104.
146 Compensation of victims of environmental degradation or implementation of communal coping strategies against the impacts of climate change for instance must be done fairly without discrimination. Section 2(4) (c), *NEMA*.
147 Section 2(4) (d), *NEMA*.

148 Section 2(4) (o), *NEMA*. Just like in any jurisdiction, the public trust doctrine is applied in South Africa to appoint the state as a trustee in the protection and preservation of the ecological integrity of South Africa. Relying on this environmental management principle, the state must plan and implement adaptation measures for the benefit of the community at large.

149 Of significance to adaptation is also the precautionary principle, which is supported by Section 2(4) (a) (vii) of *NEMA*. The precautionary principle also known as the foresight is premised on the duty of care which require that responsive actions should be taken in order to adjust or avoid environmental hazards (climate change) notwithstanding gap or uncertainty within scientific knowledge. See Section 28 of *NEMA*; Du Plessis 2004 *SAJELP* 140; J Glazewski *The Nature and Scope of Environmental Law*,In Environmental Law in South Africa. *LexisNexis South Africa*, 1–25. In the context of agriculture, an application of the precautionary would requires farmers "to take anticipatory action to avoid future risk, where the occurrence thereof is uncertain". See Strydom *Essentialia of International Environmental Law*, 78. Therefore, the precautionary principle would demand that adaptation measures be prepared in anticipation of climate change impacts being aggravated as a result of the proposed activity.

150 The application of this principle as provided by Section 2(4) (p) of *NEMA* could facilitate adaptation measures, when polluters in anticipation of penalties are forced to alter their actions, such as crops production methods which pollutes the environments. The polluter pays principle places a statutory duty of care on all people to prevent, minimise, and remedy environmental damage. See M Muriel and K Judith 2016 Climate change and insurance law. In T Humby, L J Kotzé O Rumble and A Gilder (Eds.) *Climate Change: Law and governance in South Africa*, Juta, 345.

151 See HA Strydom, ND King and FP Retief 2018 *Environmental management in South Africa,* 3rd ed. Juta, 11.

152 S Bonetti, EH Sutanudjaja, T Mabhaudhi, R Slotow and C Dalin 2022 Climate change impacts on water sustainability of South African crop production. *Environmental Resources Letter*, 17(8), 084017.

153 Glazewski 2013 *The National Environmental Management Act*, In Environmental Law in South Africa. *LexisNexis South Africa*, 1–39.

154 Section 24 of the NEMA; Government Gazette R983–985 GG 38282 of 4 December 2014. See also T Humby 2014 Law and resilience: mapping the literature. *Seattle Journal of Environmental Law*, 4, 85.

155 Appendix 1 and Appendix 2 of GN982 GG 38282 of 4 December 2014.

156 T Humby 2016 The right to development-in-environment and its ecological and developmental thresholds. *South African Journal on Human Rights*, 32(2), 219–247.

157 Rumble 2016 Water Resources Management and Climate Change. In T Humby, LJ Kotzé, O Rumble and A Gilder (Eds.) *Climate change: Law and governance in South Africa*. Juta, 239.

158 MJ Peterson 2017 Politics and law in international environmental governance. In *Research handbook on the politics of international law*. Edward Elgar Publishing, 446–447.

159 C Sadoff and M Muller 2009 *Water management, water security and climate change adaptation: Early impacts and essential responses*. Global Water Partnership, 73.

160 The nine provinces in South Africa include Limpopo, Mpumalanga, Gauteng, Northwest, Free State, Kwazulu-Natal, Northern Cape, Eastern Cape, and Western Cape. From among these provinces, it is only in Limpopo that severe drought

has not been reported. See Parliamentary Monitoring Group 2019 Drought update; Provinces' state of readiness for 2019/20 planting season. https://pmg.org.za/committee-meeting/29261/.

161 A van der Berg, A Du Plessis and M Murphree 2016 Local disaster risk reduction and management in South Africa's response to climate change In T Humby, LJ Kotzé, O Rumble and A Gilder (Eds.) *Climate change: Law and governance in South Africa*. Juta, 326.

162 See section 1 of the *DMA* for a comprehensive definition of what constitute a disaster.

163 The Western Cape is a coastal province that is also been noted for its frequent flooding which could be attributed to climate change and in addition to poor drainage systems and the clogging of waterways. The most recent flooding was recorded in June 2021. See K Dube, G Nhamo and D Chikodzi 2021 Flooding trends and their impacts on coastal communities of Western Cape Province, South Africa. *GeoJournal*, 87(4), 453–468.

164 Preamble to the *DMA*, 2002. Disaster Management Act [No. 57 of 2002]. https://www.cogta.gov.za/cgta_2016/wp-content/uploads/2016/06/DISASTER-MANAGEMENT-ACT.pdf.

165 See sections 1 and 7(2) (e), *DMA*, 2002. These approaches could also mean integrated adaptation responses that were explained under the NCCRP. See Section 5.2.2.2 above.

166 See Chapter 2 of the *DMA* 57 of 2002.

167 See long title to the *DMA* 57 of 2002; S Llosa and I Zodrow 2011 Disaster risk reduction legislation as a basis for effective adaptation. *Global Assessment Report on Disaster Risk Reduction*, 1–18.

168 Section 4(3) of the *DMA*.

169 See Section 29 of the *DMA*; T Humby 2014 Law and resilience: mapping the literature. *Journal of Environmental Law*,4, 85.

170 Section 4(1) (a)–(c) of the *DMA*.

171 Section 53(2) (a) of the *DMA*.

172 The implications for droughts and floods on agriculture has been repeatedly emphasised throughout this book, as such there is no need to revisit the subject again.

173 The new law is described as *Disaster Management Amendment Act*, Act No. 16 of 2015.

174 According to Section 1 of the *DMA* 2015 "disaster risk reduction means either a policy goal or objective, and the strategic and instrumental measures employed for—(a) anticipating future disaster risk; (b) reducing existing exposure, hazard or vulnerability; and (c) improving resilience."

175 See section 53 (1) (e) of the *DMA*, 2015.

176 See Chapter 1, Section 1.2 generally.

177 See Chapter 1, Section 1.4.2.

178 See section 1 of the *DMA*, 2015.

179 See Chapter 2, Section 2.3.1 and Chapter 3, Section 3.2.1.1.

180 Chapter 3, Section 3.2.1.1.

181 Act 15 of 2019.

182 See sections 7 and 8 of the *CTA*.

183 See schedule 2 of the *CTA*.

184 See Chapter 1, Section 1.4.

185 C Rodseth and BN Sustainability 2019 Carbon tax and agriculture. Retrieved from Confronting Climate Change https://mbtpetroleum.co.za/wp-content/uploads/2022/07/Carbon-tax-wine-and-fruit.pdf.

186 Section 1.
187 See schedule 2 of the *CTA*.
188 Direct tax are levies which farmers will be charged for emissions resulting from burning of biomass and the application of fertilisers especially urea, while indirect tax are the levies which farmers will be charged as a result of Nitrous oxide (N_2O) and the use of diesel-powered generators. See schedules 1 and 2 of the *CTA*.
189 See Section 23 of the *NEMA*, 1998.
190 A Taylor, H Davies, G Oelofse and S Roux 2016 Urban adaptation, In T Humby, L J Kotzé O Rumble and A Gilder (Eds.) *Climate Change: Law and Governance in South Africa*. Juta, 282.
191 The *Climate Change Act* 2022 has been assented after a six-year gestation. See *Climate Change Act* 22, of 2024. https://cer.org.za/wp-content/uploads/2024/07/Climate-Change-Act-22-of-2024.pdf, 151; O Rumble 2024 South Africa finally gets its *Climate Change Act* https://africanclimatewire.org/2024/07/south-africa-finally-gets-its-climate-change-act/
192 See section 35 of the *Climate Change Act*.
193 The process of promulgating a law under the *Constitution* is rigorous, and takes a lot of time. See The Parliament of the Republic of South Africa 2024. How a law is made,https://www.parliament.gov.za/how-law-made#:~:text=It%20is%20debated%20in%20the,a%20law%20of%20the%20land.
194 These principles were given as the sustainable development principle, integration principle, environmental justice principle, equitable access principle, public trust principle, and the polluter pays principle among others. See section 23 of the *NEMA*, 1998.
195 See Preamble to the *Climate Change Act* [B 9—2022].
196 For example, section 5 of the *Climate Change Act* reads: "This Act must, in relation to climate change, be read, interpreted and applied in conjunction with the National Environmental Management Act." In resolving conflict between the *Climate Change Act* with other legislation, section 6 of the *Climate Change Act* states that: "In the event of any conflict between a provision of this Act and other legislation specifically relating to climate change, the section of this Act prevails."
197 Especially schedule 1 and 2 as well as sections 7 and 8 of the *Carbon Tax Act*.
198 See Preamble to the *Climate Change Act* [B 9—2022].
199 Republic of South Africa 2024 *Climate Change Act* 22, 2024. https://cer.org.za/wp-content/uploads/2024/07/Climate-Change-Act-22-of-2024.pdf, 2.
200 Section 2, *Climate Change Act*.
201 See Bowmans 2024, South Africa: President assents to the Climate Change Bill, but no date set for commencement. https://bowmanslaw.com/insights/south-africa-president-assents-to-the-climate-change-bill-but-no-date-set-for-commencement/
202 Section 2(b).
203 See Preamble to the *Climate Change Act*.
204 Section 2(a).
205 See the PCC 2022 About the PCC. https://www.climatecommission.org.za/about.
206 See Sections 10, 11, 12, 13, and 14 of the *Climate Change Act*.
207 See Chapter 3, Section 3.3.1 and Chapter 4, Section 4.2.3.2.
208 Section 18 (1).
209 Section 18 (3).
210 Section 16 (1), (2) and 18 (3).
211 Section 19.
212 Though not relevant to the focus of this study, it should be noted that the *Climate Change Act* makes reference to agriculture as one of the national/provincial

departments that was consulted in preparing the law. See the fourth item in the Memorandum on the objects of the *Climate Chang Act*, 2024.

213 See schedule 2 to the *Climate Change Act*. Regarding the sectoral adaptation strategy for agriculture, see especially schedule 1 to the *Climate Change Act*, which must be read in conjunction to section 26 of the *Climate Change Act*.

214 This has been previously elaborated. See Chapter 2, Section 2.4.

215 See Section 5.2.1 above.

216 Schedule 4 (Part A) of the *Constitution*.

217 Refer to the discussion in Chapter 2, Section 2.2.

218 See Chapter 1, Section 1.3.3.

219 Draft Climate Smart Agriculture Strategic Framework for Agriculture, Forestry, and Fisheries https://www.gov.za/sites/default/files/gcis_document/201808/41811gen428.pdf. See Section 5.1.

220 For details about South Africa's CSA strategies, see Department of Agriculture, Land Reform and Rural Development (DALRRD) 2023 CLIMATE-SMART AGRICULTURE:Evidence-based Case Studies in South Africahttps://www.dalrrd.gov.za/images/Docs/climate-smart-agriculture-book.pdf, 1–334.

221 Department of Agriculture, Forestry, and Fisheries 2018 https://cer.org.za/wp-content/uploads/2018/07/Draft-Climate-Smart-Agriculture-Strategic-Framework.pdf, 25.

222 Department of Agriculture, Forestry, and Fisheries 2018 https://cer.org.za/wp-content/uploads/2018/07/Draft-Climate-Smart-Agriculture-Strategic-Framework.pdf, 5.

223 Department of Agriculture, Forestry, and Fisheries 2018 https://cer.org.za/wp-content/uploads/2018/07/Draft-Climate-Smart-Agriculture-Strategic-Framework.pdf, 29.

224 Department of Agriculture, Forestry, and Fisheries 2018 https://cer.org.za/wp-content/uploads/2018/07/Draft-Climate-Smart-Agriculture-Strategic-Framework.pdf, 23.

225 Department of Agriculture, Forestry, and Fisheries 2018 https://cer.org.za/wp-content/uploads/2018/07/Draft-Climate-Smart-Agriculture-Strategic-Framework.pdf, 5, 6 and 23.

226 Department of Agriculture, Forestry, and Fisheries 2018 https://cer.org.za/wp-content/uploads/2018/07/Draft-Climate-Smart-Agriculture-Strategic-Framework.pdf, 69.

227 Department of Agriculture, Forestry, and Fisheries 2018 https://cer.org.za/wp-content/uploads/2018/07/Draft-Climate-Smart-Agriculture-Strategic-Framework.pdf, 19.

228 The nexus approach seeks to strengthen inter-departmental and intradepartmental coordination. See Department of Agriculture, Forestry, and Fisheries 2018 https://cer.org.za/wp-content/uploads/2018/07/Draft-Climate-Smart-Agriculture-Strategic-Framework.pdf, 31.

229 Department of Agriculture, Forestry, and Fisheries, Notice 7 of 2013. https://www.gov.za/sites/default/files/gcis_document/201409/36063gen7.pdf.

230 These aspects were considered to be vital for any sectoral adaptation plan.

231 Department of Agriculture, Forestry, and Fisheries, Notice 7 of 2013. https://www.gov.za/sites/default/files/gcis_document/201409/36063gen7.pdf, 9.

232 Department of Agriculture, Forestry, and Fisheries, NOTICE 500 of 2015 https://www.gov.za/sites/default/files/gcis_document/201506/38851gen500.pdf, 19.

233 Department of Agriculture, Forestry, and Fisheries, NOTICE 500 of 2015 https://www.gov.za/sites/default/files/gcis_document/201506/38851gen500.pdf, 20 and 21. The benefits and impacts of these agricultural practices on the environments were discussed previously. See Chapter 1, Section 1.5.1.4.

234 Department of Agriculture, Forestry, and Fisheries, NOTICE 500 of 2015 https://www.gov.za/sites/default/files/gcis_document/201506/38851gen500.pdf,20.
235 Department of Agriculture, Forestry, and Fisheries, NOTICE 500 of 2015 https://www.gov.za/sites/default/files/gcis_document/201506/38851gen500.pdf, 20.
236 It is anticipated that this approach could also synergise the integration of sectoral adaptation especially in other sectors that supports agriculture such as land and water. See Department of Agriculture, Forestry and Fisheries, NOTICE 500 of 2015 https://www.gov.za/sites/default/files/gcis_document/201506/38851gen500.pdf, 18, 20.
237 The Paris Agreement was adopted by parties to the UNFCCC in December 2015 during COP21 session. South Africa signed the Paris Agreement on 22 April 2016. See https://www.gov.za/speeches/paris-agreement-climate-change-2-nov-2016-0000#:~:text=South%20Africa%20joins%20nations%20of,Secretary%2DGeneral%20in%20New%20York.
238 The KJWA has been discussed in Chapter 2, Section 2.4.1.
239 The most devastating years of drought recorded in South Africa were in 1981/1984, 1988/1989, 1991/92, 2004/2005 and 2014/2015. See S Mpandeli, E Nesamvuni and P Maponya 2015 Adapting to the impacts of drought by small-holder farmers in Sekhukhune District in Limpopo Province, South Africa. *Journal of Agricultural Science,* 7(2),115–124: A Mkhonza 2019 Improving the legal protection of strategic water source areas: a South African perspective. In *Law/Environment/Africa,* 453–478.
240 It has been explained earlier that agriculture relies on blue water, rainwater, and soil moisture in order to flourish, and that the availability of these forms of water is affected by droughts. See Chapter 1, Section 1.2.
241 Department of Forestry, Fisheries, and the Environment 2018. Second NAP-CDLDD (2018-2030) https://www.dffe.gov.za/sites/default/files/docs/nap_desertification_land_degradation_droughteffects.pdf, 1–48.
242 Second NAP-CDLDD (2018–2030) https:// https://www.dffe.gov.za/sites/default/files/docs/nap_desertification_land_degradation_droughteffects.pdf, 8. 8.
243 Second NAP-CDLDD (2018–2030) https:// https://www.dffe.gov.za/sites/default/files/docs/nap_desertification_land_degradation_droughteffects.pdf, 13. 13.
244 See article 9 of the UNCCD for an explanation of the guidelines on combating droughts. The process has been discussed in Chapter 4, Section 4.3.1.4.
245 Second NAP-CDLDD (2018–2030) https://www.dffe.gov.za/sites/default/files/docs/nap_desertification_land_degradation_droughteffects.pdf, 14.
246 See Chapter 1, Section 1.5.1.2 for an in-depth explanation on the use of monitoring and early warning systems, vulnerability/impact assessment.
247 Second NAP-CDLDD (2018–2030) https://www.dffe.gov.za/sites/default/files/docs/nap_desertification_land_degradation_droughteffects.pdf, 8.
248 Second NAP-CDLDD (2018–2030) https://www.dffe.gov.za/sites/default/files/docs/nap_desertification_land_degradation_droughteffects.pdf, 9.
249 See Chapter 1, Section 1.5.1.1.
250 A case for translating policies into laws has been made in several sections of this book. For example, see Chapter 2, Section 2.2; Chapter 3, Section 3.3.2; Chapter 4, Section 4.3.2.4.
251 JC Sanchez and J Roberts 2014 *Transboundary water governance: Adaptation to climate change,* Gland, Switzerland: IUCN, *229;* C Sadoff and M Muller 2009 Water management, water security and climate change adaptation: Early impacts and essential responses. *Stockholm: Global Water Partnership,* 76.
252 Section 27 of the Bill of Rights of the *Constitution,* 1996 provides that "everyone has the right to have access to (a) health care services, including reproductive health care; (b) sufficient food and water... (2) The state must take reasonable

legislative and other measures, within its available resources, to achieve the progressive realisation of each of these rights."

253 Examples of this interpretations are found in M Kidd 2009 South Africa: The development of water law. In The evolution of the law and politics of water, *Springer,*87–104; M Kidd 2016 Muddied water:(Un) cooperative governance and water management in South Africa. In *Trans-jurisdictional water law and governance*. Routledge, 157–174.1–393; J Glazewski 2016 *Marine aquaculture in South Africa: The policy and legal framework. In Aquaculture law and policy*. Edward Elgar Publishing, 386–411; CB Soyapi 2017 Water security and the right to water in Southern Africa: An overview. *Potchefstroom Electronic Law Journal,* 20; GR Backeberg 2005 Water institutional reforms in South Africa. *Water Policy,* 7(1), 107–123.

254 The 2019 water crisis in Cape Town, South Africa is an example of water scarcity resulting from temperature increase. See DW Olivier and Y Xu 2019 Making effective use of groundwater to avoid another water supply crisis in Cape Town, South Africa. *Hydrogeology Journal,* 27(3), 823–826; L Rodina 2019 Water resilience lessons from Cape Town's water crisis. *Wiley Interdisciplinary Reviews: Water,* 6(6), 1376.

255 *Water Services Act*, 1997 (108 of 1997).

256 Section 24 of the *Constitution*, 1996 provides for the right to clean and healthy environment (including water), while section 27 of the *Constitution* provides for security for all in South Africa in terms social life, health, food, and water.

257 Section 2 *WSA*.

258 These were previously discussed as part of water management strategies. See Chapter 1, Section 1.5.1.2.

259 See Section 7 of the *WSA*.

260 See Chapter 1, Section 1.4.

261 Act 36 of 1998.

262 In other words, the *NWA* is guided by some basic principles which reflects the provisions of section 24 of the *Constitution*, and it seeks to advance constitutional entrenched rights of access to water under section 27 of the *Constitution*.

263 Section 2 *National Water Act*. In practical terms, the purpose of the *NWA* simply reaffirmed the provisions of sections 24 and 27 of the *Constitution* and section 2 of the *WSA* discussed above.

264 E Van der Schyff and G Viljoen 2008 Water and the public trust doctrine – a South African perspective. *TD: The Journal for Transdisciplinary Research in Southern Africa*, 4(2), 339–354; Rumble 2016 Water resources management and climate change. In T Humby, LJ Kotzé, O Rumble and A Gilder (Eds.) *Climate change: Law and governance in South Africa*. Juta, 240.

265 Section 3 of the *NWA*.

266 See for example, Fasken 2021 *Lexology https://www.fasken.com/en/solution/clientwork/2023/03/fasken-successfully-represented-a-key-applicant-in-a-landmark*.

267 See Paragraph 5.1.2, White Paper on National Water Policy for South Africa.

268 It is believed that the PTD will promote better access to water as opposed to the previous RIparian system under the repealed *Water Act* 54 of 1956, which recognised private ownership of water with limited right of use for human needs. Similarly, the Riparian system was replaced with the PTD, because the former tied water uses to land ownership whereby water use is only allocated to a user based on public interest. See ORumble 2016 Water resources management and climate change. In T Humby, LJ Kotzé, O Rumble and A Gilder (Eds.) *Climate change: Law and governance in South Africa*. Juta, 240.

269 JH Adler 2014 Dynamic environmentalism and adaptive management: Legal obstacles and opportunities. *Journal of Law, Economics and Policy*, 11, 133.
270 ORumble 2016 Water resources management and climate change. In T Humby, LJ Kotzé, O Rumble and A Gilder (Eds.) *Climate change: Law and governance in South Africa,* Juta, 239.
271 Identified water uses are provided by section 21 of the *NWA* to include: the use of water for abstraction, storage, commercial afforestation, power generation, conservancy tanks, pit latrines and soak-aways, evaporation dams, and for recreational purposes such as water sports, fishing competitions, floating restaurants.
272 Section 43(1) *NWA*.
273 Likewise, irrigation activities exceeding allowed limits especially if not covered by the general authorisation would also require a licence. Accordingly, general authorisation outlines the conditions for specific water including the manner, limit, and the geographic area in which water use may take place. See section 39 of the *NWA*; Government Notice 399, 26 March 2004. See also M Kidd 2016 Muddied water:(Un) cooperative governance and water management in South Africa. In *Trans-jurisdictional water law and governance*. Routledge, 157–174.
274 O Rumble 2016 Water resources management and climate change. In T Humby, LJ Kotzé, O Rumble and A Gilder (Eds.) *Climate change: Law and governance in South Africa*. Juta, 239.
275 M van der Bank and J Karsten 2020 Climate change and South Africa: A critical analysis of the earthlife Africa Johannesburg and another v Minister of Energy and Others 65662/16 (2017) case and the drive for concrete climate practices. *Air, Soil and Water Research,* 13, 6.
276 Section 37 *National Water Act*.
277 See Chapter 1, Section 1.5.1.2.
278 South African National Climate Change Response Strategy 2004 https://unfccc.int/sites/default/files/sem_sup3_south_africa.pdf, 27.
279 O Rumble 2016 Water resources management and climate change. In T Humby, LJ Kotzé, O Rumble and A Gilder (Eds.) *Climate change: Law and governance in South Africa*. Juta, 217.
280 See section 39 of the *NWA* for conditions that satisfies general authorisations to use water resources.
281 Section 145.
282 See Chapter 1, Section 1.4.1.
283 See Schedule 1 to the *National Water Act*, 1998.
284 The Impala Irrigation Scheme was primarily launched in order to support 14 irrigation schemes for the agricultural sector. See Department of Water and Sanitation 2013 https://www.dws.gov.za/Documents/Other/WMPs%20Tlou%20Consulting/IMPALA_WMP_finaldraft_27%2003%2013.pdf.
285 See M Swart and NT Adams 2015, Water services provision and the Protection of Water Resources. In A Du Plessis (ed) *Environmental Law and Local Government in South Africa*. Juta:Cape Town, 445–480.
286 O Rumble 2016 Water resources management and climate change. In T Humby, LJ Kotzé, O Rumble and A Gilder (Eds.) *Climate change: Law and governance in South Africa*. Juta, 238.
287 See Chapter 1, Section 1.5.
288 Aside the *CARA*, other land use laws in South Africa include *Natives Land Act*, 27 of 1913, Expropriation Act,63 of 1975, *Deeds Registries Act*, 47 of 1937, *Land Administration Act*, 2 of 1995, and the *Land Titles Adjustment Act*, 111 of 1993. However, these laws are not relevant to this book, as such they will not be discussed. The laws are simply mentioned for the purposes of illustration.
289 Act 43 of 1983.

290 See regulations 4, 6, 7, and 8 of GN R1048 of 25 May 1984, made in terms of the *Conservation of Agricultural Resources Act* 43 of 1983: O Rumble 2016 water resources management and climate change. In T Humby, LJ Kotzé, O Rumble and A Gilder (Eds.) *Climate change: Law and governance in South Africa*. Juta, 239.
291 See, for example, M Kidd 2016 Muddied water:(Un) cooperative governance and water management in South Africa. In *Trans-jurisdictional water law and governance*. Routledge, 157–174.
292 Section 3, *CARA*.
293 Virgin soil is defined by section 1 of the *CARA* as any "land which in the opinion of the executive officer has at no time during the preceding ten years been cultivated."
294 Section 6, *CARA*.
295 See Chapter 1, paragraph 1.5.1.1.
296 Section 6(2) *CARA*.
297 Section 9(2) (i). Breaching the conditions attached to the subsidy or grant attracts some punitive measures such as fines, term of imprisonment or both. See section 20 and 23 of the *CARA*.
298 Section 8(1) *CARA*.
299 FundsforNGOs 2021 https://www2.fundsforngos.org/latest-funds-for-ngos/call-for-applications-for-south-african-farmers-support-funding/.
300 See Agri Funding: Funding Opportunitieshttps://www.greenagri.org.za/funding-and-incentives/government-and-developmental-institutes-financing/
301 T Humby 2014 Law and resilience: mapping the literature. *Seattle Journal of Environmental Law,* 4, 85.
302 See Chapter 1, Section 1.5.1.3.
303 AM Verster, JE Liang, MK Rostal, A Kemp, RF Brand, A Anyamba and van CW Huyssteen 2020 Selected wetland soil properties correlate to Rift Valley fever livestock mortalities reported in 2009–10 in central South Africa. *PLoS One,* 15(5), 326.
304 Draft Conservation of Agricultural Resources Amendment Bill, Notice 673 of 2013 https://www.gov.za/sites/default/files/gcis_document/201409/36605gen673.pdf.
305 Clause 3 GN 673 in GG 36605 of 2 July 2013.
306 The PTD and its relevance to agriculture and adaptation had been discussed above while explaining section 3 of the *NWA*, as well as in Chapter 4, Section 4.2.2.3 of this book.
307 According to the DALRRD, South Africa has a total land mass of 122,320,100 hectares, from which 101 million hectares constitute agricultural land (with grazing taking 84 million hectares and cropping 17 million hectares). Most of the agricultural lands (approximately 86%) are taken up by commercial agriculture, while subsistence agriculture makes up the remainder. See Department of Agriculture, Forestry, and Fisheries https://www.gcis.gov.za/sites/default/files/docs/resourcecentre/pocketguide/013_agriculture.pdf, 1–12.
308 See Section 5.2.3.1.
309 HA Strydom, ND King and FP Retief 2018 *Environmental Management in South Africa* 3rd ed., Juta, 339.
310 The *SPLUMA* was assented to by the President of the Republic of South Africa on 5 August 2013. The Act came into operation on 1 July 2015.
311 See preamble to the *SPLUMA*.
312 Section 3 (d) *SPLUMA*.
313 A Taylor, H Davies, G Oelofse and S Roux 2016 Urban adaptation, In T Humby, L J Kotzé, O Rumble and A Gilder (Eds.) *Climate Change: Law and Governance in South Africa*. Juta, 286.

314 Likewise, Taylor et al. maintain that climate change adaptation measures are supported by the provisions of section 12 of the *SPLUMA* and that these measures could be suitable for integration in IDPs. See A Taylor, H Davies, G Oelofse and S Roux 2016 Urban adaptation, In T Humby, L J Kotzé, O Rumble and A Gilder (Eds.) *Climate Change: Law and Governance in South Africa*. Juta, 286.
315 Section 7(b) (ii) *SPLUMA*.
316 A van der Berg, A Du Plessis and M Murphree 2016 Local disaster risk reduction and management in South Africa's response to climate change In T Humby, LJ Kotzé, O Rumble and A Gilder (Eds.) *Climate change: Law and governance in South Africa*. Juta, 338.
317 A Taylor, H Davies, G Oelofse and S Roux 2016 Urban adaptation, In T Humby, L J Kotzé, O Rumble and A Gilder (Eds.) *Climate Change: Law and Governance in South Africa*. Juta, 286.
318 A Taylor, H Davies, G Oelofse and S Roux 2016 Urban adaptation, In T Humby, L J Kotzé, O Rumble and A Gilder (Eds.) *Climate Change: Law and Governance in South Africa*. Juta, 286.
319 A Taylor, H Davies, G Oelofse and S Roux 2016 Urban adaptation, In T Humby, L J Kotzé, O Rumble and A Gilder (Eds.) *Climate Change: Law and Governance in South Africa*. Juta, 286.
320 Section 7 (d) SPLUMA.
321 South African Cities Network 2022 Spatial Planning and Land Use Management. https://www.sacities.net/spatial-planning-and-land-use-management/.
322 Government of South Africa 2022 Spatial Planning and Land Use Management White Paper. https://www.gov.za/documents/spatial-planning-and-land-use-management-white-paper.
323 A van der Berg, A Du Plessis and M Murphree 2016 Local disaster risk reduction and management I in South Africa's response to climate change. In T Humby, LJ Kotzé, O Rumble, and A Gilder (Eds.) *Climate change: Law and governance in South Africa*. Juta, 315.
324 K Lehman 2016 South Africa's climate change commitments and regulatory response potential. In T Humby, LJ Kotzé, O Rumble and A Gilder (Eds.) *Climate change: Law and governance in South Africa*. Juta, 211.
325 A Averchenkova, KE Gannon and P Curran 2019 Governance of climate change policy: A case study of South Africa. *Grantham Research Institute on Climate Change and the Environment Policy Report,* 1–40.
326 Africa Practice 2021 Introductory Political Economy Analysis of the Climate Philanthropy Space in South Africa. https://www.africapractice.com/wp-content/uploads/2021/11/Africa-Practice-Political-Economy-Analysis-for-ClimateWorks-South-Africa.pdf, 9.
327 K Ortega-Cisneros, KL Cochrane, N Rivers and WH Sauer 2021 Assessing South Africa's potential to address climate change impacts and adaptation in the fisheries sector. *Frontiers in Marine Science,* 8, 652955.
328 National Planning Commission 2021 A Review of theNational Development Plan 2030: Advancing Implementation towards a more Capable Nation. https://www.nationalplanningcommission.org.za/assets/Documents/NDP%20REVIEW.pdf, 39.
329 See RJ Scholes, M Scholes and M Lucas 2015 *Climate change: Briefings from Southern Africa*. Wits University Press, 114.
330 For example, the provisions of section 24 of the *Constitution* are repeated by the *NEMA* and the *NWA* in some instances verbatim.
331 A Du Plessis and LJ Kotzé 2014 The heat is on: Local government and climate governance in South Africa. *Journal of African Law,* 58(1),145–174.
332 See Chapter 2, Section 2.3.

333 A Taylor, H Davies, G Oelofse and S Roux 2016 Urban adaptation, In T Humby, L J Kotzé O Rumble and A Gilder (Eds.) *Climate Change: Law and Governance in South Africa*. Juta, 297.
334 L Masters and L Duff 2012 *Overcoming barriers to climate change adaptation implementation in Southern Africa*. Africa Institute of South Africa, 26.
335 For example, public comments have been received in respect of the *Conservation of Agricultural Resources Amendment Bill* since June 2013, and that of the draft CSASF were submitted since 2019. These invitations for public comments and the submissions of comments can be found in the respective documents.
336 See Chapter 1, Section 1.3.3.

References

Adler JH 2014 Dynamic environmentalism and adaptive management: Legal obstacles and opportunities. Journal of Law, Economics and Policy, 11, 133.

Averchenkova A, Gannon KE and Curran P 2019 *Governance of climate change policy: A case study of South Africa*. Centre for Climate Change Economics and Policy https://www.lse.ac.uk/granthaminstitute/wp-content/uploads/2019/06/GRI_Governance-of-climate-change-policy_SA-case-study_policy-report_40pp.pdf (accessed 25 November 2019).

Backeberg GR 2005 Water institutional reforms in South Africa. *Water Policy*, 7(1), 107–123.

Biazin B, Sterk G, Temesgen M, Abdulkedir A and Stroosnijder L 2012 Rainwater harvesting and management in rainfed agricultural systems in sub-Saharan Africa–A review. *Physics and Chemistry of the Earth, Parts A/B/C*, 47, 139–151.

Bonetti S, Sutanudjaja EH, Mabhaudhi T, Slotow R and Dalin C 2022 Aug 1 Climate change impacts on water sustainability of South African crop production. *Environmental Resources Letter*, 17(8), 084017.

Bulkeley H and Betsill M 2005 Rethinking sustainable cities: Multilevel governance and the 'urban' politics of climate change. Environmental Politics, 14(1), 42–63.

Dube K, Nhamo G and Chikodzi D 2021 Flooding trends and their impacts on coastal communities of Western Cape Province, South Africa. *GeoJournal*, 87(4), 1–16.

Du Plessis A and Kotzé LJ 2014 The heat is on: Local government and climate governance in South Africa. *Journal of African Law*, 58(1), 145–174.

Du Toit L and Glazewski J 2013 *International climate change law*. Lexis Nexis.

Eyzaguirre J and Warren FJ 2014 Adaptation: Linking research and practice. In FJ Warren and DS Lemmen (Eds.) *Canada in a changing climate: Sector perspectives on impacts and adaptation*. Government of Canada, 253–286.

Garland RM 2014 National policy response to climate change in South Africa. *South African Medical Journal*, 584–584.

Glazewski J 2005 *Environmental Law in South Africa* 2nd ed. LexisNexis Cape Town.

Glazewski J 2013, October Legal and regulatory aspects of carbon capture and storage: A developed and developing country perspective. In *Climate change: International law and global governance*. Nomos Verlagsgesellschaft mbH and Co. KG, 933–956.

Glazewski J 2016 Marine aquaculture in South Africa: The policy and legal framework. In *Aquaculture law and policy*. Edward Elgar Publishing, 386–411.

Gordon DJ and Johnson CA 2017 The orchestration of global urban climate governance: Conducting power in the post-Paris climate regime. *Environmental Politics*, 26(4), 694–714.

Harris JA 2020 The interpretation of Locke's two treatises in Britain, 1778–1956. *British Journal for the History of Philosophy*, 483–500.

Humby TL 2014 Law and resilience: Mapping the literature. Seattle Journal of Environmental Law, 4, 85.

Jegede AO and Mokoena UC 2019 Thinking into the future: Constructing social security law as climate change adaptation strategy in urban South Africa. In P B Cobbinah and M Addaney (Eds.) *The Geography of climate change adaptation in Urban Africa*. Palgrave Macmillan, 329–353.

Kidd M 2009 Important legislative developments in environmental law in South African during 2009. *IUCN Academy of Environmental Law e-Journal* 7.

Kidd M 2009 South Africa: The development of water law. In *The evolution of the law and politics of water*, Springer, 87–104.

Kidd M 2016 Muddied water:(Un) cooperative governance and water management in South Africa. In J Gray, C Holley and Rayfuse (Eds.) *Trans-jurisdictional water law and governance*. Routledge, 157–174.

Kotzé LJ 2005 *A legal framework for integrated environmental governance in South Africa and the North-West Province*. Doctoral-thesis North-West University.

Lehman K 2016 South Africa's climate change commitments and regulatory response potential. In T Humby, LJ Kotzé, O Rumble and A Gilder (Eds.) *Climate change: Law and governance in South Africa*. Juta, 189–215.

Lethoko MX 2016 Inclusion of climate change strategies in municipal Integrated Development Plans: A case from seven municipalities in Limpopo Province, South Africa. *Jàmbá: Journal of Disaster Risk Studies*, 8(3), 1–6.

Llosa S and Zodrow I 2011 Disaster risk reduction legislation as a basis for effective adaptation. Global Assessment Report on Disaster Risk Reduction, 1–18.

Locke J 1689 *Locke: Two treatises of government*. Cambridge University Press, 43–46.

Madondo M and Nkwana HM 2022 Exploring South Africa's National climate change response white paper, 2011: Challenges and prospects for public policy, 2011, Administratio Publica, 29(2), 187–206.

Madzwamuse M 2010 Climate change vulnerability and adaptation preparedness in South Africa. *Heinrich Böll Stiftung South Africa*, 16.

Mai L 2019 *Governing climate change: Global cities and transnational lawmaking*. Cambridge University Press, 1–20.

Mantlana KB, Maoela MA and Nhamo G 2021 Mapping South Africa 's nationally determined contributions to the targets of the sustainable development goals. *Natural Resources Forum*, 45(1), 3–17.

Marais M, Retief FP, Sandham LA and Cilliers DP 2015 Environmental management frameworks: Results and inferences of report quality performance in South Africa. *South African Geographical Journal= Suid-Afrikaanse Geografiese Tydskrif*, 97(1), 83–99.

Masters L and Duff L 2012 *Overcoming barriers to climate change adaptation implementation in Southern Africa. African Books Collective*, 26.

Molobela IP and Sinha P 2011 Management of water resources in South Africa: A review. *African Journal of Environmental Science and Technology*, 5(12), 993–1002.

Muriel M and Judith K. 2016 Climate change and insurance law. In T Humby, L J Kotzé O Rumble and A Gilder (Eds.) Climate Change: Law and governance in South Africa, Juta, 345.

Nel Johan and du Plessis W 2001 An evaluation of NEMA based on a generic framework for environmental framework legislation. *South African Journal of Environmental Law and Policy*, 8(1), 1–36.
Nel J and Du Plessis W 2004 Unpacking integrated environmental management-a step closer to effective co-operative governance? *Journal SA Publiekreg= SA Public Law*, 19(1), 181–190.
Olivier DW and Xu Y 2019 Making effective use of groundwater to avoid another water supply crisis in Cape Town, South Africa. *Hydrogeology Journal*, 27(3), 823–826.
Ortega-Cisneros K, Cochrane KL, Rivers N and Sauer WH 2021 Assessing South Africa's potential to address climate change impacts and adaptation in the fisheries sector. *Frontiers in Marine Science*, 8, 652955.
Peterson MJ 2017 Politics and law in international environmental governance. In *Research handbook on the politics of international law*. Edward Elgar Publishing, 443–488.
Rodina L 2019 Water resilience lessons from Cape Town's water crisis. *Wiley Interdisciplinary Reviews: Water*, 6(6), e1376.
Rodseth C and Sustainability BN 2019 Carbon tax and agriculture. Retrieved from Confronting Climate Change. https://mbtpetroleum.co.za/wp-content/uploads/2022/07/Carbon-tax-wine-and-fruit.pdf (accessed on 12 August 2022).
Rumble 2016 Water resources management and climate change. In T Humby, LJ Kotzé, O Rumble and A Gilder (Eds.) *Climate change: Law and governance in South Africa*. Juta, 217–248
Rumble 2024 South Africa finally gets its climate change Act. https://africanclimatewire.org/2024/07/south-africa-finally-gets-its-climate-change-act/. (accessed 20 August 2024).
Sadoff C and Muller M 2009 *Water management, water security and climate change adaptation: Early impacts and essential responses*. Global Water Partnership, 73.
Sanchez JC and Roberts J 2014 *Transboundary water governance: Adaptation to climate change*, Gland, Switzerland: IUCN, 23.
Scholes RJ, Scholes M and Lucas M 2015 *Climate change: Briefings from Southern Africa*. Wits University Press.
Soyapi CB 2017 Water security and the right to water in Southern Africa: An overview. *Potchefstroom Electronic Law Journal/Potchefstroomse Elektroniese Regsblad*, 20(1), 1–26.
Strydom HA, King ND and Retief FP 2018 *Environmental management in South Africa* 3rd ed. Juta. https://search.ebscohost.com/login.aspx?direct=trueand scope=siteand db=nlebkand db=nlabkand AN=2524097.
Swart JA, Van Der Windt HJ and Keulartz J 2001 Valuation of nature in conservation and restoration. *Restoration Ecology*, 9(2), 230–238.
Taylor, M. 2018 Climate-smart agriculture: what is it good for?. *The Journal of Peasant Studies*, 45(1), 89–107.
van der Bank M and Karsten J 2020 Climate change and South Africa: A critical analysis of the earthlife Africa Johannesburg and another v Minister of Energy and Others 65662/16 (2017) case and the drive for concrete climate practices. *Air, Soil and Water Research*, 13(6), 1178622119885372.
van der Berg A, Du Plessis A and Murphree M 2016 Local disaster risk reduction and management in South Africa's response to climate change In T Humby, LJ Kotzé,

O Rumble and A Gilder, (Eds.) *Climate change: Law and governance in South Africa*. Juta, 338.

Van der Schyff E and Viljoen G 2008 Water and the public trust doctrine - a South African perspective. *TD: The Journal for Transdisciplinary Research in Southern Africa*, 4(2), 339–354.

Verster AM, Liang JE, Rostal MK, Kemp A, Brand RF, Anyamba A … and van Huyssteen CW 2020 Selected wetland soil properties correlate to Rift Valley fever livestock mortalities reported in 2009–10 in central South Africa. *PLoS One*, 15(5), e0232481.

Wunsch JS 2001 Decentralization, local governance and 'recentralization'in Africa. *Public Administration and Development: The International Journal of Management Research and Practice*, 21(4), 277–288.

Ziervogel G, New M, Archer van Garderen E, Midgley G, Taylor A, Hamann R and Warburton M 2014 Climate change impacts and adaptation in South Africa. *Wiley Interdisciplinary Reviews: Climate Change*, 5(5), 605–620.

6 Conclusion

6.1 General background

By way of general observation, the findings in this book reveal that agriculture is relied upon as an important source of GDP and as a means of livelihood for the majority of the population in African countries, especially Kenya, Nigeria, and South Africa. It was also observed that the impacts of climate change, which for instance manifest as unstable temperature, droughts, desertification, water scarcity as well as poor soil fertility,[1] affect agricultural productivity and by extension the GDP of the three countries.[2] The impacts of climate change on agriculture are severe in African countries due to their geographical location, vulnerability, low resilience, and weak adaptive capacity.[3] Adaptation is therefore required in African countries, especially Kenya, Nigeria, and South Africa, so as to cope with the impacts of climate change.[4] This study advanced the hypothesis that a suitable agriculture-focused adaptation legislation framework is required in these African countries in order to enable the agriculture sector to properly adapt to the impacts of climate change.[5]

Chapter 2 of the book highlighted the laws that serve as the overarching legal foundation for climate change adaptation in a broad context, such as the UNFCCC, the Kyoto Protocol, and the Paris Agreement. In addition, the chapter also briefly discussed other agriculture-based laws, including the *Basel, Rotterdam and Stockholm (BRS) Conventions*; *International Code of Conduct for the Effective Use and Management of Fertilizers*; the *International Code of Conduct on Pesticide Management (ICCPM)*; the *Convention on Biological Biodiversity (CBD)*; the *United Nations Economic Commission for Europe (UNECE) Water Convention*; the *Revised African Convention on Conservation of Nature and Natural Resources*; the *SADC Protocol on Shared Water System*; the EU *Water Framework Directive*; the *Voluntary Guidelines for Responsible Governance of Tenure of Land, Fisheries, and Forestry in the Context of National Food Security (VGGT)*; the *Voluntary Guidelines for Sustainable Soil Management (VGSSM)*; and the *World Soil Charter*, which to some extent make provision for adaptation in the context of agriculture.[6]

Adaptation laws are also required to facilitate adaptation in a systematic, coherent, and effective way to cope with the impacts of climate

DOI: 10.4324/9781003536437-6

change on agriculture at the national level. Chapters 3, 4, and 5 of the book analysed climate-change-related laws that facilitate adaptation in Kenya, Nigeria, and South Africa, as well as agricultural-based policies and laws that have provisions which could enable adaptation in the context of agriculture. Given the gaps and challenges found in the national climate change adaptation-related laws, as well as agriculture-based laws of the three countries in enabling agriculture-focused adaptation, it was tentatively suggested that a specific agriculture-focused adaptation framework law could be created for each country.[7] This issue will be elaborated on in this concluding chapter.

The following central research question was formulated in guiding the study: to what extent do Kenyan, Nigerian, and South African laws provide for climate change adaptation in the context of the agricultural sector, and where there are gaps, how could these be addressed?

The central research question was supported by the sub-questions below, which guided the discussions in different chapters of the book:

1 What is adaptation and in what ways could it be facilitated in the context of agriculture (Chapter 1)?
2 What is the role of the law in relation to adaptation, and to what extent can the law facilitate adaptation in the context of agriculture (Chapter 2)?
3 How do Kenyan, Nigerian, and South African climate change and agriculture-related policies and laws provide for adaptation measures in the context of agriculture, and are there observable gaps in this regulatory framework (Chapters 3, 4, and 5)?
4 Considering that there are gaps in the provisions of the Kenyan, Nigerian, and South African climate change and agricultural-related policies and laws in enabling adaptation in the context of agriculture, how could they be addressed (Chapter 6)?

This chapter connects the findings in Chapters 1, 2, 3, 4, and 5 with the main research questions. A summary of the analysis and findings is presented below.

6.2 Summary of the analysis and findings

6.2.1 Climate change, agriculture, and adaptation

Chapter 1 was guided by a research sub-question: what is adaptation, and in what ways could it be facilitated? In providing answers, the chapter explored the concept of adaptation in a broad context and in the context of agriculture. The chapter commenced with background information on how agricultural practices partly contribute to climate change, and how climate change impacts agriculture. In practical terms, GHG-intensive crops and livestock production methods were noted as drivers of climate change, while climate change (which results in short-term, medium-term, and/or long-term ecological changes) affects water cycles and temperature and, by extension, impairs crop yields and animal productivity.[8]

The chapter identified adaptation as a strategy which allows human and natural systems to cope with the impacts of climate change.[9] In addition, the chapter classified climate change adaptation relative to location, needs, and circumstances.[10] Furthermore, the chapter highlighted that adaptation primarily revolves around vulnerability reduction, boosting of resilience, and the enhancement of adaptive capacity. In brief, vulnerability reduction encapsulates all measures aimed at ameliorating the risk of climate change impacts on human and natural systems.[11] Boosting of resilience in the context of climate change adaptation was explained as the application of measures that could assist human and natural systems to absorb or overcome the shocks of climate change impacts such as flooding and droughts.[12] The chapter described adaptive capacity in the context of climate change as measures or strategies for enhancing the ability of infrastructure involved in driving the process of adaptation, including legal and institutional frameworks.[13]

The chapter also discussed the forms and levels of adaptation. In that regard, the implementation of adaptation was described as occurring at the individual level, communal level, and institutional level through short-term, medium-term, and/or long-term plans.[14] The forms and levels of adaptation were linked to the concept of mainstreaming and how it could facilitate low-carbon and climate-resilient development (including agricultural development) in vulnerable societies.[15] While discussing the benefits of mainstreaming, the chapter also identified the challenges of adaptation in a broad context. Accordingly, ecological, economic, social, and governance (including legal) constraints are the main challenges to effective adaptation in vulnerable societies – especially in Kenya, Nigeria, and South Africa.[16]

Chapter 1 also analysed adaptation in the context of agriculture, focusing on the need for adaptation, the forms of adaptation, and the levels of adaptation. The analysis revealed that anticipatory and planned adaptation is the most suitable type of adaptation for agriculture. Anticipatory and planned adaptation in the context of agriculture could be implemented as on-farm strategies, technological solutions, or insurance mechanisms.[17] Collectively, these strategies were described as CSA strategies,[18] and it was noted that they could be implemented through short-term, medium-term, and/or long-term plans.[19]

The chapter concluded by highlighting the various challenges against adaptation in the agriculture sector, which are partly derived from socio-economic factors, infrastructural factors, and organisational instability.[20] The chapter argued that the challenges for adaptation in a generalised context and in the context of agriculture could be addressed, possibly by an effective regulatory mechanism embedded in policies and law.[21]

6.2.2 Climate change adaptation, agriculture, and the Law

The link between climate change adaptation and the law was made in Chapter 2. The chapter was built on the question: what is the role of the law in relation to adaptation, and to what extent can the law facilitate adaptation in the context of agriculture?

Through discussions on the meaning and objectives of the law in regulating human conduct, the chapter found that law could also be utilised in guiding humans to alter their behaviour and specific activities (such as crop rotation and livestock production methods) in order to cope with the impacts of climate change. To illustrate how the law could promote adaptation, the chapter first explained the law in the context of adaptation as a manifestation of hard laws (treaties, conventions, protocols, agreements, regulations, directives, rules, and Acts of parliaments) and soft laws (pre-legislation policies and guidelines) that could be used in coordinating institutional and governance structures.[22] In the same context, a law was described as an instrument for outlining the roles and responsibilities of actors, for setting standards and targets, and for translating commitments with respect to climate change actions.[23]

The findings in the chapter revealed that the main objectives and principles of adaptation law as established by the UNFCCC, Kyoto Protocol, and the Paris Agreement are intended to support adaptive capacity, increase resilience, and reduce the vulnerability of human and natural systems against the impacts of climate change.[24] The UNFCCC, for instance, supports the objectives of adaptation through a set of guiding principles that oblige member states to foster cooperation in promoting low-carbon and climate-resilient development through adaptation.[25] Similar provisions supporting the objectives of adaptation exist under the Kyoto Protocol and the Paris Agreement (which replaced the Kyoto Protocol).[26] For example, the Kyoto Protocol places obligations on both developed and developing parties to promote adaptation by requiring them to integrate climate change concerns into their social and economic planning.[27] The provisions of the Paris Agreement are broader than the UNFCCC and the Kyoto Protocol in promoting the objectives of adaptation, as it conceptualises adaptation based on local, subnational, national, regional, and international dimensions.[28]

Moving from a broad to a narrower focus, the chapter found that law also has a role in facilitating adaptation in the context of agriculture, for instance, by guiding farmers in adjusting their crops and livestock production activities, and by enabling the resources that support agriculture such as water and soil to adapt to the impacts of climate change. Law also supports agriculture-related disasters and risk-management plans, especially through insurance services.[29] In this regard, it was observed by way of inference that the provisions of article 4 of the UNFCCC, article 4(1) of the Kyoto Protocol, and article 7 of the Paris Agreement provide the context for adaptation measures in the agriculture sector.[30]

In addition to an analysis of the UNFCCC, the Kyoto Protocol, and the Paris Agreement, the chapter also analysed other legal instruments that are relevant to agriculture such as the *BRS Conventions*, the *CBD*, *Agenda 21*, the *UNECE Water Convention*, the *SADC Protocol on Shared Water System*, the *EU Water Framework Directive*, and the *Revised African Convention on Conservation of Nature and Natural Resources*. Other agriculture-based laws that were analysed in the chapter include the *VGGT*, *VGSSM*, and the *World Soil Charter*. All these legal instruments seem to support the objectives of adaptation in the context of agriculture.

Based on the above understanding of the role of the law in facilitating adaptation in a broad context and in the context of agriculture, the study moved on to the next aspect of the inquiry in Chapters 3, 4, and 5, namely, how climate change and agriculture-related policies and laws in Kenya, Nigeria, and South Africa provide for adaptation measures in the context of agriculture.

6.2.3 Assessment of the Kenyan, Nigerian, and South African regulatory frameworks for climate change adaptation and agriculture

How do Kenyan, Nigerian, and South African climate change and agriculture-related policies and laws provide for adaptation measures in the context of agriculture, and are there observable gaps in this regulatory framework?

By signing and ratifying the UNFCCC, Kyoto Protocol, and the Paris Agreement, the three focus countries (Kenya, Nigeria, and South Africa) are committed (though at a lesser level compared to developed countries) to adapting to the impacts of climate change.[31] The commitments of these countries, as discussed in the various chapters, are expressed through various policies and laws.

6.2.3.1 Regulatory frameworks for adaptation: Kenya, Nigeria, and South Africa

For a start, the *Constitutions* of the three countries serve as the basis for the formulation of policies and the promulgation of laws that promote adaptation. Although nominally and textually different, the provisions of the three *Constitutions* on environmental governance seek to serve the same purpose. Namely, protecting the environment from harmful impacts such as climate change and, to a greater or lesser extent, creating an enabling regulatory environment to design and implement adaptation measures. In this regard, there are climate-change-related policies and laws in the three countries that contain some provisions that enable adaptation, either directly or indirectly.

Article 42 of the *Constitution* of Kenya provides for the right to a clean and healthy environment in the country.[32] An interpretation of the provisions of article 42 implies that adaptation is one of the measures for promoting the right to a clean and healthy environment. The provisions of article 42 are further supported by the provisions of article 69 of the *Constitution*, which obliges the State (as well as other relevant stakeholders in environmental governance) to equitably utilise and conserve important components of the environment, especially land and water. Undoubtedly, land and water are vital in supporting agriculture, and adaptation remains one of the means of utilising and conserving land and water.[33]

There are various climate-change-related policies and laws in Kenya which derive their validity pursuant to the provisions of article 42 of the *Constitution*. The policies include the National Environment Action Plan (NEAP), National Climate Change Response Strategy (NCCRS), National Climate Change Action Plan (NCCAP), and the National Adaptation Plan (NAP), while the laws consist of the *Environment Management and Coordination Act (EMCA)* and the *Climate Change Act*.

The NEAP has been described as a mechanism for promoting the formulation of other policies and strategies and the establishment of institutions for environmental governance, including adaptation measures in Kenya. Based on the recommendations in the NEAP, climate change policies such as the NCCRS, NCCAP, and the NAP were formulated, and laws including the *EMCA* and *Climate Change Act* were enacted. The NCCRS, NCCAP, and the NAP were noted to facilitate adaptation in the context of agriculture by, for instance, supporting the production of drought- and pest-resistance crops and animal varieties or promoting the production of crops that are resilient to extreme weather events.[34] These climate change policies were also observed to have some provisions that promote adaptive water strategies, which are important elements of adaptation in the context of agriculture.

The *EMCA* has been explained as the framework of the environmental law of Kenya that provides strategies such as Environmental Impact Assessment (EIA) and Strategic Environment Assessment (SIA), which could be used in addressing the negative impacts of climate change especially on agriculture.[35] The *Climate Change Act* also provides for mainstreaming of climate change adaptation strategies with development plans in a bid to promote low-carbon and climate-resilient development pathways in the country.[36] Though not directly provided for, adaptation in the context of agriculture can be inferred from the provisions of the Kenyan *Climate Change Act*.

In Nigeria, section 20 of the *Constitution* (CFRN) provides that "the State shall protect and improve the environment and safeguard the water, air and land, forest and wildlife of Nigeria".[37] Therefore, environmental protection entails taking all measures that will also allow the environment to adapt against natural and human-induced hazards such as climate change. Based on the description of agriculture and the types of resources that are required in supporting agriculture,[38] it was concluded that section 20 of the CFRN is broad enough to encompass agriculture-focused adaptation.[39]

The provisions of section 20 of the CFRN on environmental protection lay the foundation for adaptation governance in Nigeria. Climate change policies such as the National Adaptation Strategy and Plan of Action for Climate Change in Nigeria (NASPA-CCN), Nigeria Climate Change Policy Response and Strategy (NCCPRS), National Policy on Environmental Conservation (NPEC), and the NAP were developed by the Federal Government, and environmental-based laws with provision on adaptation especially the *National*

Environmental Standards and Regulations Enforcement Agency (NESREA) Act and the *Climate Change Act* were enacted pursuant to the provisions of section 20 of the CFRN.[40] A critical analysis of these Nigerian climate change policies revealed that they have some provisions which could enable agriculture to adapt to the impacts of climate change. By using a multi-sectoral approach, the NASPA-CCN, NCCPRS, NPEC, and the NAP facilitate adaptation strategies for agriculture and water at the community level and national level either as short-term, medium-term, and/or long-term measures.[41]

The *NESREA Act* was discussed as a framework of environmental legislation which serves almost the same purpose as the *EMCA* of Kenya. In other words, the *NESREA Act* is a law enacted to ensure that environmental standards in Nigeria are complied with in a general sense.[42] The standards as provided by sections 7, 20, 21, and 23 of the *NESREA Act* were identified as vital for adaptation in agriculture because the standards are intended to be applied in also promoting adaptive land management practices and adaptive water uses, which were noted as vital resources that support agriculture.[43]

The Nigerian *Climate Change Act* was enacted to give legal impetus to Nigerian climate change policies that provide for adaptation, especially the NCCPRS and the NAP.[44] Agriculture-focused adaptation is recognised by the *Climate Change Act*, though not in explicit terms. Section 27 of the Act provides for conservation tillage, restorative agriculture, and regrowing clear-cut forests as part of nature-based solutions which, in simple terms, qualify as forms of adaptation strategies in the agriculture sector.[45]

In South Africa, sections 24 of the *Constitution* provides for the right to a healthy environment and the need to promote the conservation of natural resources through reasonable legislative and other measures.[46] In the context of this book, natural resources are those resources (such as land and water) that are required to support agriculture. Therefore, sections 24 of the *Constitution* also (implicitly) relates to agriculture-focused adaptation.

Pursuant to the provisions of section 24 of the *Constitution*, climate change policies such as the NCCRS, National Climate Change Response Policy White Paper (NCCRP), the National Climate Change Adaptation Strategy (NCCAS), the National Development Plan (NDP), and the National Determined Contributions (NDCs) were developed. Section 24 also serves as the constitutional basis for enacting environmental-based laws, including the *National Environmental Management Act (NEMA)*, *Disaster Management Act (DMA)*, the *Carbon Tax Act (CTA)*, and a *Climate Change Act*, which is about to be promulgated.[47] These climate change policies and environmental-based laws underscore the need for adaptation in critical sectors of the economy, particularly agriculture. For example, the NCCRS adopts the vulnerability approach in identifying the appropriate adaptation strategies against the impacts of climate change for critical sectors of the economy of South Africa, including agriculture.[48]

Adaptation under the NCCRP takes a collaborative approach with cross-sectoral linkages.[49] In other words, the NCCRP identifies water, health, agriculture, and human settlements as sectors of the economy that are vulnerable to climate change, and consolidates adaptation actions in those sectors through a process known as the Integrated Development Plan (IDP).[50] In the same instance, the NCCRP requires all government departments (especially those departments that focus on climate change governance) to review and align their regulatory mechanisms and climate action plans (including adaptation) with the ambitions of the NCCRP.[51]

The NCCAS is a climate change governance policy that was developed in order to operationalise and consolidate existing climate-change-related policies, such as the NCCRS, with new policies, particularly the NAP. The goals of the NCCAS were highlighted as to support government departments and agencies such as the Department of Forestry, Fisheries, and the Environment and Department of Water and Sanitation at all levels, in the planning and implementation of adaptation measures and actions. These departments have specific mandates for managing resources required in support of agriculture (especially land and water) that were identified as more susceptible to the impacts of climate change. The development of water resources infrastructure (such as boosting the storage capacity of surface water) and the development of integrated water catchment techniques are a few illustrations of agricultural adaptation measures which the NCCAS supports.[52]

The role of the NDC of South Africa in enabling adaptation in the context of agriculture is observable through the programmes and measures that it promotes. For example, the NDC promotes the resilience of agriculture against climate-change-induced floods through water conservation techniques and land restoration methods.[53]

Therefore, most of the climate change policies and environmental-based laws in the three countries discussed above are suitable, to a lesser or greater extent, in facilitating adaptation measures in the context of agriculture.

6.2.3.2 *Kenyan, Nigerian, and South African regulatory frameworks for agriculture*

As was established in Chapter 2, adaptation in the context of agriculture is also facilitated by some agriculture-based policies and laws.[54] The agriculture-based policies were discussed to provide the context for the discussion of agriculture-based policies and laws in Kenya (Chapter 3), Nigeria (Chapter 4), and South Africa (Chapter 5). In the respective chapters, the *Constitution* of each country was affirmed as the basis for agriculture-based policies and laws.

In Kenya, the provisions of articles 42, 60, 69, and 72 of the *Constitution* of Kenya were discussed as the basis for the development of policies which support crop and livestock production, as well as policies for the management of land and droughts, and a policy for the provision of agriculture-based insurance.[55] These policies are exemplified by the Kenyan CSA Strategy, the National Land Policy, a draft National Livestock Policy, and the

National Agricultural Insurance Policy.[56] These agricultural-based policies are supported by several agriculture-related laws, which include the *Water Act*, the *National Drought Management Authority (NDMA) Act*, the *Kenyan Agriculture and Livestock Research Act (KARLA)*, the *Crops Act*, the *Land Act*, and the *Agriculture and Food Authority (AFA) Act*.[57]

The Kenyan CSA Strategy is a platform for guiding all government agencies and other actors in Kenya on how to promote measures that could address the vulnerabilities of agriculture to changing temperatures and poor precipitation, among others.[58] The National Land Policy promotes agricultural adaptation through soil conservation measures by, for instance, encouraging equitable land allocation in order to avoid overgrazing or over-cultivation practices.[59] These strategies could allow lands that have been depleted by climate change impacts (such as droughts or erosion) to recover.

The draft National Livestock Policy of Kenya provides directions on how to implement measures that could reduce the vulnerability of livestock and resources that support livestock production such as water, soil, and fodder to the impacts of climate change. In this regard, the draft National Livestock Policy recommends climate-friendly livestock breeding techniques such as animal-fattening methods and drought- and pest-resistant fodder.[60] In a bid to promote agriculture-based insurance as an adaptation strategy against climate-change-induced risks such as floods and droughts, the National Agricultural Insurance Policy provides farmers with indemnity-based insurance and index-based insurance covers.[61]

In addition to the agricultural policies discussed above, the following agriculture-based laws of Kenya have provisions that enable adaptation measures in agriculture. The *Water Act*, for instance, contains provisions which emphasise the need for water conservation techniques, and the protection of water catchment areas through an integrated water strategy plan.[62] These measures were discussed as suitable adaptive water management strategies.[63] Furthermore, the *NDMA Act* has provisions on early warning systems which seek to enable farmers to proactively adapt against anticipated climate change-induced droughts or soil erosion.[64] The *KARLA* recognises and recommends livestock migration as an important adaptation strategy against the impacts of climate change.[65] The draft National Livestock Policy of Kenya is part of the recommendation of the *KARLA*. In addition to the *KARLA*, the *Crops Act* also supports adaptation (though not in explicit terms) by making provisions for soil and water conservation measures under Section 6, and crop insurance (among other measures) under Section 34(1).[66] Similarly, the *Crops Act* makes provisions for the control of climate-change-induced invasive weeds, the salination and saltification of the soil, and the control of grazing activities (as relevant adaptation strategies).[67]

The *Land Act* supports adaptive land management strategies through its provisions on "justified measures" that have been interpreted to mean actions that could moderate GHG-intensive crops and livestock production activities on agricultural lands in Kenya. Examples of "justified actions" were inferred from the provisions of sections 11(2) and 48(1) of the *Land Act*.[68] The *AFA Act* complements the provisions of the *Land Act* on adaptive land management and further recommends the adoption of measures such as conservation tillage and cover cropping in order to allow agricultural lands to adapt to climate change impacts.[69] All of these measures potentially qualify as adaptation strategies based on the broad framing of adaptation in earlier parts of this book.[70]

In Nigeria, section 20 of the CFRN makes provision for the protection of the environment, including air, land, and water which are specifically relevant to agriculture.[71] Section 20 is therefore broadly interpreted to encompass agriculture-focused adaptation. In other words, pursuant to the provisions of section 20, the Federal Government developed some agriculture-related policies to, among other things, facilitate adaptation in the context of agriculture. Prominent among these adaptation policies are the Nigerian Agricultural Policy, the National Agricultural Resilience Framework (NARF), the National Water Policy, the National Drought Plan, and the National Policy on Erosion and Flood Control (NPEFC).[72] A critical analysis of these policies was carried out in Chapter 4 of this book, and it was observed that most of the policies contain the types of provisions that could enable adaptation in the context of agriculture.

The National Agricultural Policy, for instance, has provisions which enable agriculture-focused adaptation by aiming to protect land from droughts, erosion, and flooding as well as to promote adequate water supply for irrigation purposes. The policy also recommends the adoption of special livestock breeds that can withstand the impacts of climate change. Most importantly, the policy recognises and promotes a mandatory insurance scheme as an adaptation strategy against climate-based risks.[73] The NARF further provides guidance on how capacity building, monitoring and evaluation, and resource mobilisation in relation to adaptation measures in agriculture can be implemented.[74] In a related development, the National Water Policy of Nigeria was developed in order to promote strategies that can control risks and vulnerabilities (including climate-change-induced risks) on water resources through integrated water resources management techniques.[75] The National Drought Plan also promotes adaptive water management strategies by, for example, recommending the use of meteorological and hydrological information facilities as anticipatory adaptation methods against droughts, the salination of water, water scarcity, and water stress.[76]

Complementary to the National Drought Plan, the NPEFC also exists to support adaptation measures against droughts. In addition, the NPEFC also has provisions which enable farmers to adapt to erosion and floods, both of which are aggravated by climate change. In this respect, the NPEFC recommends forecasting, monitoring, and control mechanisms.[77]

Chapter 4 also discussed laws that were enacted pursuant to the provisions of section 20 of the CFRN to support the above-mentioned policies. For example, the *Water Resources Act* and the *River Basin Development*

Authority (RBDA) *Act* are laws which support the National Water Policy, while the National Drought Plan and the NPEFC are backed by the *Land Use Act*.[78] In facilitating adaptation in the context of agriculture, the *Water Resources Act* provides for technical, social, and commercial standards for water,[79] while the *RBDA Act* has provisions for strategies that could boost surface water and groundwater. The *RBDA Act* also contains provisions for the enhancement of water storage schemes for irrigation purposes.[80] The *Land Use Act* adopts the Public Trust Doctrine (PTD) as part of fair and effective use of land, and to promote less GHG-intensive agricultural activities.[81]

Other agriculture-based laws such as the *National Environmental (Soil Erosion and Flood Control) (NESEFC) Regulations* and the *Nigerian Agricultural Insurance Corporation (NAIC) Act* also have provisions that could enable agriculture to adapt to the impacts of climate change. For example, the *NESEFC Regulations* provide for Erosion Control Plan measures such as the installation of rock beams, sandbags, and the construction of slopes and the planting of cover crops.[82] As a means of cushioning the effect of climate-change-related risk on agriculture, the *NAIC Act* provides for agricultural insurance schemes and other forms of financial support (credits) to farmers.[83]

As for South Africa, section 24 of the *Constitution* has also been interpreted to serve as the basis for the regulatory framework for agriculture. As stated above, natural resources consisting of land and water are to be conserved and protected by reasonable legislative and other measures.[84] These legislative measures are exemplified by agriculture-based policies and laws. The draft Climate-Smart Agriculture Strategic Framework (CSASF) and the Climate Change Sector Plan for Agriculture, Forestry, and Fisheries (CCSPAFF) are the two main agriculture-based policies in the country.[85] Resources which support agriculture, such as water and land, are also guided by policies, and supported by additional laws based on the notion of PTD. In this regard, the NCCRP provides direction on the allocation and use of water resources in the country, while the *Water Service Act (WSA)* and the *National Water Act (NWA)* offer a comprehensive approach to (adaptive) water management strategies.[86]

Drought management in South Africa is facilitated by the Second National Action Programme for South Africa to Combat Desertification, Land Degradation and the effects of Droughts (Second NAP-CDLDD). Land as an important resource of agriculture is regulated by the *Conservation of Agricultural Resources Act (CARA)* and the *Spatial Planning and Land Use Management Act (SPLUMA)*.[87] Insurance for agriculture in South Africa against climate-change-induced risk is already adequately encapsulated in climate change adaptation policies such as the NCCRP and the NCCAS.[88]

A detailed review of the South African agriculture-based policies reveals that most of them contain provisions that could enable adaptation in the context of agriculture. For example, the draft CSAF has made provisions for the planting of early-maturing and drought-resistant crops and the breeding of resilient livestock breeds, in addition to irrigation techniques and conservation agriculture practices.[89] Apart from the draft CSAF, the CCSPAFF also promotes knowledge and public awareness on how to adjust GHG-intensive agricultural practices, such as the application of industrial fertilisers and organic agriculture methods.[90] For connected purposes, the Second NAP-CDLDD has put in place monitoring measures such as early warning systems, vulnerability/impact assessment, and watershed protection measures to enable farmers to proactively adapt to droughts.[91]

South African agriculture-related laws discussed in Chapter 5 also have provisions that are relevant to adaptation. The *WSA* (under section 2) has provided for water resource management and conservation such as the protection of aquifers, rainwater-harvesting methods, and the construction of strategic boreholes.[92] As a means of promoting adaptive water-management strategies, the *WSA* also prohibits activities that could introduce industrial effluents or aggravate climate-change-related effluents into water systems.[93] The *NWA* makes provision for licenses that promote adaptive water management. In addition, the *NWA* promote adaptive water management through its provisions on controlled water activities and provisions that categorise water according to its uses.[94] The *CARA* has provisions that recommend the cultivation of virgin lands, as an adaptation strategy for lands affected by the impacts of climate change. Other adaptation measures provided for by the *CARA* include the irrigation of lands with less soil moisture content, as well as the prevention of waterlogs and water run-off to control the escalation of soil erosion.[95] In addition, financial supports such as grants and subsidies are provided by the *CARA*,[96] and these might qualify as or at least support adaptation strategies on the basis of previous discussions in this book.[97] The *SPLUMA* is another law closely related to agriculture and also has provisions that could enable adaptation for agriculture. Through its spatial sustainability and spatial resilience principles, the *SPLUMA* promotes adaptive land management strategies previously described as droughts and/or flood control measures, and control of overgrazing and over-cultivation.[98]

6.3 Regulatory gaps

The analysis in Chapters 3, 4, and 5 demonstrated that Kenya, Nigeria, and South Africa are committed to adapting to the impacts of climate change on critical sectors of the economy, especially agriculture, through laws and policies (in addition to existing action plans and strategies). Based on the analysis carried out on climate change and agriculture-related policies and laws in the three focus countries, it has been observed that their regulatory frameworks and sectoral approach appear to contain relevant provisions that could enable agriculture to adapt to the impacts of climate change. However, there are also gaps in the climate-change-related and agriculture policies and laws in each country as far as adaptation in the context of agriculture is concerned.

In Kenya, the gaps that were observed in the climate change and agriculture-related policies and laws include a lack of specific provisions on some important strategies that could enable agriculture to adapt to the impacts of climate change. For example, there are no provisions in the NEAP on irrigation and desertification management.[99] There is also no provision for institutional support in the NCCAP.[100] The NCCAP also presented a poor understanding of the vulnerabilities of climate change on agriculture, in addition to a lack of coordinated programmes such as water conservation schemes for boosting the resilience of agriculture against the impacts of climate-change-related droughts.[101]

The two laws of Kenya that were considered to have provisions on adaptation are the *EMCA* and the *Climate Change Act*. The major gap that was discovered in the *EMCA*, in the context of agriculture-focused adaptation, has to do with the lack of specific provisions on the need to employ trained personnel to ensure monitoring and compliance of EIAs and SEAs on agricultural projects.[102] The implication of this on adaptation being that EIAs and SEAs remain just mere requirements that are unenforceable. Gaps were also identified in the *Climate Change Act*. First, as a generic law, agriculture is not (expected to be) specifically mentioned in the Act. Therefore, agriculture-focused adaptation could only be inferred from some provisions of section 13 of the *Climate Change Act*.[103] Also, the *Climate Change Act* does not have provisions for synchronising scientific solutions with legal solutions to climate change.[104] Provisions on how to translate scientific data into readable information are required in implementing adaptation strategies as provided by the Act. In the same regard, though the *Climate Change Act* makes provision for mainstreaming, institutional and sectoral compartmentalisation were noted as challenges for mainstreaming climate change issues with agricultural development.[105]

In the same vein, some of Kenya's agricultural laws, especially the *Land Act*, do not contain provisions on how adaptive land practices could be implemented.[106] In the same context, the provisions of the *AFA Act* are not clear in linking the agencies and actors responsible for promoting adaptation actions.[107] By implication, adaptation measures in the context of agriculture are weakened. Although section 22 of the *AFA Act* provides for adaptation measures against climate change resulting from livestock production activities, but the provisions of section 22 are not clear on the number of cattle herds that can be kept on agricultural lands.[108] This omission could make it difficult to comply with grazing authorisations in the Act, and for the environment to be safeguarded from livestock-induced climate change. Furthermore, Kenya's organisational framework for adaptation seems diffused and uncoordinated, leading to duplication of activities. For example, the mandates of the National Climate Change Council (NCCC) and other institutions tasked with implementing adaptation strategies in the context of agriculture (such as the Agriculture and Food Authority, Water Resources Authority, NDMA) are not properly aligned.[109]

In Nigeria, some climate change policies, especially the NCCPRS, have been found to be too broad in generally facilitating adaptation initiatives in agriculture.[110] It has also been observed that some Nigerian adaptation policies (especially the NAP) are designed as short-term climate change adaptation approaches which could likely lead to maladaptation. This is different from Kenya's NAP, where the NAP is conceived as a long-term adaptation approach, which has been considered cost-effective and could promote socio-economic development.[111] Furthermore, Nigeria's NAP does not directly provide the required guidance and types of adaptation suitable for most critical sectors of the economy impacted by climate change, including the agricultural sector.[112] Adaptation for agriculture could only be indirectly inferred from the provisions of the NAP. Equally, Nigeria's NAP failed to stimulate the required collaboration among stakeholders, and it neither integrates nor complements adaptation efforts provided by policies in other sectors such as agriculture, forestry, and water resources.[113]

The *Climate Change Act* is a law that broadly serves as the basis for coordinating long-term climate change plans and actions including adaptation across all sectors of the economy in Nigeria.[114] The Act seeks to identify sectors of the economy vulnerable to climate change impacts, while striving to enhance the resilience, and to increase the adaptive capacity, of such sectors against the impacts of climate change.[115] Only relatively few gaps were identified in the *Climate Change Act* in facilitating adaptation either broadly or in the specific context of agriculture. Noticeably, the provisions of the *Climate Change Act* do not facilitate vertical and horizontal cooperation between the federal and state levels of government as well as civil societies, but it rather makes provision for inter-ministerial cooperation at the federal level.[116] In short, the Act does not provide guidelines or a framework for collaboration amongst the various stakeholders. The implication of this shortfall for adaptation in the context of agriculture is that there would be poor mainstreaming of climate change issues with agricultural development projects.

Another gap in Nigeria's agriculture regulatory framework is noticeable in the *NESEFC Regulations*, which exempt livestock grazing activities from its provisions.[117] Livestock grazing has been noted as a driver for climate change and soil erosion when it takes place in poorly allocated or space-constrained land.[118] Without provisions controlling cattle grazing, the *NESEFC Regulation* could be considered to be insufficient as far as adaptation is concerned.

The *Nigerian Agricultural Insurance Corporation Act* (NAIC), for its part, excludes non-commercial farmers from those who can benefit from insurance as a result of (especially climate-change-related) risk on agricultural products. As stated in previous discussions, most Nigerian farmers are poor subsistence farmers.[119] Excluding this category of farmers from agricultural insurance would defeat the purpose of the *NAIC Act* in supporting agricultural adaptation in Nigeria. In other words, without agricultural insurance, poor subsistence farmers found in drought-prone areas of northern Nigeria, for instance, would not be able to adapt to crops or livestock losses ensuing from climate-induced droughts.[120]

With regard to South Africa, the regulatory framework for adaptation appears to be dominated by policies rather than by laws. Comprehensive as they are, some of the policies were noted to contain gaps in facilitating adaptation for agriculture. For example, the NCCRP (as the focal point of climate change policies in South Africa) is mostly concerned with mitigation rather than adaptation.[121] Therefore, the policy could not be expected to contain more specific provisions targeting agriculture adaptation. In a similar vein, the NCCRP does not have provisions that are necessary in stimulating cross-sectoral mainstreaming of climate change issues between, for instance, the energy sector and agriculture.[122]

The NDP has provisions that seem to support agriculture-focused adaptation, but it fails to include budgetary processes and finance for adaptation.[123] Without these, adaptation measures, whether in a broad context or in the context of agriculture, can arguably not be successfully implemented. Additionally, for failing to consider underused lands in communal areas, the NDP's projection with respect to the potential of irrigation is noted to be unrealistic.

Furthermore, there are no specific climate change adaptation laws in South Africa; as a result, adaptation is derived from environmental-related laws such as the *NEMA*, *DMA*, and *CTA*. These laws have been presented and discussed as general framework laws that provide basic principles relevant to driving adaptation processes. However, there are observable gaps in these laws as far as agriculture-focused adaptation is concerned. The *DMA*, for instance, does not have provisions that elaborate upon how the impacts of climate-change-related disasters such as droughts and floods could be adapted to.[124] Likewise, the *CTA* (at the moment) also excludes commercialised agricultural practices from a carbon tax.[125] This implies that farmers responsible for carbon-intensive farming practices might not be compelled to alter or adjust their farming-related activities.

South African agriculture-based policies and laws from which adaptation is derived also contain some gaps. The CCSPAFF, for example, was formulated by South Africa before the signing of the Paris Agreement in 2015, and its provisions on adaptation, therefore, cannot be expected to be in line with the guidelines of the Paris Agreement, especially the Koronivia Joint Work on Agriculture (KJWA) of the Paris Agreement, which provides guiding norms on agricultural adaptation strategies.[126] The *CARA* is another example of an agriculture-based law with gaps in promoting agricultural adaptation. The lack of provisions in the *CARA* on the need to include skilled personnel in the implementation of soil conservation strategies has been identified as an example of such a gap.[127]

Based on the above observations, it is clear that there are some instances where the regulatory frameworks for climate change and agriculture in Kenya, Nigeria, and South Africa do well in facilitating agriculture-focused adaptation, but their provisions also contain gaps that could prevent agriculture from effectively adapting to the impacts of climate change.

6.4 Recommendations

Considering that there are gaps in the provisions of the Kenyan, Nigerian, and South African climate change and agricultural-related policies and laws in enabling adaptation in the context of agriculture, how could they be addressed?

The gaps identified in the regulatory frameworks for agriculture and adaptation in the three countries of focus are solvable – at least in part – by way of legal reforms. Three broad recommendations are offered in this regard. First, as a short-term and immediate solution, the national governments of Kenya, Nigeria, and South Africa should make regulations to cater for details that were not expected to be provided in some of their climate-change-related laws and agriculture-based laws. Examples of these details include specific on-farm adaptation strategies such as adaptive water management strategies, soil conservation methods, and so on. In addition, the provisions of some climate-change-related and agricultural policies of Kenya, Nigeria, and South Africa could be amended or altered in order to solve the specific gaps that were identified in them. Second, legislative amendment is required in the three countries to consolidate all relevant national climate-change-related policies and laws with agricultural policies and laws that have relevant provisions on adaptation as medium-term solutions. Third, specific framework law for agriculture-focused adaptation, incorporating essential agricultural adaptation strategies, could perhaps be enacted as long-term solutions to the regulatory gaps in Kenya, Nigeria, and South Africa.[128] These recommendations are elaborated in more detail below, and they could be implemented independently of each other and/or concurrently, depending on the needs and urgency of the problems caused by the regulatory gaps.

6.4.1 Short-term solutions

A review of climate-change-related and agriculture-based policies and laws in Kenya, Nigeria, and South Africa could be carried out to add specific provisions that could enable agriculture-focused adaptation where they appear to be missing. In addition, regulations could also be made to add specificities that appear to be missing in the provisions of some generic climate-change-related and/or agriculture-based laws in the three countries of focus. As already pointed out, regulations are made pursuant to oversight functions by government agencies and departments tasked with enforcing a law, in order to cover specificities that are usually missing in generic laws.[129]

In Kenya, specific provisions on irrigation and desertification management could be added to the NEAP. Similarly, the NCCAP could be improved by the addition of provisions on the need for (and direction on how to implement) water conservation scheme programmes. A regulation with specific provisions on the roles of trained personnel in ensuring monitoring and compliance of EIAs and SEAs could be made as a means of improving the *EMCA*. The provisions of the Kenyan *Climate Change Act* – especially section 13, from which agricultural adaptation is inferred – could also benefit from an added regulation with a provision on how to synchronise scientific solutions such as agricultural adaptation strategies with legal solutions. Likewise, the addition of a Regulation to the *Land Act* with provisions on how to implement adaptive land management practices could solve the gap that was identified in the law. The *AFA Act* could also be improved by regulations with provisions (among others) specifying the number of cattle herds which can be allowed to graze on agricultural lands, as an immediate solution to the gap that was identified in, especially, section 22 of the law.

In Nigeria, the provisions of the NCCPRS on adaptation (as a replacement for the NASPA-CCN) were observed as being too broad in focus. Thus, the NCCPRS could be revised to accommodate specific contents relevant to agricultural adaptation such as water conservation techniques, irrigation strategies, and how to promote drought-resistant crops and climate-smart livestock breeding methods. The NAP could also be revised to serve as a long-term adaptation approach and to explicitly include other critical sectors of the economy which were identified as vulnerable to climate change impacts, especially agriculture. In addition, revising the NAP could possibly solve the gap in its lack of facilitation of stakeholders' collaboration and lack of integration with other sectoral policies with content on adaptation, particularly agriculture and water resources.

The immediate solution to the gaps that exist in the *Climate Change Act* is to amend the provisions of section 2 of the Act which provides for interministerial collaboration at the federal level. Amending section 2 of the *Climate Change Act* could possibly further promote mainstreaming efforts through both vertical and horizontal collaborations between different stakeholders in Nigeria's climate change adaptation governance, including the agricultural sector. The provisions of the *NESEFC Regulations* could be amended to also include livestock grazing activities, which are currently exempted. Provisions controlling livestock grazing would allow agricultural lands to adapt to soil erosion and methane emissions, among others.

In order to fully promote adaptation against the impacts of climate change on agricultural practices, the provisions of the *NAIC Act* should be amended to also include non-commercial and/or poor subsistence farmers as beneficiaries of the agricultural insurance scheme. Including this category of farmers in the agricultural insurance scheme would enhance their adaptive capacity, boost their resilience, and reduce their vulnerability to the impacts of climate-change-related losses and risks.

The same solutions that have been recommended in addressing specific regulatory gaps in Kenya and Nigeria could also work well for South Africa. For example, more specific provisions on adaptation, especially in the context of agriculture, could be added to the NCCRP. In the same instance, the NCCRP could benefit from an added provision on cross-sectoral mainstreaming of climate change issues, especially between energy, water, and agriculture. For its part, the NDP could be improved to fully accommodate important aspects of adaptation particularly budgeting and finance, which were reported to have been omitted.

Gaps in climate-change-related laws such as the *DMA* and *CTA* could also be possibly solved by a few modifications. The gaps in the *DMA*, for instance, are solvable by a regulation containing elaborate provisions on the required adaptation measures for climate-change-related disasters such as droughts and floodings, and how they could be implemented. Probably too, some current gaps in the *CTA* such as the exclusion of commercial-based farmers from carbon tax could be addressed by a regulation modifying provisions of section 7 and 8 of the law in preparation for the second phase of implementation of the *CTA* as from 2023.

On a final note, since the *CARA* is tabled for amendments through the draft *Conservation of Agricultural Resources Amendment Bill*, it is recommended that subsequent amendments to the Bill should be done to consider and incorporate some of the most recent guiding norms for agricultural adaptation practices especially those provided by the KJWA of the Paris Agreement. Simply because the KJWA norms were developed much later than the *CARA* and the draft *Conservation of Agricultural Resources Amendment Bill*.

6.4.2 *Medium-term solutions*

Most of the climate-change-related policies in Kenya, Nigeria, and South Africa, which appear to have contents related to adaptation in the agriculture sector that were discussed above, could be merged, consolidated, or amalgamated with agriculture-based policies in the respective countries. For example, in Kenya, the NCCRS, NAP, and the NCCAP could be merged with agriculture-based policies, especially the National Land Policy, the draft National Livestock Policy, and the draft National Agricultural Insurance Policy. In the same instance, agriculture-based laws with similar goals such as the *NDMA Act*, *Crops Act* and the *AFA Act* could be consolidated and possibly implemented jointly.

In Nigeria, the NCCPRS have broad, elaborate, and comprehensive provisions on agriculture-focused adaptation, which could be consolidated with agriculture-based policies such as the Nigerian Agricultural Policy, the NARF, the National Drought Plan, and the NPEFC. Likewise, agriculture-based laws with similar objectives and related mandates such as the *RBDA Act* and the *NESEFC Regulations* could be amalgamated.

In the case of South Africa, the NCCRS, the NCCRP, and the NCCAS are climate-change-related policies that are relevant in facilitating adaptation in the context of agriculture, though with few gaps. Consequently, these policies can be consolidated with other agriculture-based policies that seem to fully support adaptation, particularly the Second NAP-CDLDD.

Some examples may serve to illustrate the benefits of merging climate-change-related policies with agriculture-based policies; as well as in consolidating climate-change-related laws with agriculture-based laws of Kenya,

Nigeria, and South Africa. For a start, it would be possible to have an increased coordination of adaptation activities for agriculture in Kenya, Nigeria, and South Africa by way of consolidation. Merging climate-change-related policies with agriculture-based policies, or consolidating climate-change-related laws with agriculture-based laws, could be an ideal way of overcoming the institutional challenges identified in Kenya, and the lack of coordination of adaptation measures and duplication of functions observed in Nigeria and South Africa. The amalgamation of climate change adaptation policies that have provisions on agriculture with agricultural-based policies could provide an opportunity for balancing competing national development goals (especially those goals that are envisaged to be achieved through agriculture) with climate and resilient development in Kenya, Nigeria, and South Africa. By extension, consolidation could pave the way for a seamless development of a specific national framework law on agricultural adaptation in Kenya, Nigeria, and South Africa.

6.4.3 Long-term solutions

Given the role of agriculture in the economic growth and development of Kenya, Nigeria, and South Africa, and its susceptibility to the impacts of climate change, adaptation in agriculture deserves a specific framework law. Merely inferring agriculture-focused adaptation from other laws which were not originally intended to regulate adaptation, or laws which are either too broad or with insufficient provisions on adaptation might lead to a situation where aspects of adaptation are not regulated. The proposed framework law for agriculture-focused adaptation could for instance be named "the *Agricultural Adaptation Act*" and could be similar but better than existing environmental-related framework laws such as the *EMCA* of Kenya, the *NESREA Act* of Nigeria, and the *NEMA* of South Africa.

In anticipation, the new framework law should incorporate aspects of adaptation strategies for crop production, livestock production, water management, land management, and agricultural insurance provided by some climate-change-related and agriculture-based policies and laws that currently exist in Kenya, Nigeria, and South Africa. The envisaged specific agriculture-focused adaptation framework law should also have provisions on mandatory agricultural adaptation strategies, with compliance, enforcement, and sanctions specified in clear terms.

6.5 Future research

The possibility of climate change events in Kenya, Nigeria, and South Africa spilling over to, and impacting the agricultural practices of other East African, West African, and Southern African countries, respectively, since they share geographical borders, are issues that are connected to this study, but they were not addressed.[130] In this context, it becomes imperative to study the development of a sub-regional legal framework for climate change with provisions that could enable agriculture in East African, West African, and Southern African countries to adapt against the spillover effects of climate change impacts. Likewise, this study did not compare the regulatory frameworks for climate change adaptation and agriculture of Kenya, Nigeria, and South Africa with other developing countries in Asia (such as Thailand) or Mexico in Latin America. These are developing countries that have a similar degree of susceptibility to climate change and weak adaptive capacity as that seen in African countries.

Therefore, either of these proposals will serve as the main themes for future research projects. A proposal for the development of a sub-regional legal framework for climate change with provisions on agriculture-focused adaptation for East Africa (EAC), West Africa (ECOWAS), and Southern Africa (SADC); or a proposal for a comparative study of agriculture-focused adaptation governance in African countries with other developing countries in Asia and Latin America. Hence these issues are briefly presented as the following research questions:

- How and to what extent could climate change in Kenya, Nigeria, and South Africa have an impact on agricultural practices in other countries in East Africa, West Africa, and Southern Africa?
- In view of the possibility of a spillover of climate change impacts from Kenya, Nigeria, and South Africa to other countries in East Africa, West Africa, and Southern Africa, respectively, would it be viable to develop a sub-regional climate change legal framework with provisions on agriculture-focused adaptation for East Africa, West Africa, and Southern Africa?
- How does agriculture-focused governance in African countries compare with agriculture-focused governance in other developing countries in Asia and Latin America?

Notes

1 Chapter 1, Section 1.2.
2 Chapter 1, Sections 1.3.1, 1.3.2, 1.3.3.
3 Chapter 1, Section 1.2.
4 Chapter 1, Section 1.2.
5 Chapter 2, Section 1.6.
6 Chapter 2, Sections 2.3, 2.4.
7 Chapter 3, Section 3.5; Chapter 4, Section 4.6; Chapter 5, Section 5.6.
8 Chapter 1, Section 1.2.
9 Chapter 1, Section 1.2.
10 Chapter 1, Section 1.4.1.
11 Chapter 1, Section 1.4.2.
12 Chapter 1, Section 1.4.2.
13 Chapter 1, Section 1.4.2.
14 Chapter 1, Section 1.4.3.
15 Chapter 1, Section 1.4.4.
16 Chapter 1, Section 1.4.5.

17 Chapter 1, Section 1.5.1.
18 Chapter 1, Section 1.5.1.4.
19 Chapter 1, Section 1.5.2.
20 Chapter 1, Section 1.5.3.
21 Chapter 1, Section 1.5.3.
22 Chapter 2, Section 2.2.
23 Chapter 2, Section 2.2.
24 Chapter 2, Section 2.3.
25 Article 4(1) UNFCCC.
26 It was previously explained in Chapter 2, Section 2.3.3, that the Paris Agreement improves on, and replaces the Kyoto Protocol. Hence, the Kyoto Protocol is simply referred in reaffirmation of its contextual role in the development of adaptation law.
27 Article 10(b) (i), Kyoto Protocol.
28 Article 7 Paris Agreement.
29 Chapter 2, Section 2.4.2.3.
30 Chapter 2, Section 2.3 generally.
31 Chapter 3, Section 3.2; Chapter 4, Section 4.2; Chapter 5, Section 5.2.
32 Chapter 3, Section 3.2.1.
33 Chapter 1, Sections, 1.2; 1.5.
34 Chapter 3, Section 3.2.2.
35 Chapter 3, Section 3.2.1.1.
36 Chapter 3, Section 3.2.1.2.
37 Section 20, CFRN 1999; Mondaq 2022 https://www.mondaq.com/nigeria/energy-law/53804/environment-law-in-nigeria.
38 Chapter 1, Section 1.2.
39 Chapter 4, Section 4.2.1.
40 See generally, Chapter 4, Section 4.2.
41 Chapter 4, Section 4.2.2.
42 Chapter 4, Section 4.2.3.1.
43 See mostly, Chapter 1, Section 1.5.1.
44 Chapter 4, Sections; 4.2.2.2; 4.2.2.4.
45 Chapter 4, Section 4.2.3.2.
46 Chapter 5, Section 5.2.1.
47 Chapter 5, Section 5.2.1.
48 National Climate Change Response White Paper GN 757/Government Gazette No 34695 https://www.gov.za/sites/default/files/gcis_document/201409/nationalclimatechangeresponsewhitepaper0.pdf, 16–24.
49 Chapter 5, Section 5.2.2.2.
50 Chapter 5, Section 5.2.2.2.
51 Chapter 5, Section 5.2.2.2.
52 Chapter 5, Section 5.2.2.3.
53 Chapter 5, Section 5.2.2.5.
54 See generally, Chapter 2, Section 2.4.
55 See Chapter 3, Section 3.2.1.
56 Chapter 3, Section 3.3.2.
57 Chapter 3, Section 3.3.3.
58 Chapter 3, Section 3.3.1.
59 Chapter 3, Section 3.3.2.1.
60 Chapter 3, Section 3.3.2.2.
61 Chapter 3, Section 3.3.2.3.
62 Chapter 3, Section 3.3.3.1.
63 Chapter 1, Section 1.5.1.2.

64 Chapter 3, Section 3.3.3.2.
65 Chapter 3, Section 3.3.3.3.
66 Chapter 3, Section 3.3.3.4.
67 Chapter 3, Section 3.3.3.4.
68 Chapter 3, Section 3.3.3.5.
69 Chapter 3, Section 3.3.3.6.
70 Chapter 1, Section 1.5.1.
71 Chapter 4, Section 4.2.1.
72 Chapter 4, Section 4.3.1.
73 Chapter 4, Section 4.3.1.1.
74 Chapter 4, Section 4.3.1.2.
75 Chapter 4, Section 4.3.1.3.
76 Chapter 4, Section 4.3.1.4.
77 Chapter 4, Section 4.3.1.5.
78 Chapter 4, Section 4.3.2.3.
79 Chapter 4, Section 4.3.2.1.
80 Chapter 4, Section 4.3.2.2.
81 Chapter 4, Section 4.3.2.3.
82 Chapter 4, Section 4.3.2.4.
83 Chapter 4, Section 4.3.2.5.
84 Chapter 5, Section 5.2.1.
85 Chapter 5, Sections 5.3.1.1; 5.3.1.2.
86 Chapter 5, Sections 5.3.2.1; 5.3.2.2.
87 Chapter 5, Sections 5.3.2.3; 5.3.2.4.
88 Chapter 5, Sections 5.2.2.2; 5.2.2.3.
89 Chapter 5, Section 5.3.1.1.
90 Chapter 5, Section 5.3.1.2.
91 Chapter 5, Section 5.3.1.3.
92 Chapter 5, Section 5.3.2.1.
93 Chapter 5, Section 5.3.2.2.
94 Chapter 5, Section 5.3.2.2.
95 Chapter 5, Section 5.3.2.3.
96 Chapter 5, Section 5.3.2.3.
97 Chapter 1, Section 1.5.1.3.
98 Chapter 5, Section 5.3.2.4.
99 Chapter 3, Section 3.2.2.1.
100 Chapter 3, Section 3.2.2.3.
101 Chapter 3, Section 3.2.2.3.
102 Chapter 3, Section 3.2.1.1.
103 Chapter 3, Section 3.2.1.2.
104 Chapter 3, Section 3.2.1.2.
105 Chapter 3, Section 3.2.1.2.
106 See Chapter 3, Section 3.3.3.5.
107 Chapter 3, Section 3.3.3.6.
108 Chapter 3, Section 3.3.3.6.
109 Chapter 3, Section 3.5.
110 See Chapter 5, paragraph 5.5.
111 Chapter 3, Section 3.2.2.4.
112 Chapter 4, Section 4.2.2.4.
113 Chapter 4, Section 4.2.2.4.
114 Chapter 4, Section 4.2.3.2.
115 Section 27, *Climate Change Act*.
116 See Chapter 4, Section 4.2.3.2.

117 Chapter 4, Section 4.3.2.4.
118 Chapter 2, Section 2.4.1.2.
119 Chapter 4, Section 4.3.2.5.
120 See Chapter 1, Section 1.3.2.
121 Chapter 5, Section 5.2.2.2.
122 Chapter 5, Section 5.2.2.2.
123 Chapter5, Section 5.2.2.4.
124 Chapter 5, Section 5.2.3.2.
125 Chapter 5, Section 5.2.3.3.
126 The KJWA norms were introduced and discussed in Chapter 2, Section 2.4.1, while the extent of their application in the context of South African agricultural adaptation governance was explained in Chapter 5, Section 5.3.1.2.
127 Chapter 5, Section 5.3.2.3.
128 Enacting a new framework law is recommended as a long-term solution, considering the delay involved in, for instance passing the *Climate Change Act* of Nigeria into Law, and the recently assented South African *Climate Change Act*. See Chapter 4, Section 4.2.3.2; Chapter 5, Section 5.2.3.4.
129 See Chapter 2, Section 2.2.
130 It is possible for the impacts of climate change in Kenya to spill over and affect agricultural practices in other countries in East Africa, especially Burundi, Democratic Republic of Congo, Rwanda, South Sudan, Tanzania, and Uganda. As well, climate change in Nigeria could spill over and impact the agricultural sector of other West African countries, including Benin, Burkina Faso, Cabo Verde, Côte d'Ivoire, the Gambia, Ghana, Guinea, Guinea Bissau, Liberia, Mali, Niger, Senegal, Sierra Leone, and Togo. Similarly, climate change in South Africa could have spillover effects on agricultural practices in Angola, Botswana, Eswatini, Lesotho, Mozambique, Namibia, Zambia, and Zimbabwe.

Index

Note: Page numbers followed by "n" denote endnotes.

Adamawa State 142–143, 167n128
adaptive capacity 5, 8, 9, 14, 15, 18, 57, 58, 67, 101, 132–136, 141–143, 150, 152, 184, 185, 195, 231, 232, 242, 245
Adler R 67
Agrarian Insurance Law 71
Agricultural Adaptation Act 247
agricultural adaptation laws 62–64, 72n2, 72n6; chemical-use laws 65–66; Conservation of Agricultural Resources Act (CARA) 179, 202–204, 240, 243; controlling agricultural production 64–66; Crops Act 108–109; fertiliser application laws 65–66; insurance law 71–72; Kenya 105–111; Kenyan Agriculture and Livestock Research Act (KALRA) 108; Land Act (2012) 109–110; Land Use Act (1978) 130, 154–155; land-use laws 69–71; livestock production laws 66; National Drought Management Authority Act (NDMA) 107; National Environmental (Soil Erosion and Flood Control) Regulations (NESEFC Regulations) 130, 155–156, 242, 245; National Water Act (NWA) 199–201, 221n262, 222n271, 240; Nigeria 151–157; Nigerian Agricultural Insurance Corporation (NAIC) Act 130, 156–157; promoting agricultural production 67–72; River Basins Development Authority (RBDA) 130, 148, 153–154; role of 62; soil-conservation laws 69–71; Spatial Planning and Land Use Management Act (SPLUMA) 204–205; Water Act (2016) 106–107; water-management laws 67–69; Water Resources Act 130, 152–153; Water Services Act (WSA) 198–199; *see also individual entries*
Agricultural and Fishery Disaster Insurance Act (AFDIA) 71
agricultural-based policies 178, 194; Climate Change Sector Plan for Agriculture, Forestry, and Fisheries (CCSPAFF) 178, 196–197, 243; draft Climate-Smart Agriculture Strategic Framework (CSASF) 195–196; draft National Agricultural Insurance Policy 104–105; draft National Livestock Policy of Kenya 103–104; economic growth 180; Kenya 101–105; National Agricultural Resilience Framework (NARF) 146–147; National Drought Plan 130, 149–150; National Land Policy 103; National Policy on Erosion and Flood Control (NPEFC) 130, 151, 238; National Water Policy 130, 147–149; Nigeria 143–151; Nigerian Agricultural Policy (2001) 130, 144–146; Second National Action Programme for South Africa to Combat Desertification, Land Degradation and the effects of Drought (NAP-CDLDD) 197–198; *see also individual entries*
agricultural insurance law 71–72
agriculture 231–233; challenges for 20; climate change impacts 1, 20–21; climate-smart agriculture (CSA) 16–17; contributions to climate

change 2–3; economy, African countries 1; financial management techniques 15–16; insurance 15–16; Kenya 3–4; Nigeria 4–5; on-farm adaptation strategies 14; planning and implementation 17–18; South Africa 5–6; technological solutions 14–15
Agriculture and Food Authority (AFA) Act 110–111, 245
agro-forestry techniques 18
agro-processing 147
Ajani E 19
Alder 200
alternative fallow 14, 35–36n181
Anchor Borrowers' programme 16, 38n213
Angola 251n130
anticipatory adaptation 7, 8
Arcidiacono C 19
Ardakanian R 12
autonomous adaptation 7

Baker I 12
Bali Action Plan (2007) 77n72
Bank M 200
Basel Convention on the Control of Transboundary Movement of Hazardous Wastes and their Disposal (1989) 65
Basel, Rotterdam and Stockholm (BRS) Conventions 65, 229
Benin 251n130
Benue 172n221
Berlin Rules on Water Resources (2004) 68
Billiri 133
Blesh J 13
blue water 15, 23n21, 69
Bodansky D 59
Borno State 142–143, 167n128, 169n169
Botswana 121n121, 125n220, 126n236, 251n130
Brunnée J 59
Burkina Faso 10, 39n226, 251n130
Burundi 251n130

Cabo Verde 251n130
Cancun Adaptation Framework (CAF) 59, 76n53, 136
Cancun Agreements 59
Cape Town 221n254
Carbon Budget 141
Carbon Tax Act (CTA) 178, 191–192, 246
catchment areas 106, 111, 124n185, 188, 237
Cattle Grazing Act 66
civil rule 152, 160n23
chemical-use laws 65–66
Clean Development Mechanism 77n66
climate change 230–231; adaptation 231–233; agriculture 1–3, 20–21; causes 1; hotspots 23n23, 30n103; human-induced 21n1; impacts of 1; livestock, impacts on 24n25
Climate Change Act (2016, Kenya) 99–101, 112, 117n55, 121n116
Climate Change Act (2021, Nigeria) 130, 140–141, 158, 166n112, 235, 245
Climate Change Act (2024, South Africa) 178, 192–194, 218n196
climate change adaptation laws 56–58, 139, 179, 229–230; Carbon Tax Act (CTA) 178, 191–192, 246; Climate Change Act (2016) 99–101; Climate Change Act (2021) 130; Climate Change Act (2024) 178, 192–194; climate-smart agriculture (CSA) 142–143; directives and standards, actors 57; Disaster Management Act 178, 190–191; Environment Management and Coordination Act (EMCA) 97–99; gaps and challenges 230; Kenya 97–101; Kyoto Protocol 61; National Environmental Management Act (NEMA) 97–98, 178, 188–190, 215n144, 215n145, 216n148–150; National Environmental Standards and Regulations Enforcement Agency (NESREA) Act 130, 139–140; Paris Agreement 61–62; planning 57; principles of 58–62; regional and national levels 58; South Africa 188–194; United Nations Framework Convention on Climate Change (UNFCCC) 59–60; *see also individual entries*
climate change adaptation policies 232–233; Kenya 93–97; National Adaptation Plan (NAP) 91, 96–97, 130, 136–137; National Adaptation Strategy and Plan of Action for Climate Change in Nigeria (NASPA-CCN) 130, 132–133; National Climate Change Action Plan

(NCCAP) 91, 95–96, 117n56, 141, 166n118, 234, 244; National Climate Change Adaptation Strategy (NCCAS) 178, 186, 206, 235–236; National Climate Change Response Policy White Paper (NCCRP) 178, 184–186, 205, 235, 243, 245; National Climate Change Response Strategy (NCCRS) 91, 94–95, 116n35, 183–184, 234–236; National Development Plan (NDP) 187; National Environment Action Plan (NEAP) 91, 93, 234; Nationally Determined Contributions (NDCs) 96, 131, 137–138, 164n82, 164n91, 178, 179, 187–188, 235–236; National Policy on Environmental Conservation (NPEC) 130, 135; Nigeria Climate Change Policy Response and Strategy (NCCPRS) 130, 133–135, 162n56, 245, 246; *see also individual entries*
climate change adaptations 1, 6–7, 232–233; agriculture, context of 13–20, 21–22n5; challenges for 12–13; ecological constraints 13; economic constraints 12–13; forms of 7–8; governance of 6; mainstreaming 11–12; need for 8–10; social constraints 12; stages of 10–11
Climate Change and Air Quality Management (CCAQM) Agency 181
Climate Change Sector Plan for Agriculture, Forestry, and Fisheries (CCSPAFF) 178, 196–197, 243
climate-smart agriculture (CSA) strategy 16–17, 66, 72, 102–103, 142–143, 185
cloning 15
coal mining 5
coal mining-related emissions 26–27n56
commercial-based farming method 5
communal knowledge-sharing 10
community-based adaptation (CBA) 18
community-based measures 191
Conferences of the Parties (COPs) 59, 75n51
Conservation of Agricultural Resources Act (CARA) 179, 202–204, 240, 243
Conservation of Agricultural Resources Amendment Act 179
Conservation of Agricultural Resources Amendment Bill 246
conservation tillage 14
Constitution of Kenya 91–93, 115n16, 120n108; article 42 92; article 60 92; article 69 92
Constitution of South Africa (1996) 178, 181–183, 189, 194, 198, 205, 207, 209n32, 210n51, 233; climate change governance 180; international instruments 182; Section 24 182; Section 27 182–183
Constitution of the Federal Republic of Nigeria (CFRN, 1999) 130–132, 152, 234–235, 238; section 4(2) 172n210; section 12 (1) 160n29; section 20 160n25; section 62 161n31
Contiades X 59
Convention on Biological Biodiversity (CBD) 66, 229
Convention on the Protection and Use of Transboundary Watercourses and International Lakes (1992) 68
Côte d'Ivoire 251n130
cover crops 14
crop diversification 7
Crops Act 108–109, 237
Cross River 172n221

Dashe 133
Decrees 152
Delta River 172n221
Democratic Republic of Congo 34n163, 251n130
Department of Climate Change 131, 147
Department for Environmental Affairs 180
Department of Environment, Forestry and Fisheries (DEFF) 181
Department of Forestry, Fisheries and the Environment (DFFE) 179, 186, 188
Department of Water and Sanitation; the Department of Agriculture, Land Reform and Rural Development (DALRRD) 181, 186, 189, 201, 223n307
direct tax 218n188
Disaster Management Act (DMA) 178, 190–191
draft Climate-Smart Agriculture Strategic Framework (CSASF) 178, 195–196, 239–240
draft Conservation of Agricultural Resources Amendment Bill (2013) 203, 206
draft National Agricultural Insurance Policy 104–105

draft National Livestock Policy of Kenya 103–104
draft South African Development Commission (SADC) 186
Duff L 181
Du Plessis A 180, 182, 185, 190, 204, 205

early warning systems 184
Eastern Cape 203, 216n160
East Africa 17, 18, 24n24, 247–248, 251n130
ecolabels 73n10
ecological constraints 13
economic constraints 12–13
ecosystem-based water management approach 82n155
El Nino-Southern Oscillation 15
enteric fermentation 23n14
environmental governance principles 188–189
Environmental Impact Assessments (EIAs) 98, 99, 112, 189–190, 234
environmental management principles 189
Environmental (Soil Erosion and Flood Control) Regulations, Lagos 151
environmental stewardship 22, 56, 73n11
Environment Management and Coordination Act (EMCA) 91, 97–99, 120n102, 234
Eriksen SH 6
Erosion Control Plan Measures 155
Eswatini 251n130
European Union (EU) Water Framework Directive (2000) 67, 69
EU Water Framework Directive 229

Falgore 133
fertilisers: application laws 65–66; industrial 2, 22n7
financial management techniques 15–16
flood defence system 10
flooding 31n127, 31n129
Food and Agricultural Organization (FAO) 57, 65, 73n14, 101, 143; international code of conduct 80n119; modern CSA practices 143
Fotiadou A 59
Free State 216n160
Fufore 142
FundsforNGOs 203

Gambia 251n130
Gauteng 216n160
Ghana 22n6, 38n208, 251n130
GHG-intensive farming activities 98
Gilpin J 14
Glasgow 140, 165n109
Glatzle A 3
global anthropogenic emissions 22n7
Global Climate Observing System Report 137
Green Climate Fund (GCF) 60, 77n66
greenhouse gases (GHGs) emissions 1, 2, 4, 25n37, 141, 142, 147, 154, 179; reduction targets 60
gross domestic product (GDP) 1, 158, 229
Guinea 251n130
Guinea Bissau 251n130
Guyuk 142

Hadejia-Chad 172n221
hard laws 56
Härtel I 63
Hettiarachchi H 12
Hollo EJ 59
Horn of Africa 39n228, 107
human systems 6, 9, 28n72, 57
Humby T 189, 203

Imo-Anambra 172n221
Impala Irrigation Scheme 222n284
indemnity-based insurance 15, 38n208
index-based insurance 15–16, 38n208
indiscriminate bush burning 4
industrial fertilisers 2, 22n7, 145
Insurance Act (1984) 109
insurance system 15–16, 64
integrated development plans (IDPs) 180–181, 209n36
integrated resource planning 184
Intergovernmental Committee on Disaster Management (ICDM) 190
Intergovernmental Panel on Climate Change (IPCC) 7, 11, 13, 20, 57, 191
Inter-Ministerial Committee on Climate Change (IMCCC) 181
Inter-Ministerial Committee on Climate Change (ICCC) 134
International Code of Conduct for the Effective Use and Management of Fertilizers 65, 229
International Code of Conduct on Pesticide Management (ICCPM) 65, 229

international legal framework: agricultural adaptation 64; fertiliser application and chemical-use laws 65; water-management laws 67
invasive alien species 38n210
irrigation: activities 222n273; services 153–154, 187
Iza R 67

Jigawa 169n169
Johannesburg 27n55, 208n11, 222n275

Kano 169n169
Karsten J 200
Katsina 169n169
Kenya: agricultural income 24n32; agriculture 3–4; agriculture, regulatory framework 101–111, 236–240; broad recommendations 243–247; challenges 114; climate change adaptation, regulatory framework 90–101, 233–236; Constitution of Kenya 91–93; CSA Strategy 237; future research projects 247–248; gaps, policy and legal frameworks 111–113, 240–243; impacts of climate change 24n33; insurance law 125n209; long-term solutions 247; medium-term solutions 246–247; migration 40n232; NAP 114n6, 118n75; policies and laws 90; regulatory frameworks, agriculture 236–240; regulatory frameworks, climate 233–236; short-term solutions 244–246; social cohesion 25n36
Kenya Climate Smart Agriculture Strategy (2017–2026) 126n234
Kenyan Agriculture and Livestock Research Act (KALRA) 108, 237
Kenyan Land Act 71
Kenyan National Water Master Plan 123n178
Kenyan Shillings 3
Kilifi District 99
Klassen S 14
Koronivia Joint Work on Agriculture (KJWA) 64, 243, 251n126
Kotzé LJ 57, 180
Kulovesi K 59
Kwaikong 133
Kwazulu-Natal 216n160
Kyoto Protocol 61, 62, 64, 71, 72, 77n66, 130, 232–233, 249n26; article 10(c) 77n73; Kenya 91; Nigeria 131; South Africa 182; United States 28n76
Kyrgyz Law on Soil Fertility and Protection of Agricultural Lands 71

Lagos State government of Nigeria 151
Lake Shores Regulations (2009) 156
Land Act (2012) 109–110, 237–238
land tenure 78n92; rights 83n169
land topography 35n180
Land Use Act (1978) 130, 154–155, 172n228, 172n230
land-use laws 69–71
laws: adaptation planning and strategies 57; agricultural adaptation laws 62–72; climate change adaptation 56–58; directives and standards, actors 57; Kyoto Protocol 61; meaning of 55, 56; Paris Agreement 61–62; principles, climate change adaptation laws 58–62; regional and national levels 58; regulating human conduct 232–233; types 56; United Nations Framework Convention on Climate Change (UNFCCC) 59–60
Least Developed Countries Fund (LDCF) 60; *see also individual entries*
Lehman K 185
Lesotho 251n130
Li S 13
Liberia 251n130
Limpopo 209n37, 216n160, 220n239
livestock production laws 66
livestock-rearing 2–3
long-term adaptation plans 11, 19
long-term solutions 247

Malawi 38n208, 41n250
Mali 10, 124n195, 251n130
Mandryk M 12
Mantlana KB 188
manure 2, 22n11
Masters L 181
medium-term adaptation plans 11, 19
medium-term solutions 246–247
Mehling M 59
Mendelsohn R 1
micro-insurance mechanism 185
migration 10, 17, 31n127, 31n129, 40n232
military rule 152

Millennium Development Goals (MDGs) 132, 161n32
Ministry of Environment and Natural Resources, Kenya 117n56
Ministry for Agriculture 12, 140
Ministry for Water Resources 12
Mozambique 10, 31n127, 31n129, 251n130
Mpumalanga Province 26n55, 216n160
Mubi South 142
Muller M 190
Multilateral Environmental Agreements (MEAs) 6, 56, 65, 132
multi-sectoral approach 162n50
multi-sectoral measures and actions 6
Murphree M 182, 185, 190, 204, 205
Mwebaza R 57

Namibia 251n130
Nanda V 60
National Adaptation Plan of Actions (NAPAs) 41n245, 59
National Adaptation Plans (NAPs) 18, 41n245, 91, 96–97, 114n6, 118n75, 130, 136–137, 157–158, 245
National Adaptation Strategy and Plan of Action for Climate Change in Nigeria (NASPA-CCN) 130, 132–133
National Agricultural Policy 238
National Agricultural Resilience Framework (NARF) 146–147, 238
National Agricultural Value Chain Development Project 16
National Assembly 152, 160n29, 161n31, 166n110, 172n210, 174n263
National Climate Change Action Plan (NCCAP) 91, 95–96, 117n56, 141, 166n118, 234, 244
National Climate Change Adaptation Strategy (NCCAS) 178, 186, 206, 235–236
National Climate Change Council (NCCC) 95, 193
National Climate Change Response Policy White Paper (NCCRP) 178, 184–186, 205, 235, 243, 245
National Climate Change Response Strategy (NCCRS) 91, 94–95, 116n35, 178, 183–184, 234–236
National Council on Climate Change (NCC) 141
National Development Plan (NDP) 178, 187
National Drought Management Authority Act (NDMA) 107, 113, 237, 241
National Drought Plan (2018) 130, 149–150
National Environment Action Plan (NEAP) 91, 93, 234
National Environmental Management Act (NEMA) 97–98, 178, 188–190, 215n144, 215n145, 216n148–150, 235
National Environmental (Soil Erosion and Flood Control) Regulations (NESEFC Regulations) 130, 155–156, 242, 245
National Environmental Standards and Regulations Enforcement Agency (NESREA) Act 130, 139–140, 153, 165n105, 235
National Land Policy (2009) 103, 237
national legal framework: agricultural insurance law, Korea 71; Azerbaijan law 71; land-use and soil-conservation laws 71; Ugandan cattle grazing 66
national level adaptation policies: National Climate Change Adaptation Strategy (NCCAS) 178, 186, 206, 235–236; National Climate Change Response Policy White Paper (NCCRP) 178, 184–186, 205, 235, 243, 245; National Climate Change Response Strategy (NCCRS) 116n35, 178, 183–184, 234–236; National Development Plan (NDP) 187; Nationally Determined Contributions (NDCs) 96, 131, 164n82, 164n91, 178, 179, 187–188, 235–236; South Africa 179–180
National Livestock Policy 237
Nationally Determined Contributions (NDCs) 96, 131, 137–138, 164n82, 164n91, 178, 179, 187–188, 235–236
National Policy on Environmental Conservation (NPEC) 130, 135
National Policy on Erosion and Flood Control (NPEFC, 2006) 130, 151, 238
National Water Act (NWA) 179, 199–201, 221n262, 222n271, 240
National Water Policy 130, 147–149
National Water Resources Policy (NWRP) 148–149

National Water Resource Strategy (NWRS) 106, 185, 201
natural systems 1, 6, 7, 9, 11, 28n73, 101, 136
Niger Delta Region 174n263
Nigeria: Agricultural Policy 144–146, 168n148; agriculture 4–5; broad recommendations 243–247; climate change adaptation, regulatory framework 131–143; deficiencies, policy and legal frameworks 157–158, 240–243; droughts 26n48; first national policy, environmental conservation 162n59; future research projects 247–248; GDP 158; livestock grazing reserves 168n148; long-term solutions 247; medium-term solutions 246–247; policies and laws 130–131; Presidential Constitution 160n23; rainfall patterns 25–26n47; regulatory frameworks, agriculture 143–157, 236–240; regulatory frameworks, climate 233–236; short-term solutions 244–246
Nigeria Climate Change Policy Response and Strategy (NCCPRS) 130, 133–135, 162n56, 245, 246
Nigerian Agricultural Insurance Corporation (NAIC) Act 130, 156–157, 239, 242, 245
Nigerian Agricultural Policy (2001) 130, 144–146, 168n148
Nigerian Agricultural Resilience Framework (NARF) 130, 146–147
non-governmental organisations (NGOs) 57
Northern Cape 216n160
Northwest South Africa 216n160

off-farm adaptation 35n177
Ogun Basin 172n221
on-farm adaptation strategies 14
organisational stability 20
Osun Basin 172n221
Owena Basin 172n221

parliament 97, 117n56, 119n81, 120n108, 203
Paris Agreement 61–64, 71, 72, 130, 232–233; article 7 78n79, 78n85; Kenya 91; Nigeria 131; Nigeria's NDCs 138; South Africa 182
Parry ML 7
pastoralists, assets of 39n230
pathway 21, 25n37, 34n161, 95, 100, 113, 117n57, 136, 158, 160n26, 193, 234
Pelling M 6
pesticides 36n185
planned adaptation 7–8
Presidential Climate Commission (PCC) 179–180, 193
Presidential Employment Stimulus Initiative (PESI) 16
prevention of significant harm principle 67–68
principle of reasonable and equitable use 67–68
private sectors 60
Pro-Resilience Action (PROACT) 142–143, 167n127
public trust doctrine (PTD) 154, 199–200, 203, 221n268
Puthucherril TG 77n73

quasi-legal instruments 56

rainwaterharvesting technologies 184
Rajamani L 59
regional legal framework, land-use and soil-conservation laws 69–70
resilience 9
resource mobilisation 146–147
Revised African Convention on Conservation of Nature and Natural Resources (RACNNR, 2003) 67–70, 165n106, 229
risk assessment exercise 32n136
River Basin Development Authority Act (RBDA) 130, 144, 148, 153–154, 168n144, 238–239
Roosevelt 69
Rotterdam Convention on the Prior Informed Consent Procedure for Certain Hazardous Chemicals and Pesticides in International Trade (1998) 65
Rumble O 57, 200, 201
Rwanda 251n130

Sadoff C 190
Scotland 140, 165n109
Second National Action Programme for South Africa to Combat Desertification, Land Degradation and the effects of Drought (NAP-CDLDD) 178, 197–198
Senegal 251n130

shifting cultivation 4, 25n45
short-term adaptation plans 11, 19
short-term solutions 244–246
Sierra Leone 251n130
social constraints 12
socio-economic factors 20
socio-economic systems 62
soft laws 56
soil-conservation laws 69–71
soil moisture 23n22
Sokoto 169n169
Sokoto-Rima 172n221
Song 142
South Africa: adaptation, regulatory framework 179–194; agriculture 5–6; agriculture-based policies 178–179, 194–198; agriculture, regulatory framework 194–205, 236–240; annual precipitation rate 26n54; broad recommendations 243–247; Carbon Tax Act (CTA) 178, 191–192, 246; Climate Change Act 178, 192–194; climate change adaptation laws 188–194; climate change-related policies 178–188, 233–236; coal mining-related emissions 26–27n56; future research projects 247–248; impacts of climate change, adapting to 178; long-term solutions 247; medium-term solutions 246–247; net zero GHG emissions 179; regulatory frameworks, climate 233–236; short-term solutions 244–246; weaknesses, regulatory frameworks 205–206, 240–243
South African Development Commission (SADC) Protocol on Shared Water System (1995) 67–69
South Sudan 251n130
spatial planning 204
Spatial Planning and Land Use Management Act (SPLUMA) 179, 204–205, 239–240
spatial resilience 205
Special Climate Change Fund (SCCF) 60, 77n64
special climate change unit (SCCU) 131
sprinkler 39n223
stakeholders 150
Stein R 67
Stockholm Convention on Persistent Organic Pollutants (2001) 65
Strategic Environment Assessment (SEA) 98, 99, 112, 191, 234
Strydom HA 189
susceptibility 183, 190, 206, 213n104, 247, 248

Takaungu 99
Tanzania 38n208, 41n250, 124n195, 251n130
Taylor A 180, 192, 224n314
technological solutions 14–15, 155
technology transfer 79n108
Togo 251n130

Uganda 40n232, 84n183, 251n130
unconventional tillage farming methods 2, 22n8
UN Convention on the Law of the Non-Navigational Uses of International Watercourses (1997) 68
United Nations Conference on Environment and Development (UNCED) 93, 115n27
United Nations Convention to Combat Desertification (UNCCD) 149, 170–171n191, 197
United Nations Economic Commission for Europe Convention on the Protection and Use of Transboundary Watercourses and International Lakes (UNECE Water Convention of 1992) 67, 68, 229
United Nations Environmental Program (UNEP) 57, 66
United Nations Framework Convention on Climate Change (UNFCCC) 7, 59–60, 63, 64, 71, 72, 75n50–75n51, 76n53, 76n59, 76n61, 77n62, 130, 165n109, 232–233; adaptation plans 76n59; adequacy, adaptation 76n61; Kenya 91; Nigeria 131, 161n30; Nigeria's NAP 136; South Africa 179, 182
United States dollars (USD) 3, 5
Uyombo villages 99

Van der Berg A 182, 185, 190, 204, 205
Vision 20:2020 147
Vision 2030, Second Medium Term Plan 3, 104
Voluntary Guidelines for Responsible Governance of Tenure (VGGT) of Land, Fisheries and Forestry in the

Context of National Food Security (2012) 69–70, 229
Voluntary Guidelines for Sustainable Soil Management (VGSSM) 69–70, 229
vulnerability 8–9, 21

Water Act (2016) 106–107
water conservation techniques 184
Water Framework Directive, EU 69
water-management laws 67–69
water-management techniques 153, 165n107
water resource management 198
Water Resources Act (1993) 130, 144, 152–153, 172n214
Water Resources Management Reform Program (WRMRP) 148
Water Services Act (WSA) 178–179, 198–199, 240
Watershed, Mountainous, Hilly and Catchment Areas Regulations (2009) 156
West Africa 32n136, 247, 248, 251n130
Western Cape 28n68, 216n160, 217n163
Wetlands, Riverbanks, Lake Shores, and Sea Shore Management Regulations 119n98, 120n100, 156
Wolf J 13
World Soil Charter (2015) 69–70, 229
World Summit on Sustainable Development (WSSD) 208n11

Yobe State 167n128

Zambia 35n175, 251n130
Zamfara 169n169
Zimbabwe 41n250, 251n130

Kolibri 360 GmbH
Ferdinand-Jühlke-Straße 7
99095 Erfurt, Deutschland
produktsicherheit@kolibri360.de